ROADS AND ECOLOGICAL INFRASTRUCTURE

Wildlife Management and Conservation

Paul R. Krausman, Series Editor

Roads and Ecological Infrastructure

Concepts and Applications for Small Animals

EDITED BY

KIMBERLY M. ANDREWS
PRIYA NANJAPPA
SETH P. D. RILEY

 Published in Association with *THE WILDLIFE SOCIETY*

JOHNS HOPKINS UNIVERSITY PRESS | BALTIMORE

© 2015 Johns Hopkins University Press
All rights reserved. Published 2015
Printed in the United States of America on acid-free paper
9 8 7 6 5 4 3 2 1

Johns Hopkins University Press
2715 North Charles Street
Baltimore, Maryland 21218-4363
www.press.jhu.edu

Library of Congress Cataloging-in-Publication Data

Roads and ecological infrastructure : concepts and
applications for small animals / edited by Kimberly M.
Andrews, Priya Nanjappa, and Seth P. D. Riley.
      pages cm. — (Wildlife management and conservation)
   Includes bibliographical references and index.
   ISBN 978-1-4214-1639-7 (hardcover : alk. paper) — ISBN
1-4214-1639-5 (hardcover : alk. paper) — ISBN 978-1-4214-
1640-3 (electronic) — ISBN 1-4214-1640-9 (electronic)
1. Wildlife crossings. 2. Animals—Effect of roads on.
3. Wildlife conservation. I. Andrews, Kimberly M., 1978–
editor. II. Nanjappa, Priya, editor. III. Riley, Seth P. D., editor.
IV. Wildlife Society.
   SK356.W54R63 2015
   333.95'4—dc23   2014024719

A catalog record for this book is available from the British
Library.

Special discounts are available for bulk purchases of this book.
For more information, please contact Special Sales at 410-516-
6936 or specialsales@press.jhu.edu.

Johns Hopkins University Press uses environmentally
friendly book materials, including recycled text paper that
is composed of at least 30 percent post-consumer waste,
whenever possible.

The editors dedicate this book to their spouses, parents, siblings, Kimberly's technicians and students (having gone from graduate student to faculty herself during the course of this book), Priya's daughter (conceived, birthed, and grown to a three-year-old during the development and production of this book), and in loving memory of Seth's rats and his cat Cleo, who passed for reasons unrelated to this book. Mostly, we cannot forget the many smashed frogs, snakes, turtles, rodents, and even gators, among other below-the-bumper critters both present and future, who gave and continue to give their lives toward the scientific advances we have presented here, and who continue to remind us of the proverbial long road we have to go in keeping them, and us, safe.

ecological infrastructure (n). /ĕk′ə-lŏj′ĭ-kəl, ˈinfrə͵strək chər/
The basic habitat components and their connections necessary for species survival, and for natural populations, communities, and ecosystems to function properly.

Contents

Contributors

Kimberly M. Andrews, University of Georgia and Jekyll Island Authority

Steven P. Brady, Yale University

Alvin R. Breisch, Roosevelt Wild Life Station at SUNY–College of Environmental Science and Forestry

Brennan Caverhill, Toronto Zoo

Barbara Charry, Maine Audubon

Patricia Cramer, Utah State University

Brian A. Crawford, University of Georgia

Bryan C. Eads, Indiana–Purdue University Fort Wayne

Lenore Fahrig, Carleton University

Mark Fitzsimmons, Albany County Office of Natural Resources (retired)

Cathryn H. Greenberg, USDA Forest Service

Kari E. Gunson, Eco-Kare International

Lindsey Hayter, Indiana–Purdue University Fort Wayne

Scott D. Jackson, University of Massachusetts

Sandra L. Jacobson, USDA Forest Service

Jochen A. G. Jaeger, Concordia University Montreal,

Nancy E. Karraker, University of Rhode Island

Bruce A. Kingsbury, Indiana–Purdue University Fort Wayne

Julia Kintsch, ECO-resolutions

Tom A. Langen, Clarkson University

Thomas E. S. Langton, Herpetofauna Consultants International

Jeffrey E. Lovich, US Geological Survey

David M. Marsh, Washington and Lee University

David A. Mifsud, Herpetological Resource and Management, LLC

Kevin Moody, Federal Highway Administration

Priya Nanjappa, Association of Fish and Wildlife Agencies

David J. Paulson, Massachusetts Division of Fisheries and Wildlife

Seth P. D. Riley, National Park Service

William Ruediger, Wildlife Consulting Resources

Raymond M. Sauvajot, National Park Service

Fraser Shilling, University of California, Davis

Namrata Shrestha, Toronto and Region Conservation Authority

Paul R. Sievert, USGS Massachusetts Cooperative Fish and Wildlife Research Unit

Daniel J. Smith, University of Central Florida

Stephen Tonjes, Florida Department of Transportation

Rodney van der Ree, University of Melbourne

Patricia A. White, TransWild Alliance

Derek T. Yorks, University of Massachusetts

Foreword

When it comes to roads, the proverbial elephant in the room is much smaller than an elephant. Research has shown that many, possibly most, of the vertebrate species heavily affected by roads and traffic are small animals, especially amphibians and reptiles. Why then is the public, particularly the North American public, so blissfully unaware of the threat of roads to small animals? High profile road mitigation measures, like the iconic wildlife overpasses in Banff National Park, have focused public concern on the impacts of roads on large animals such as grizzly bears and elk. But effects of roads on small animals remain largely below the radar for several reasons. Collisions with small animals do not typically endanger humans or damage vehicles, and road-killed small animals often go unnoticed just because they are small. In addition, small animal population numbers are seldom tracked, so we don't often know whether roads are causing declines. And, many of these taxa, such as snakes, are simply less popular than the larger, more charismatic species. Thus, road mitigation measures aimed at small animals frequently are viewed with a mixture of bemusement, skepticism, and downright hostility over money "wasted" on saving a few toads, turtles, or rodents. This needs to change, especially given that road mortality is an important contributor to the endangerment of amphibians and reptiles in particular.

So, the time is ripe for a book summarizing the wealth of research on the effects of roads and traffic on small animals. Andrews, Nanjappa, and Riley have pulled together an impressive collection of authors who lay out in detail the wide array of information on how roads affect small animals and how we can miti-

gate these effects. The volume's accessible style and practical approach will ensure that its messages reach not only researchers but also government transportation and environment agencies, nongovernment environmental groups, and individuals interested in understanding the conservation impacts of the human-built environment.

Perhaps the strongest contribution of the volume is that by bringing together this information, important themes emerge that are otherwise lost in a literature distributed over a wide range of research journals and agency reports. I highlight three examples. The first is the role of "complementary" habitats in modulating road impacts. The small animals most affected by roads are those that need to move among different kinds of habitats to complete different phases of their life cycles (e.g., breeding, dispersing, and overwintering phases). When these animals have to cross one or more roads to find resources for the next life history stage, the toll in road mortality can be horrendous. This means that road agencies should implement mitigation measures for small animal species that have multiple habitat needs when the habitats are separated by roads. A second emerging theme is that places where the most roadkill occurs ("roadkill hotspots") are not necessarily the best sites for road mitigation. In fact, the best places for mitigation are sometimes the places where roadkill is low, if the *ecological infrastructure* is intact. In such sites the accumulation of past roadkill has reduced the populations so much that current roadkill numbers are low, although the per capita roadkill rate remains high. Mitigation at these roadkill hotspots is most effective because it can lead to recovery of a depressed population. A third emerging theme is the idea that the most cost-effective way to mitigate road effects on small animals is to take advantage of the opportunities presented in the thousands of road repair projects undertaken daily on the world's road networks. Each of these projects is an opportunity to reduce the negative effects of the road with a relatively small marginal increase in the cost of the road repair project. For example, barriers can be erected to reduce the likelihood of animals entering the road and being killed. When road work involves repairs to culverts and bridges, these can be done in ways that improve their use as wildlife passages.

Nearly as important as the information contained in this volume is the information not contained, revealing lacunae in our knowledge and thus future research needs. I highlight two examples. First is the question of which road effect—mortality, traffic disturbance, or movement barrier—is the most important, and therefore most in need of mitigation. While reductions in traffic on roads would simultaneously mitigate most road effects, reducing traffic is often not feasible. Rather, mitigation usually focuses on the particular road effect thought to be most important. But if the wrong road effect is mitigated, the mitigation will be ineffective at best. For example, if the most important road effect for a given small animal population is mortality, but the presumed effect is a movement barrier, then mitigation to encourage animal movement across roads may lead to higher mortality and thus exacerbate the road impact. Estimating the relative importance of road effects is not easy. To wit, road-killed small animal carcasses often go unnoticed due to their small size, and because they are quickly devoured by scavengers or disintegrated by passing vehicles. Thus, an important area of future research will be to determine the relative importance of different road effects on small animals. A second challenge for future research will be to evaluate the effectiveness of road mitigations. This statement may seem surprising to readers who have attended road ecology research meetings. These are typically replete with photo evidence of animals using wildlife passages of various shapes and dimensions, retrofitted variously with, for example, shelves, natural substrates, or openings for overhead light. However, showing that small animals use a passage is only the very first step in proving that the passage actually mitigates the effect of the road on small animal populations. Studies of populations, both before and after mitigation and in comparison to control sites, are the next, crucial step in road mitigation research.

Richard Forman referred to the global road network as the "giant embracing us," but to a small animal the road network may be more like the fabled 800-pound gorilla, stomping about in complete ignorance of the 8-ounce animals under its feet. This new volume provides the information needed to start reining in that gorilla.

Lenore Fahrig

Acknowledgments

The editors wish to thank Dr. Vincent Burke, Executive Editor, Johns Hopkins University Press (JHUP), for his guidance and advice during the development of this book. Similarly, we thank Catherine Goldstead, Sara Cleary, Jennifer Malat, Laura Ewen, Kathryn Marguy, and Courtney Bond of JHUP for their answers to our many questions along the way and for keeping us on track. We appreciate the comments provided by two anonymous peer reviewers provided by JHUP, which greatly enhanced our manuscript. We also sought internal reviews by our authors of one another's chapters; and for certain chapters, we benefitted from external comments provided by Alex Levy and Elke Wind. Bill Branch, Margaret Griep, and Joe Mitchell participated in early discussion and planning of the book and contributed to some of the concepts. Kevin Moody offered guidance on the initial proposal for the book, and Marlys Osterhues and Daniel Buford assisted us in adjusting content for our target audiences. We are indebted to our excellent copy editor, Sheila Ann Dean, whose meticulousness we relished. We were pleased to use the indexing services of Susan Harris. We are extremely appreciative of the impressive assistance provided by Katie Mascovich, whose professionalism and attention to detail was critical and invaluable in the final throes of assembling our manuscript. Additional support early in our book development process was provided by the US Forest Service (Agreement No. 10-PA-11083150-023), the Amphibian and Reptile Conservancy, the American Society of Ichthyologists and Herpetologists, and Resource Management Services, LLC; these funds were critical in allowing us both to assemble the amazing cadre of authors included here, and to advance progress on manuscript preparation. The concept for this book was conceived by K. M. Andrews during regional meetings of Partners in Amphibian and Reptile Conservation (PARC), a network whose breadth and depth further enhanced our collection of contributing authors; the PARC Roads Task Force was key to identifying some of the issues and challenges that we attempted to address in this volume. While we tried to include as many projects and contributors as possible in the field of road ecology, we would be remiss not to acknowledge the numerous graduate students, wildlife experts, road ecologists, and transportation planners whose innovative efforts have advanced our understanding to the point that we could write this book. We also raise our glasses to the future professionals who will maintain forward progress.

We are grateful to the following organizations that provided both financial and in-kind contributions to JHUP for this book:

ROADS AND ECOLOGICAL INFRASTRUCTURE

Kimberly M. Andrews,
Priya Nanjappa,
and Seth P. D. Riley

Introduction

Why Did We Create This Book?

The need for the material presented in this book has existed since the establishment of the field of road ecology. With the increase in road studies over the past four decades, this synthesis is timely given the recent rising interest in how to reduce the ecological effects of public infrastructure on small animals. We have elected to focus here on "low profile" wildlife taxa that have received less attention in the scientific literature and at scientific conferences, in transportation planning and guidance documents, and, perhaps most importantly, among public and governmental sectors of all societies around the world. Specifically, these taxa are amphibians, reptiles, and small mammals; references to other taxa are included where they provide guiding examples that apply to small animal species.

We refer to small animals as low profile on roads for several reasons. Literally, small animals are less visually obvious to drivers and therefore present a reduced risk to human safety and property. Also, while they are more vulnerable to being struck on roads, it can be difficult to determine when their populations are declining because of their low detectability and covert behavior. Lastly, many of these taxa, such as snakes, are not popular in today's society and do not raise public concern to the same extent as large vertebrates such as deer when there are obvious road conflicts. For these reasons, assessments of road effects on small vertebrates have demonstrated a pressing issue; yet these taxa receive less attention in the transportation planning process, even where quantified mortality rates exceed those of other vertebrate species. Some species are long lived with low reproductive potential relative to many large vertebrates that reproduce annually and have higher offspring survivorship (Figure I.1). Therefore, the continued loss of reproductive-age adults due to roadkill can have severe ecological consequences that either cannot be reversed or will take multiple generations to reverse.

Increasingly, small animal populations are being studied by the scientific and wildlife management community, and more and more of these animals are determined to be at risk because of habitat fragmentation (i.e., prevention of movement and greater isolation) as well as because of climate change. Thus, concepts and considerations of safety for these animals go beyond roadkill impacts and can mean ensuring population and metapopulation health.

Designing roads and mitigation measures to avoid wildlife conflict is possible, but expert recommendations are needed. Transportation planning cycles and decision-making processes are, necessarily, focused on public infrastructure needs, but we have room for growth in the integration of ecological considerations. Adapted from Merriam-Webster, *public infrastructure* is defined as *the basic equipment and structures (roads, bridges, and buildings) that are needed for a country, region, or community to function properly*. However, there must be consideration given to *ecological infrastructure*, which we hereby define as *the basic habitat components and their connections necessary for species survival, and for natural populations, communities, and ecosystems to function properly*. In this definition, our reference to

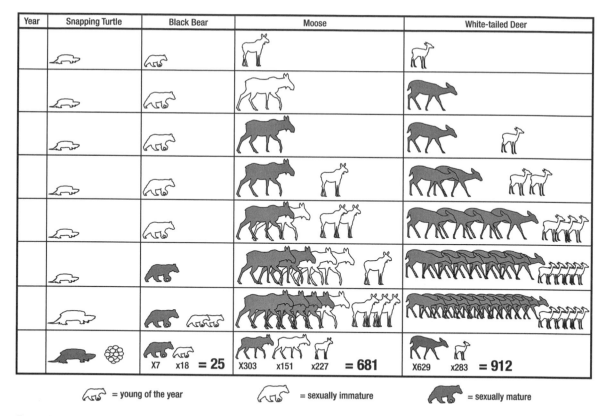

Year	Snapping Turtle	Black Bear	Moose	White-tailed Deer
		X7 x18 = 25	X303 x151 x227 = 681	X629 x283 = 912

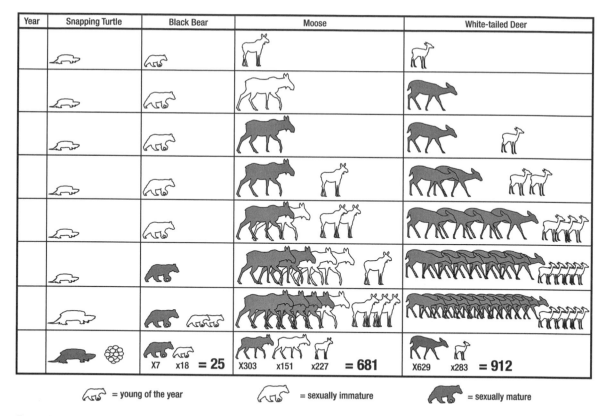 = young of the year   = sexually immature   = sexually mature

Figure I.1. Comparison of reproductive potential of snapping turtles to other large animals. An adult snapping turtle may lay up to 1,400 eggs before a single offspring reaches sexual maturity; this may not occur until year 50. Chart recreated and caption adapted from an original idea by George Kolenosky; modified in 2005 by the Ontario Ministry of Natural Resources Black Bear Technical Team and further modified in 2008 by the Ontario Multi-Species Turtle Recovery Team to include snapping turtles. *Credit: © Victor Young Illustration.*

"species" applies to plants, animals, and people. While public infrastructure is hard and static, ecological infrastructure is unpredictable and dynamic. By integrating these habitat components and allowing these processes to continue as naturally as possible, we can proactively ensure that public infrastructure will do minimal harm to the existing ecological infrastructure. This book addresses how and why to do so via expert guidance and recommendations of concepts and applications for both planning and mitigation relevant to small animals.

Ecological infrastructure existed long before people roamed the earth, while public infrastructure is a relatively modern advancement that often conflicts with the natural environment and ecological processes. Similarly, ever since people have been traveling along roads, conflicts between their movements and those of wildlife have arisen, and the science of road ecology is

another, more modern advancement to address these challenges. Just as ecological and public infrastructure do not function as independent components, small animals are components within wildlife communities and broader landscapes. Therefore, the focus of our management efforts must shift from managing these components independently to managing these integrated systems.

In a world laden with anthropogenic change, not only do we have to be adaptive, but we are challenging wildlife to adjust on increasingly short time frames. The effects of these abrupt changes could be reduced by instituting flexible (adaptive) mitigation systems that have long-term viability. Ideally, these mitigation systems are also accompanied by regular monitoring, allowing us to identify when passage structures or other mitigation measures are not working and to then adjust management plans to meet our original objectives. This challenge should inspire us and not paralyze our planning process.

In fact, it was under this premise that K. M. Andrews and P. Nanjappa began discussing the need for an ecosystem-based synthesis of knowledge applicable to small animals and roads. In addition to the urgent need for novel approaches, the information available in this arena was largely inaccessible and unpublished (and in some cases stored only in the brains of a few transportation practitioners). To extract some of their expertise, in 2010 we organized a symposium at the Joint Meetings of Ichthyologists and Herpetologists where a diverse panel of experts in road ecology, wildlife biology, and transportation policy convened to present on the knowns, unknowns, and desired future conditions. The symposium served as the kickoff to development of this book, and was the first of many collaborative planning sessions that fleshed out our skeleton of ideas and allowed us to shape the material to accommodate the diversity of challenges for our readers. We recognized that to reach the suite of professionals who are essential to the transportation process, we had to create a book that speaks to a varied readership. Through the time invested in numerous hours on conference calls and at computers, the team planned and wrote this book as a labor of love. We believe that this has resulted in a truly unique and useful product.

This book is a call to action. It is a call to researchers, engineers, landscape planners, politicians, and laypersons to tear down the artificial walls that often exist between these groups and to collaborate in order to address basic and shared goals; a call to show a greater willingness to share data and future research goals; and a call to remind all of us that no one wants to see species harmed to the point of endangerment, which is why we need to work together proactively.

Are You Talking to Me?

Yes, we are, and to *every one* of you! This book is for wildlife biologists specializing in our target taxa, but it is also for other biologists who can apply our models in their research or management systems and, in turn, can teach us their models that have application in ours. This book is equally for professionals and nonbiologists working in transportation arenas—federal, state, county, and municipal planners, those in the private sectors that influence land use, engineers drafting the blueprints, and the many, many people that it takes to execute transportation plans on the ground. The result of those plans allows us to access the places that we need to go for our livelihood and well-being, including travel to work, to grocery stores, farmers markets, or restaurants for food for our family, and to those favorite spots where we can relax and have fun. Parts of this book will also appeal to the general public who simply like wildlife and want to support conservation. Road planning is influenced by every member of every society who relies on public infrastructure for their day-to-day living needs. Do you drive a car on federal, state, county, or private roads? Do you pay taxes? Do you vote? Do you own or rent a home and thereby influence local land-use patterns? Then this book has relevance to you.

What Does This Book Provide?

Along with an introductory history on small animal road ecology, we present information on the natural history of small animals, with specific focus on amphibians, reptiles, and small mammals. These are the taxa for which we are lacking the greatest amount of information relative to roads. In particular, we have summarized what makes these animals vulnerable to roads and road-related impacts. This book presents direct effects of roads on small animals; it also presents effects on habitat quality, and thus, indirect effects on small animals. We orient the reader to the importance of public engagement along with transportation planning processes and funding sources. We offer summaries of mitigation planning and design guidance and identify areas where more research (ecological and engineering) is needed. Further, we employ an ecosystem perspective that allows for species groupings based on similarity in habitat needs or ecological specializations (e.g., fully aquatic species vs. aquatic breeders, or fully terrestrial species vs. terrestrial breeders). We describe why and when to retrofit crossings, how to plan ahead for construction and maintenance phases, and provide the tools to design monitoring programs and to incorporate adaptive management approaches.

We present current approaches for responding to and managing road effects on small animals but caution the reader that the recommendations and solutions offered are not the only way. In fact, one of the objectives of this book is to identify challenges and opportunities for improvement within the current approaches in order to inspire innovative lines of inquiry and application. As the field of road ecology evolves, new techniques will become available, and we intend for our readers to be on the front lines of developing pioneering approaches. There will always be ecological and social variability among landscapes and cultures, and the examples and concepts presented here should be adapted to accommodate that variation. We emphasize that each planning and mitigation project should be unique and customized for a particular location, target animals, or objective. In short, we are writing this book to encourage growth, including more research, monitoring, and quantification of road effects and the efficacy of mitigation measures.

This book is intended to be a navigable resource manual or guidance document rather than an academic discussion. We mostly focus on North America but intermittently present international examples when pertinent to understanding the current state of our knowledge. We further acknowledge some inherent geographic biases to the eastern United States, given the higher density of roads, and the greater number of small animal road studies and mitigation projects accomplished. Information is presented in the form of Chapters, Case Studies, Boxes, and Practical Examples. We have divided material into chapters and sections that can be read consecutively cover-to-cover, or independently, based on what is most relevant to your role or needs in the transportation realm. We present the customizable road map below to help you find the shortest distance between A and B in your journey through this book. Choose your own adventure.

Roads and Small Animals

What is the history of road ecology and small animals? What are the effects of roads on small animals and why do they matter for sustaining populations?

- How did the field of road ecology with small animals begin? How has it evolved and what challenges were faced over time relative to those faced today? How have these challenges varied across the globe? → Chapter 1 (History)

- Why do animals move? How do movements vary between aquatic and terrestrial species? What biotic and abiotic factors drive movement? What aspects of small animal natural history predispose them to road conflicts? How do roads affect the movements of individuals, populations, and metapopulations? → Chapter 2 (Natural History)

- What are the different consequences of roadkill and barrier effects? How do they compare in severity relative to each other? How do we study these effects? What are the ecological costs of habitat loss and landscape change? → Chapter 3 (Direct Effects)

- How are microhabitats surrounding roads altered? What are the different types of contaminants associated with roads? How do noise and light pollution from roads affect wildlife? How can roads and roadsides be managed to reduce negative effects? What are secondary results of human access to landscapes via roads? How is animal behavior affected by roads? → Chapter 4 (Habitat Effects)

Planning and Design

What is the role of the public in transportation projects? How do agencies operate in terms of the planning process and funding? How do you identify when and where mitigation measures are needed? How do you then design appropriate mitigation features and their placement?

- How important is public support? What are some examples of tools for interfacing with the public? How can you foster interagency collaborations? What is the best way to engage citizen scientists and political officials? With citizen assistance, how do you set data protocols and manage quality? → Chapter 5 (Public Engagement)

- Which transportation and resource agencies are involved in which processes of building and maintaining roads? When do the best opportunities occur for wildlife experts to engage? What information is used in planning roads, and what types of information is most useful when provided? How does that information vary with the different phases of road planning? How does the planning and management of different types of roads (public, county, municipal, industrial, residential) vary? → Chapter 6 (Planning and Design)

- Who controls the funding for roads? Where is funding secured for roads and associated mitigation efforts before they are built? After they are built? What road features are funded through which mechanisms? When are the best opportunities to engage in the process? → Chapter 7 (Funding)
- Which spatial scales are relevant for planning? What tools are used to proactively plan mitigation designs? What factors should be considered in placing roads? What should be included in baseline surveys and data collection? What are the best ways to quantify hotspots of mortality? → Chapter 8 (Mitigation Planning)
- What techniques are commonly employed for the mitigation of road effects? How do mitigation approaches vary among habitat types? Which habitat features are most important to consider? Which design features are most important to consider? → Chapter 9 (Mitigation Techniques)
- Which situations warrant integrating or modifying a passage structure? How is retrofit design influenced by species considerations, including mobility? Which landscape features should be considered when designing retrofits? Which external and internal structure design features influence the likelihood of passage use by animals? What is the process of enhancing existing structures? → Chapter 10 (Modifying Structures [i.e., retrofitting])

Construction, Maintenance, and Monitoring

How do you develop construction contracts and ensure long-term maintenance schedules? How do you implement monitoring objectives in the pre- and post-construction phases? How do you measure the success of mitigation efforts and manage objectives adaptively? Which topics and research needs are emerging as leading issues?

- What are the primary steps in the construction process? What needs to be considered when negotiating a maintenance contract? Who is responsible for maintaining wildlife passage structures? Who funds construction and maintenance phases? What are some of the greatest challenges in insuring proper construction and maintenance operations? → Chapter 11 (Construction and Maintenance)

- What are the primary elements of a monitoring project? How are they different before and after construction? How is a study designed to produce quantifiable and comparable results of structure effectiveness? What are the best ways to monitor mortality reductions? What are the best ways to monitor improvements in habitat connectivity at individual and population levels? Why is it important for management plans to be adaptive? → Chapter 12 (Monitoring and Adaptive Management)
- What are the emerging topics in road ecology for small animals? What ecological and social constraints will continue to present challenges? What research and monitoring needs should be prioritized to allow our knowledge base to necessarily progress for designing responsible data-based management plans? → Chapter 13 (Road Ahead)

How Can This Book Help You Help Small Animals?

Making a difference in the field of conservation is harder than ever. You can no longer simply be an expert in a given arena; it is now necessary to wear many hats and to apply these areas of expertise in an interdisciplinary fashion. The transportation process is multivariate, and one must be able to converse in the languages of transportation planning, wildlife biology, environmental engineering, and landscape planning. Communications and collaborations should be inclusive of local and state officials and members of the general public.

Specifically, the need for far-reaching communications on wildlife-road conflicts is most critical. Generally, people are unaware of the importance of small animals in ecosystems and the relevance of population declines related to road impacts. Furthermore, the media often misrepresents the science, often because we, as scientists, have not provided the proper communication and translation tools. The public must be convinced of why small animals are important—by you. Learning to speak simply and pertinently about irreversible environmental change (population extirpation, species extinction, landscape-level habitat fragmentation) and the loss of "free" services provided to us by our ecosystems is critical for making global-level impacts.

In order to obtain the resources necessary for effective conservation, we must have the public on our side. If we have public support, political support can follow and can garner the momentum needed for high-level prioritization of these largely fixable problems. Consistent momentum will establish a standard place for these measures in decision-making and planning processes. Thus, the public is key in making it possible for us to get ahead. Compromising project time frames and budget bottom lines may be inevitable. However, by using informed approaches, forging partnerships to share and maximize resources, and analyzing existing data for new objectives, we can weather these resource trade-offs without compromising the effectiveness of public or ecological infrastructure.

Proactively ensuring that ecological infrastructure is both integrated and maintained throughout the process offers a better and cheaper way of conducting business. The costs and regulatory burden of adding mitigation measures as a result of concerns about threatened, endangered, or declining species, as well as the costs of species recovery, far exceed the costs of proactive planning. Further, negative media attention, either in support of or against a mitigation project, can greatly affect the likelihood of funding. It is our intent for this book to provide the information and justifications needed to facilitate the proactive measures and collaborative partnerships that will better address these low profile animals, and thereby minimize high profile problems.

1

Thomas E. S. Langton

A History of Small Animal Road Ecology

1.1. Introduction

Small vertebrate animals have received an increasing amount of attention in road ecology. This is evidenced by a growing number of case studies that use a variety of measures to prevent the mortality of animals when crossing busy roads. The simple and ultimate goal of building small-scale structures and other activities is to ensure population persistence. For the purposes of this book, small animals have a normal standing height (to top of head from ground level) of less than one meter. Most are shorter, less than 0.2 m, including anurans (frogs and toads), caudates (salamanders), chelonians (freshwater turtles, terrapins, and tortoises), squamates (lizards and snakes), and small animals. Small animals tend to receive attention only when they are considered to be declining, usually through the regulatory means of formal statutory listing, or when they are incorporated as part of a mitigation project geared toward large animals. Efforts by wildlife professionals have provided a basic foundation on which future work can build, fostering broader and more consistent applications as a standard part of landscape planning and remediation.

One traditional approach to small animal road crossings has been to install small underpasses (Chapter 9). Wildlife barriers, such as fencing, are commonly used to divert or funnel animals to dual-purpose (wildlife and drainage) and single-purpose (wildlife) passage structures. These multi-component passage systems have been used since the 1960s, particularly in northern Europe, and include those for the Eurasian badger (*Meles meles*), which had experienced high collision frequency over expansive stretches of roads. By virtue of variable dispersal cues and interspecific and intertaxonomic variation in behavior, the degree of acceptance and usage of such systems varies greatly among wildlife. Many small animals can readily extend, jump, or climb vertically one meter or more over obstacles. This agility presents a challenge in preventing their movement onto the road when building effective, low-maintenance, but long-lasting fences to exclude or direct their movement.

The need for a greater understanding of which techniques effectively mitigate which ecological effects is now apparent. Additionally, recent studies have more frequently established measurable success criteria. Calls for the development of better small animal protection measures on roads have increased since the 1980s, in particular those measures that employ comparative designs and rely on evidence-based studies (e.g., amphibians, Schmidt and Zumbach 2008; turtles, Section 9.5.3). These approaches need to guide future road-crossing specifications, followed by standardized performance measures that can be tested against predetermined objectives.

1.2. The Engineering Contribution

For as long as people have been moving from place to place, simple earthworks and bridges have been built to make journeys easier. The threat to animals on roads was mentioned early in the progression of transportation systems, including the mortality of common toads (*Bufo bufo*) from horse-drawn carriages in England (Hudson 1919). Manufacturing of the combustion engine spurred massive growth in the car industry, from under a million vehicles to tens of millions by the 1940s. Transportation convenience and subsequent growth in motor vehicle production was further enabled by extensive construction of tarmac-surfaced, or paved, roads. Amphibian mortality, along with that of other small vertebrates, began to be consistently noted in the published literature in association with this burst in transportation activity (e.g., Savage 1935).

In order for transportation systems to progress, engineering technology had to accomplish the feat of managing the flow and direction of a large volume of traffic around rivers, streams, swamps, and other aquatic features. Channels and embankments to divert water or prevent flows from degrading a road have facilitated the movement of people, livestock, and wagons through the ages. Drainage pipework with brick and stonework voids made from terracotta (fired clay), lead, wood stave, and even leather predate Roman times in Europe. Trenches were dug and then lined with gravel and loose stones or pebbles using ancient techniques to further enable surface water infiltration and diversion. Measures to enable the drainage of water from and around road surfaces, particularly on sloping ground, included catchwater drains, shallow trenches set back from road edges, and side drains (American Association of State Highway and Transportation Officials 2007) to reduce scouring, erosion, and water logging. In more recent times, preformed tube and box materials to carry water under linear transport routes have been composed of bitumen, plastic pipes with coated (galvanized) steel, and concrete.

As culverts and bridges increased in size for vehicle use, concerns over safe passage for large animals under roads became more widespread. Provisions for small animals drew little attention because of generally low public and scientific interest and modest levels of participation in environmental protection efforts for these wildlife groups. Further, animal-vehicle conflicts on roads were not as extensive or as widespread due to the predominance of low rural road traffic densities. However, this negligence began to

change in the 1960s when awareness of, and interest in, mortality issues grew alongside rapidly increasing vehicle numbers and collisions. Protective measures for small animals have subsequently developed over the past 50 years, particularly with technological advances in the culvert materials used for water drainage and with bridge designs incorporating reinforced concrete.

From the 1930s to the 1960s, in larger, more progressive continents, such as Europe, North America, and Australia, a concerted interest developed in reducing the conflict between dispersing wildlife and road traffic, primarily on the basis of driver safety. As travel speeds and volumes increased steeply from the 1940s, the need to fence domestic animals and large wild vertebrates grew because of rising numbers of vehicle collisions and human fatalities (e.g., from deer and elk [*Odocoileus* and *Cervus* spp.] and kangaroos [*Macrophus* spp.]). The increase in road building that accompanies residential, commercial, and industrial development changed roadside habitats throughout the 1970s.

Some of the coordinated measures to reduce animal mortality on roads have been present for almost 100 years, beginning around the start of the mass motor vehicle manufacturing industry. Starting in the late nineteenth century, some of these initial measures included simple exclusion fencing constructed with steel or wood posts and wire fencing. These fences aimed to keep large wild and domestic animals off roads. Additionally, construction of bridges and raised roads along rivers, streams, floodplain edges, and seasonally flooded ground have been effective in accommodating surface water flow and flooding events. These overpasses have enabled animal passage during both wet (e.g., seasonal flow) and dry (e.g., drought) conditions and do not present a barrier to wildlife movement or contribute further to landscape fragmentation. In the 1960s, as barrier fencing was shown to effectively direct large mammals to bridges (i.e., overpasses) and drainage underpasses, this design was incorporated more routinely by several government and state highway departments. Construction of wildlife-crossing features on roads and the use of other collision reduction measures continued more consistently in the 1970s (e.g., in North America, Bissonette and Cramer 2008; in Europe, Iuell et al. 2003).

1.3. The Evolution of Techniques to Mitigate Ecological Effects of Public Infrastructure

1.3.1. Road Warning Signs

The first government-funded road warning signs for amphibian migration were perhaps at Freiburg, in Baden-Württemberg, Germany, in the early 1960s, where homemade and occasionally official signs began to be used for animals other than deer and domestic animals, such as for European otter (*Lutra lutra*) and Eurasian badger. Signs became much more popular with the general increase in interest in the environment in the 1980s. For example in Manitoba, Canada, warning signs were installed in the 1980s for red-sided garter snakes (*Thamnophis sirtalis parietalis*), which are killed in large numbers on roads following mass emergence from hibernation dens. Signs alerting motorists of turtle and tortoise crossings along with other wildlife species have continued to become more popular in both Canada and the United States. Sign designs use wording, symbology, or both to indicate the particular wildlife hazard; the designs vary greatly among and within countries, with variations even among neighboring areas (Section 9.2.1). However, using signage alone may not adequately alter driver behavior to mitigate population-level impacts.

1.3.2. Access Limitations and Human Intervention

Traffic calming and temporary road closure actions can be effective options where wildlife crossings are short term or seasonal (Section 9.2.4), particularly where extreme mortality events may only occur one to two times per year (e.g., Coelho et al. 2012). Additionally, local citizens can be organized to physically assist in moving animals across the roads in instances of mass movements over a concentrated time period (Chapter 5). In Switzerland, the first collective organization of drift fences and pitfall bucket rescues of migrating amphibians was recorded in 1968 (Ryser and Grossenbacher 1989). Similar citizen efforts occurred around the same time in Germany and Austria, later began in the Netherlands in the mid-1970s, and were more formalized in the United Kingdom in 1985. Sporadic efforts were also made in several other countries.

1.3.3. Habitat Compensation and Creation

Habitat compensation and creation were first considered in response to road development in the mid-1970s in central Europe when it became apparent that in some circumstances tunnel and fence systems were not always sufficient to maintain species movement across the road. For example, early tunnel systems were not effective with European newts (*Triturus* spp.; Ryser and Grossenbacher 1989), which rejected the dry or otherwise unsuitable internal environment. In the Länder (state) of Lower Saxony, Podloucky (1989) described a generalized pond shape for amphibians of approximately 50 × 20 × 1.5 m, stressing the importance of its orientation to sunlight in order to maximize warm conditions for spawning and larval development. During the 1980s, the construction of these alternative spawning ponds away from a new road became a preferred strategy to building road underpasses in Lower Saxony, due to uncertainty over tunnel effectiveness (Podloucky 1989). While the general idea of habitat compensation and management became a more typical consideration in road projects, it often did not fully take into account the long-term isolation effects of road fragmentation on adjacent wildlife habitats.

1.3.4. Passages and Barriers
One-Way Systems

Some of the earliest underpass systems were created in the late 1960s in Europe, including one-way passage systems. One-way systems were developed as a basic concept for controlling the direction of under-road movement of amphibians. Underpasses were designed to take a large proportion of an adult breeding population under a road and then facilitate their safe return later in the season along with the new cohort. This technique was typically applied near large lakes or meres where terrestrial and aquatic habitats were almost completely partitioned by the road.

In this design, animals fall down a pitfall entrance to enter an underpass where they have to travel to the other side to escape (Berthoud and Müller 1986; Müller and Berthoud 1996). The system acts as a compulsory transfer of animals from one side of the road to the other. One-way systems were first tested for amphibians at Neeracherried, near Zurich, Switzerland, in 1969 (FFPS 1988). One-way systems were further installed at 15 of the first 25 (60%) amphibian tunnel sites in Switzerland (Ryser and Grossenbacher 1989), including the Turlersee and Etang de Sepay sites (Berthoud and Müller 1986; Müller and Berthoud 1996). In the 1980s, open-ended passages with no trapping element (noncompulsory) were designed and used in experimental studies (Dexel 1989) in Germany. Here, individuals moving toward the road were presented with a behavioral choice of entering the underpass from either direction.

Barrier and Guide Fencing

Fencing can be installed as a single management technique simply to prevent animals from entering the road (barrier fencing). Alternatively, it can be installed with passages as a measure to direct animals to the safe crossing either over or under the road (guide fencing). In many instances, it is ideal if fences are coupled with passages to both reduce mortality (excluding animals from roads) and decrease the fragmenting effects of roads (connecting habitats across roads), which may be a greater problem for populations than mortality. Much early attention toward wildlife crossing systems was focused on underpass suitability and not as much on barrier fencing, which is an equally important feature of multi-component passage systems. One of the earlier uses of barriers in addition to passages was the amphibian project at Neeracherried, Switzerland (FFPS 1988). This project also demonstrated the importance of post-construction monitoring to assess structure performance. In 1970, monitoring by the Schaffhausen Society for Scientific Research Monitoring revealed that fences should be angled in order to direct a maximum number of individuals (FFPS 1988; see also Chapter 9).

Typical fence designs have used a solid material with an overhang to prevent trespass from climbers and some jumpers (see also Table 10.2). Ryser and Grossenbacher (1989) observed that some amphibians, such as common toads, moved away from fences that were made from clear plastic. Later studies with mesh fences showed that when some animals, including chelonians, see light through a barrier they may keep trying to move forward, which is energetically costly (e.g., Hagood and Bartles 2008). Some freshwater turtles and tortoises can climb large-gauge mesh fences by using them as a ladder. Despite this periodic occurrence, mesh barrier

fencing has been shown to be highly effective at reducing mortality in tortoises and other vertebrates in the west Mojave desert of the southwestern United States (Boarman and Sazaki 1996).

Fences should be durable and easily maintained with no obstructions, or the entire system may become compromised and rendered useless. The main problem in open and harsh environments with shifting soil zones is the cost of installing permanent fencing where sunlight duration and intensity are high and many materials such as metals and plastics rapidly degrade due to UV exposure. Ryser and Grossenbacher (1989) also gave some preliminary recommendations for system design including the placement of small deflection fences, or "swallowtails," close to tunnel entrances to deflect moving animals into a tunnel entrance. In 1994 in Germany, after the development of ACO brand tunnels, the company Maibach Vul GmbH began selling metal wildlife fences and an L-shaped guide wall. In 1996, a second company, Volkmann and Rossbach GmbH (V&R), also began competing with the ACO product for the permanent fence system market using similar materials to Maibach's. Some examples of typical amphibian barriers (wall and fence structures) are shown in Figure 1.1.

Passage Structure Materials

In Germany, a review of performance results was conducted on 13 passage systems and 3 adapted road drainage culverts for their use by amphibians. All systems were constructed of concrete except for one made of steel (Stolz and Podloucky 1983). The steel passage was considered unsuccessful due to its high conductivity and resulting cold temperatures, and because of a lack of evidence of its use by amphibians. Other experimental wildlife tunnels installed in the 1990s were made from cast or preformed concrete, and were often round or rectangular culverts mass produced for drainage purposes. In many countries, structures larger than precast units require inspection and maintenance and therefore are more expensive and less attractive to engineers.

Another early lesson learned was the need to manipulate the amount of light and moisture within the passage to reduce resulting injury or mortality of amphibians through exhaustion and dehydration. With one-way systems, some studies showed that light levels at the bottom of the entrance to the pitfall area at night

Figure 1.1. Examples of materials, configurations, and heights of barrier fencing for small animals: (a) semi-permanent plastic fence with UV inhibitors can withstand sunlight for several years in the open but suffer from expansion, which causes distortion and damages joints; (b) permanent metal fence with overhang but no base plate, Cheshire, England; (c) permanent "stop grid" to enable vehicles access across passage and amphibian guide wall. These barriers restrict movement of many amphibians that cannot cross the surface and instead fall down into the safe zone; (d) a woven plastic fence with overhang connecting to a surface tunnel, Quebec, Canada. *Credits: Thomas E. S. Langton (a, b, c); Dinu Filip, ACO Canada (d).*

were often greater than within the passage. Hence, animals remained in the moonlit area rather than proceeding through the passage structure; this issue was later remedied by shielding the entrance area to dim the light levels at night (Krikowski 1989). Further, the dryness and absorbency of the surfaces of the concrete pipes were reported to aggravate dehydration effects.

Alternative tunnel types were developed as a result of these early experiments, notably in Germany, and were subject to experimentation with a range of materials and structures both in controlled conditions and in the field (Dexel 1989; Brehm 1989). Passage designs of different dimensions and construction material were tested, including those with more inert, nonab-

sorbent, and heat-stable characteristics than concrete and metal. Designs to allow greater wetness and a more natural tunnel floor substrate were chosen to mimic natural habitat conditions. Passages with slots at the road surface allowed water to drain in, keeping underpasses wet. This design feature was incorporated into one of the first modular wildlife passages, which were made from polymer concrete, with a small stone aggregate or recycled glass component that bonds with polymer resins, making it stronger than concrete. These were tested in situ at road locations in Europe and the United States on a trial basis in the late 1980s. A larger, 500 mm span passage for small animals was introduced shortly thereafter in the early 1990s. Amphibians appeared to benefit the most from the inert nonabsorbent nature of polymer concrete.

Passage Structure Placement

Standards for constructing underpasses were published in Switzerland in the mid-1970s (Honegger 1977) and Germany in 2000 (FMTBHRCDRT 2000). More wildlife passage designs were being considered in the early 1990s, particularly in Europe; this was in response to a growing demand for structures with dimensions of at least 1 m width that had been suggested by the central European (mainly Germany, Austria, and Switzerland) Highway Guidelines (i.e., *Merkblatts*) for amphibian tunnels. Beginning in the 1990s, larger tunnel sizes were recommended. It then became clear that burying and fitting large underpasses into the ground was not feasible in many circumstances, particularly with retrofitted passages. Further, sinking the structures below the ground surfaces caused conflicts with expensive water, telephone, and electricity lines, usually buried at 1–2 m. Brehm (1989) was the first to place a passage at ground level with a 200 mm wide structure at Methorst pond in Schleswig-Holstein. Surface underpasses were initially favored where an embankment on one side was nonexistent and the other side had a steep incline, such as on a hillside. This configuration represents around 30% of localities where wildlife passages are built in mountainous central Europe.

Passage Structure Design

Slots in surface passages allow light, rain, and air circulation to maintain an interior environment more consistent with the outside, removing some of the problems of cold, dry, and dark passages (Chapter 9). In addition to slotted passages, unslotted (solid top) versions have been designed for situations where the middle of a road rises, or "crowns," or in locations where rainfall and road runoff levels are so high that the wildlife tunnel may become flooded and slots are not advantageous. A structure may also be partially slotted (often in the verge or median); this controls the amount of water passing into a tunnel. On rainy nights, moderating excessive surface water flow is particularly important on sloping ground. Passages have also been constructed using modular rectangular-shaped box culverts, which can be durable enough to place close to the road surface.

Around the early 1990s, production began at three major suppliers in Germany to produce "stilt" tunnel passages. With stilt designs, the three sides of an inverted preformed box shape are fixed onto foundation plinths (to provide strength), allowing any substrate to be placed on the passage floor as a walking platform. More recently, durable paints and inert tiles have been designed to reduce any abrasive or caustic surface conditions at the bottom and sides of concrete passages. Use of these materials also reduces the formation of ridges along the surface floor that result from water or wind movement of substrates through the passage; the ridges can interfere with rapid movement of animals through underpasses.

A diversity of design shapes, sizes, and simulated microhabitat features became available, albeit with minimal field validation and monitoring (Figure 1.2). By 1997, the commercial value of amphibian migratory system solutions in central Europe commanded a competitive marketplace. At the same time, the lack of evidence that larger tunnels performed better began to encourage questions regarding the validity of the 1 m threshold of Federal Ministry of Transport guidelines (FMTBHRCDRT 2000). Wildlife passages under a 1 m span became worthy of reevaluation. While small, narrow, dry, concrete pipes were clearly outdated and to be discouraged, the statement of "bigger is better" was replaced with "how big is big enough?"

Lessons Learned with Passage Structure Design

The size and shape of fences have been a challenge to builders for many years, and new and more complex designs are being constantly tested. Experimental testing is complicated due to the broad suite of environmental and engineering factors that influence structure effectiveness (Figure 1.3). The durability and lifespan of fencing is an important factor in determining its effectiveness as a barrier. Resistance to environmental conditions varies both with materials and climate factors, such as temperature and exposure to sunlight. Roadside factors, such as salt applications to roads, may also influence corrosion of materials such as metals.

Problems with poor design and construction details of tunnel and fence systems were pervasive due to lack of monitoring and experimental testing in the early decades of wildlife passage installation; for example, there were thought to be over 20 nonfunctioning

Figure 1.2. Examples of different passage structure designs, shapes, sizes, and microhabitat features: (a) traditional concrete wall and drainage pipe tunnel for amphibians in Berezinsky Protected Area, Belarus; (b) interior of a 10 m near-surface precast concrete stilt tunnel in Nudow, near Potsdam, Berlin, Germany, where newts (*Triturus cristatus*) hibernate in leaf litter along tunnel base; (c) concrete circular tunnel (1,200 mm in diameter—tunnel lengths were shortened by cutting into the embankment) with dirt floors and concrete guide walls (800 mm) for a population of amphibians estimated at 1,000 grass frogs (*Rana temporaria*) and 15,000–20,000 common toads (*Bufo bufo*) in Møre og Romsdal, western Norway; (d) experimental "smart" tunnels with managed levels of heat, light, and moisture in Germany; (e) box culverts across a flat ditch, near Basingstoke, England; (f) steel arch underpass (5 m × 2 m) in Peterborough, Engand, placed for what may be the largest population of newts in Europe. *Credits: Tom Kirschy (a); Thomas E. S. Langton (b, e, f); Oddvar Olsen (c); Anne Martine, ARCADIS Nederland BV (d).*

a  b

Figure 1.3. An underpass (a) built in 2009 on the N310 road next to the Elspeetshe Heide (heathland) in Gelderland Province, Netherlands, which is also used to transfer sheep. Cut tree branches provide a "soft" cover to encourage small animal movement toward the underpass. The wooden "split-rail" cage walkway (b) provides a sheep-proof zone, where layers of old wood and cut branches are placed as a shelter that is warmed by part-day sun exposure. Viviparous lizards (*Zootoca vivipara*) have been observed basking on the wood rails and branches. *Credit: Thomas E. S. Langton.*

systems in Lower Saxony alone (Stolz and Podloucky 1983; Ryser and Grossenbacher 1989). Function had been compromised for reasons such as cracks in fencing and overgrown vegetation encroaching on fences that aid climbing of arboreal species and prevent lateral movement of animals along the base of fences. While we are aware of these maintenance issues, they are still common problems with many structures due to the pervasive pattern of a lack of support for monitoring and maintenance following passage installation (Chapter 12).

Attempts at testing amphibian passages in the United States were delayed relative to other continents and did not come to fruition until 1990. Drainage culverts designed for water passage are often round or elliptical and made from steel, plastic, or concrete and can be retrofitted and linked up to wildlife fencing to provide a crossing facility. This system was first tested in the United States for the declining Houston toad (*Anaxyrus houstonensis*). Results from surveys of State Highway 21 in northeast Bastrop, Texas, recommended an extensive design with a construction price estimate of $630,000 USD (Brown and Mesrobian 2005). Following extensive political backlash opposing such expenditure, $50,000 USD was spent on barriers leading to the existing metal drainage culverts. In addition, an elliptical corrugated metal pipe was haphazardly placed without fencing for western toads (*A. boreas*)

at an urban road junction in the town of Davis, near Sacramento, California. Neither of these projects was designed or implemented with a sound scientific objective or realistic mitigation strategy and neither invested in the necessary public outreach or political support. Therefore, the projects suffered from frivolous and critical interventions by the public and were persecuted in the media. This poor planning and negative public and political support undermined the concept of amphibian crossings in North America for several years, delaying progress in road effect mitigation for small animals. These projects also served as strong examples of the absolute importance of public and political involvement in road projects beyond the basic biological and environmental engineering planning.

1.4. Corridor and Defragmentation Initiatives

Road building is an inextricable part of land-use planning, and the movement to proactively design "environmentally friendly" roads and buildings that reduce environmental consequences are sometimes referred to as green infrastructure (GI). Multispecies "umbrellas" for wildlife protection have also been discussed in various places to varying degrees for some time. In Australasia, it is recognized that "flagship" or "focal species" preservation is only one part of a suite of measures needed to prevent biodiversity collapse from landscape-wide challenges (Lambeck 1997). There are already moves toward multispecies and wider landscape plans and programs based on examples from California (e.g., Regan et al. 2008) and Europe, which has a number of green corridor projects, such as The European Green Belt initiative (Terry et al. 2006). One of the most extensive multispecies approaches under implementation is the defragmentation program that aims to restore connectivity; this initiative is coordinated by the Federal Ministry for the Environment, Nature Conservation, and Nuclear Safety in Germany (FMENCNS 2012). This defragmentation program was adopted by the Federal Cabinet in February 2012 to better develop the basic principles of GI networks, including both terrestrial and freshwater components, and to maintain adequate levels of ecological infrastructure around transportation corridors. The key components are (1) a map of the country showing existing and proposed principal corridors, called "multiuse habitat corridors for people and nature," and (2) a goal of reducing land development by the transport and housing industries from around 120 hectares per day to under 40 hectares per day, approaching a 70% decrease by 2020 (FMENCNS 2012). In early 2009, an economic stimulus package resulted in the design and construction of 15 green bridges on main roads in Germany and an additional 3 at the planning stage. In addition, over 20 other initiatives have been undertaken, including a dozen small wildlife passage projects (FMENCNS 2012).

In other countries, the need for partial or transcontinental GI projects is recognized, and many examples are underway in an increasing number of locations. A GI strategy was approved by the European Union in May 2013. Similar initiatives to those in Germany are underway in Austria and Switzerland (e.g., Trocmé 2006). In France, the "Trame Verte, Trame Bleue" (Green and Blue Infrastructure) is a major national initiative incorporating regional protective wildlife corridors into planning policy (Bonnin 2008). In 2003, the Carpathian Convention was established by the Czech Republic, Hungary, Poland, Romania, Serbia, Slovakia, and Ukraine; the representatives met to address roadbuilding threats to the environment and discuss how to foster sustainable development in the large border region between the former Soviet Union boundary with Western Europe and the Carpathian Mountain area. Projects such as "The Safety Net of the Wildcat" (Felis silvestris), managed by Bund/Friends of the Earth Germany (Franck 2012), are beginning to better define and create political awareness for the required GI networks, even if the linear transportation defragmentation measures are not yet fully developed or built. Poland has a green corridor scheme with much potential, given its large areas of relatively unspoiled habitats (Jędrzejewski et al. 2009).

In many countries legal battles still occur and road routes are not selected based on ecological criteria. The Czech Republic has started to assess landscape fragmentation (Anděl et al. 2005) and GI and transport corridor defragmentation projects, but these have not been implemented despite the concepts being fully developed. The reality of their implementation hinges on the adequate policy, legislation, or programs necessary to introduce proposed changes on the ground.

Examples of studies with multispecies foci are rare. In the District of Segeberg in the Länder of Schleswig-Holstein, Germany, a green bridge has been built as a result of impacts on species listed in the Federal Nature Conservation Act, which is aimed at preventing significant interference in ecosystem functions and to ensure biodiversity protection (Reck et al. 2011). Target species include invertebrates (e.g., beetles, bush crickets), snakes, amphibians, small mammals, and larger animals such as red deer (Cervus elaphus). Habitat corridors were enhanced on both sides of the road for a distance of several kilometers to encourage use of the built structure. Additionally, a bridge was constructed over the river and associated ponds to increase the number of species that could cross (e.g., European otters, Reck et al. 2011).

Multispecies approaches have also been implemented in the Netherlands, with the construction of several green bridges, including the Woeste Hoeve and Terlet overpasses that were built in the late 1980s. While these were designed principally for red deer, they also accommodated wild boar (Sus scrofa), and a range of small- and medium-sized vertebrates, including red fox (Vulpes vulpes), roe deer (Capreolus capreolus), badgers (Meles meles), and domestic cats (Felis catus). The bridges are also used by a range of herpetofauna (Renard et al. 2008). There has been further construction of 25 green bridges since 2004 in the Netherlands

(Creemers and Struijk 2012) that were principally designed for amphibians, reptiles, and mammals. One of the monitored green bridges on the A2 motorway is the Goene Woud or Green Forest Overpass built in 2005 that included a chain of ponds on the overpass (van der Grift et al. 2010) and was used by at least seven species of amphibians. The 27 European Union countries have committed themselves recently (European Commission 2011) to achieve no further loss of biodiversity across the European Union area after 2020 and to restore biodiversity where possible; however, because the targets lacked realistic delivery mechanisms, the goals of this project did not come to fruition.

Within the United States, multispecies crossing structures were perhaps first installed along approximately 135 kilometers of road when "Alligator Alley" (Burghard 1969), a part of Interstate 75 / State Road 93 and State Road 84, was widened beginning in 1986 (Foster and Humphrey 1995) into a four-lane highway across a very large expanse of Everglades wetlands. The crossings are now part of the "Florida Wildlife Corridor" project. As part of a $77 million project, it was primarily intended to accommodate passage of the declining Florida panther (*Puma concolor coryi*), but a range of crossing structures were built, including 36 medium-sized wildlife underpasses (2.4 m tall, 21.3 m wide, 30.5 m long). Tall chain link fences (~3 m) funnel wildlife toward the passages (Evink 2002; USFWS 2012).

The concept of GI programs and pancontinental ecological networks (e.g., in Europe, Bonnin et al. 2007) offers a comprehensive future approach for defragmentation measures that support the surrounding ecological (green) infrastructure. Transport-focused defragmentation initiatives such as those promoted in the European Union (EU Green Infrastructure Strategy and White Paper on Transport) have ambitious goals and challenging 2020 and 2050 targets that represent radical changes to transportation modes as well as the potential for biodiversity restoration.

Current legislation is moving in the right direction, but there are challenges requiring long-term commitments to defragmentation, particularly in regard to roadside habitat management. At Heikamp in the Netherlands, a heathland defragmentation project funded as a compensatory action following destruction of sand lizard (*Lacerta agilis*) habitat from development,

Figure 1.4. A heathland defragmentation project was funded in 2012 as a compensatory action for sand lizards (*Lacerta agilis*) on the Koningsweg (N311), Gelderland Province at Heikamp, the Netherlands, and includes the construction of an underpass for a two-lane road. *Credit: Thomas E. S. Langton.*

enabled the construction of a passage under a two-lane road and cycleway; this passage was aimed at allowing sand lizards and other heathland animals to move between two areas (Figure 1.4). Once constructed, the bordering habitat will need managing in order to connect the fragments. Managing roadside habitats as attractive environments for multiple species in spatial proximity to passages has emerged as an important objective.

1.5. Variation among Species and Landscapes

A review of small tunnel design for amphibian dispersal was conducted at the end of the 1980s at a small conference in Rendsburg, Germany (Langton 1989). Subsequently, awareness increased regarding the difficulty of retroactively maintaining amphibian dispersal patterns once hard infrastructure is built. This challenge is largely due to the dynamic and variable nature of dispersal patterns of not only amphibians but of all wildlife species. Even in situations unaltered by anthropogenic processes, wildlife distribution and dispersal patterns vary among communities, taxa, and species. These patterns may also differ within species among habitats, populations, and individuals, and individual patterns may vary seasonally, as influenced by health and reproductive status. Differing fauna assemblages can have different ecological needs in various climatic zones. For example, a large lake in alpine Europe serves as breeding habitat for six species of amphibians that travel seasonally up to several kilometers. This lake is very different from a tropical, seasonally flooded wetland where 20 amphibian species vary in distribution and density on a regular basis and exhibit small, localized home ranges because they don't have to move as far for the resources needed to survive. This stochasticity and variation are both the challenge and charm of natural systems because they are necessary for wildlife to adapt to changes in the environment. Even where regular patterns are present, we usually lack the field observations and data to understand the mechanisms, a deficit that is pervasive with most small vertebrate species.

This challenge is confounded even further as the movement patterns and impacts of human society and culture also vary over time. Economic fluctuations drive which resources people acquire and how they obtain them. As transportation is one mechanism through which human society derives growth and acquires resources, and resource acquisition varies with society's economic prosperity, this somewhat unpredictable feedback loop influences the number of vehicles and the consequent degree or intensity of wildlife conflicts with roads. This, in turn, affects how we propose to mitigate the effects of public infrastructure on ecological infrastructure. A crossing system that is spatially static cannot adjust without great cost to the dynamic nature of wildlife habitats and human society, and must be designed for a long life and expected changes from both variables.

1.6. Political Challenges and Transportation Planning

Minimizing consequences to biodiversity at the landscape scale when planning roads is rarely a primary objective in most countries around the world. Road construction necessarily reflects short-term human economic demands along with long-term economic growth and resource acquisition needs. Cost models identify the most efficient route from A to B based primarily upon journey time and fuel use. When considerations of ecological infrastructure have been in-

corporated in road building, it has traditionally been a reactive compromise rather than proactive prevention; such incorporation has perhaps accepted as inevitable the major impacts of road fragmentation, pollution, and confounding effects from, for example, climate change. Generally, it has not focused upon mitigation objectives beyond simple mortality reduction. Further, mainstream mitigation practices that typically target a single species may be ineffective in overall biodiversity retention and protection and in contributing significantly to preservation of the economic value of a healthy ecosystem and the services it provides to our society (MEA 2005).

The cost of GI construction can be costly for governments, road builders, and taxpayers if comprehensive wildlife and habitat compensation measures are incorporated. To date, legislation and compliance-based action do not address fully and directly the ecological expense incurred by road construction and maintenance. Therefore, concern for avoidable environmental damage and the pressure to modify plans is largely voluntary and tends only to be sophisticated within global regions with smaller, more politically stable, and wealthy countries (e.g., Europe and North America). Mitigation of effects on wildlife is much less likely in larger countries with new or unstable political conditions or in rapidly growing economies such as India and China, which promote rapid economic growth even if the public gains are unsustainable and short lived.

Slow rates of incorporating small animal passage designs in road-building practices have also resulted in insufficient monitoring of existing designs and little feedback on their effectiveness (e.g., van der Ree et al. 2007; van der Grift et al. 2013). In many instances, it appears that mortality was reduced, which is likely attributed to the mitigation features. However, without data to put those mortality levels in context, it is not certain whether the decreases are attributed to greater connectivity and safe passage or whether they are artifacts of decreased population levels adjacent to the road due to chronic mortality. Stolz and Podloucky (1983) appraised two amphibian tunnel systems and discovered that the majority of the tunnels were not functioning as intended. The authors caution that the incorporation of wildlife tunnel and fence systems should not be used as a justification for supporting road development. This partly reflected a view that single-species solutions, even if possible, are insufficient to fully address the complexity of road construction impacts upon the broader wildlife community.

Rapidly increasing development uses new technology to enable faster and cheaper infrastructure growth but often does not set limits and safeguards against degradation of natural habitats and wildlife communities. Although spatial planning in designated reserve areas such as national parks, wilderness areas, and smaller-scale nature reserves incorporates these planning measures proactively, these reserves still experience barrier and fragmentation effects from roads due to high traffic volumes in peak seasons. They even experience these effects at low traffic volumes and on unpaved roads for particularly sensitive species. As human access is the causal effect of road fragmentation and we cannot restrict human access as a standard practice, how do we build roads that do not harm our ecosystems? Resolution of these conflicts will require scientific experimentation and collaboration among multiple sectors of society in order to learn how to build environmentally sustainable transportation systems.

1.7. Setting Future Goals for Standards and Guidelines for Small Animal Road Ecology

The development of standards and guidelines within and among countries has generally received relatively little attention until recently. Initial guidelines have been produced (e.g., Iuell et al. 2003), and the move toward more detailed specifications for a wide range of structures will likely follow. While many earlier experiences often look like mistakes and poor planning, the lessons learned offer opportunities for more realistic and justified plans and protocols based on improved science and engineering.

Road authorities throughout the world are responsible for promoting the development of standards and guidelines, and the Infra Eco Network Europe (IENE) and International Conference on Ecology and Transportation (ICOET) movements in Europe and North America are now important for everything from policy development to practical applications. However, standards and guidelines are glaringly absent from many of the countries currently building huge numbers of new roads or paving dirt tracks into roads or highways. In

parts of India, China, Africa, and Central and South America, road construction is proceeding at unsustainable rates and often without basic safety measures for people, much less wildlife. It is essential that the global road ecology community address this concern over the next 5 to 10 years. Most important, central governmental organizations and international scientific collaborations must instigate standardized wildlife construction and maintenance measures, such as those promoted by the American Association of State Highway and Transportation Officials (AASHTO 2007). This approach would mean new training for transportation and construction personnel to recognize and protect existing wildlife barriers and to incorporate them into regular routine maintenance, in addition to engineers more routinely adding them to designs for new roads. The Rauischholzhausen Agenda for Road Ecology (Roedenbeck et al. 2007) and the Before-After-Controlled-Impact (BACI) experimental design are valuable approaches to consider for road ecology studies; they may offer better ways to assess small passage effectiveness. An international network of information sharing can be built to help inform the effectiveness of the variety of systems now installed around the world.

In Europe, a network of interested nongovernmental bodies has formed an informal network to better promote road ecology issues relating to amphibians and reptiles (European Network for the Protection of Amphibians and Reptiles from Transport Systems [ENPARTS]). ENPARTS had an initial meeting in Peterborough, United Kingdom, in 2011 and a second meeting in Berlin in 2012. The network is seeking to share information and experiences to better inform the transportation sector and to promote collaboration across Europe. Organizations such as RAVON (Reptile, Amphibian and Fish Conservation Netherlands,), NABU (Nature and Biodiversity Conservation Union, Germany), Froglife (United Kingdom), and Karch (Koordinationsstelle für Amphibien-und Reptilienschutz in der Schweiz, Switzerland) are assessing the potential for collaborative ventures that extend beyond national boundaries. The information handbook (Iuell et al. 2003) produced by the European Union's research exercise COST 341 (Cooperation in the Field of Scientific and Technical Research) is already being updated and revised, an indication of the fast growth in this knowledge base. There is little reason that setting guidelines for protection of small animals cannot grow rapidly and begin to tackle some of the great challenges facing wildlife on transportation corridors in the near future.

LITERATURE CITED

AASHTO (American Association of State Highway and Transportation Officials). 2007. Highway drainage guidelines. AASHTO, Washington, DC, USA.

Anděl, P., I. Gorčicová, V. Hlaváč, L. Miko, and H. Andělová. 2005. Assessment of landscape fragmentation caused by traffic. Systematic Guide. Agency for Nature Conservation and Landscape Protection of the Czech Republic, Prague, Czech Republic.

Berthoud, G., and S. Müller. 1986. Protection des Batraciens le Long Des Routes. Mandat De Recherche No. 26/74, Commission Des Recherches en Matiere de Consruction Des Routes. Department Federale Des Transports, Des Communications et de l'energie, Lausanne, Switzerland. [In French.]

Bissonette, J. A., and P. C. Cramer. 2008. Evaluation of the use and effectiveness of wildlife crossings. NCHRP Report No. 615. National Cooperative Highway Research Program, Transportation Research Board, National Academies Press, Washington, DC, USA. www.trb.org/Main/Public/Blurbs /160108.aspx. Accessed 25 April 2014.

Boarman, W. I., and M. Sazaki. 1996. Highway mortality in desert tortoises and small vertebrates: success of barrier fences and culverts. Pages 169–173 in Transportation and wildlife: reducing wildlife mortality and improving wildlife passageways across transportation corridors. G. Evink, D. Zeigler, P. Garrett, and J. Berry, editors. US Department of Transportation, Federal Highway Administration, Washington, DC, USA.

Bonnin, M. 2008. Les corridors écologiques, Vers un troisième temps du droit de la conservation de la nature, collection Droit du patrimoine culturel et naturel. L'Harmattan, Paris, France. [In French.]

Bonnin, M., A. Bruszik, B. Delbaere, H. Lethier, D. Richard, S. Rientjes, G. van Uden, and A. Terry. 2007. The Pan-European Ecological Network: taking stock. Nature and Environment, No. 146. Council of Europe, Strasbourg, France.

Brehm, K. 1989. The acceptance of 0.2-metre tunnels by amphibians during their migration to the breeding site. Pages 29–42 in Amphibians and roads. Proceedings of the toad tunnel conference, Rendsburg, Federal Republic of Germany. T.E.S. Langton, editor. ACO Polymer Products, Shefford, UK.

Brown, L. E., and A. Mesrobian. 2005. Houston toads and Texas politics in amphibian declines: the conservation status of United States species. Pages 150–167 in Amphibian declines: The conservation status of United States species. Michael Lannoo, editor. University of California Press, Berkeley and Los Angeles, California, USA.

Burghard, A. 1969. Alligator Alley: Florida's most controversial highway. Lanman, Washington, DC, USA.

Coelho I. P., F. Z. Teixeira, P. Colombo, A. V. Coelho, and A. Kindel. 2012. Anuran road-kills neighboring a peri-urban reserve in the Atlantic Forest, Brazil. Journal of Environmental Management 112:17–26.

Creemers, R.C.M., and R. Struijk. 2012. Tunnel systems and fence maintenance, evaluation and perspectives in citizens science. Abstract a66 in Proceedings of the 2012 Infra Eco Network Europe (IENE) international conference, Potsdam, Germany.

Dexel, R. 1989. Investigations into the protection of migrant amphibians from the threats from road traffic in the Federal Republic of Germany—a summary. Pages 43–49 in Amphibians and roads. Proceedings of the toad tunnel conference, Rendsburg, Federal Republic of Germany. T.E.S. Langton, editor. ACO Polymer Products, Shefford, UK.

European Commission. 2011. Our life insurance, our natural capital: an EU biodiversity strategy to 2020 communication from the Commission to the European Parliament, the Council, the Economic and Social Committee, and the Committee of the Regions. Com(2011) 244 final, {SEC(2011) 540 final}, {SEC(2011) 541 final}. European Commission, Brussels, Belgium.

Evink, G. L. 2002. Interaction between roadways and wildlife ecology: a synthesis of highway practice. National Cooperative Highway Research Program, Synthesis 305. Transportation Research Board, National Academies Press, Washington, DC, USA.

FFPS (Fauna and Flora Preservation Society). 1988. Toads on roads. Fauna and Flora Preservation Society, London, UK.

FMENCNS (Federal Ministry for the Environment, Nature Conservation, and Nuclear Safety). 2012. Federal Defragmentation Programme: fundamental principles—fields of action—cooperation. FMENCNS Public Relations Division, Berlin, Germany.

FMTBHRCDRT (Federal Ministry of Transport, Building and Housing, Road Construction Department, Road Traffic), eds. 2000. Instruction sheet for protecting amphibians on roads (MAMS). Forschungsgesellschaft für Straßen- und Verkehrswesen (FSGV; Research Association for Roads and Transport [FSGV]), Cologne, Germany. [In German.]

Foster, M. L., and S. R. Humphrey. 1995. Use of highway underpasses by Florida panthers and other wildlife. Wildlife Society Bulletin 23:95–100.

Franck, N. 2012. Safety net for the wildcat. Friends of the Earth Germany / BUND, Berlin, Germany.

Hagood, S., and M. J. Bartles. 2008. Use of existing culverts by eastern box turtles (Terrapene c. carolina) to safely navigate roads. Pages 169–170 in Urban herpetology. J. C. Mitchell, R. E. Jung Brown, and B. Bartholomew, editors. Herpetological Conservation No. 3. Society for the Study of Amphibians and Reptiles, Salt Lake City, Utah, USA.

Honegger, R. 1977. Unknown . . . unloved . . . threatened. Naturopa 27:13–18.

Hudson W. H. 1919. The book of a naturalist. Hodder and Stoughton, London, UK.

Iuell, B., G. J. Bekker, R. Cuperus, J. Dufek, G. Fry, C. Hicks, V. Hlavac, et al. 2003. COST 341: Habitat fragmentation due to transportation infrastructure; Wildlife and traffic: a European handbook for identifying conflicts and designing solutions. European Co-operation in the Field of Scientific and Technical Research, Brussels, Belgium.

Jędrzejewski W., S. Nowak, R. Kurek, R. W. Mysłajek, K. Stachura, B. Zawadzka, and M. Pchałek. 2009. Animals and roads: methods of mitigating the negative impact of roads on wildlife. Mammal Research Institute, Polish Academy of Sciences, Białowieza, Poland.

Krikowski, L. 1989. The "light and dark zones": two examples of tunnel and fence systems. Pages 89–91 in Amphibians and roads. Proceedings of the toad tunnel conference, Rendsburg, Federal Republic of Germany. T.E.S. Langton, editor. ACO Polymer Products, Shefford, UK.

Lambeck, R. J. 1997. Focal species: a multi-species umbrella for nature conservation. Conservation Biology 11:849–856.

Langton, T.E.S. 1989. Tunnels and temperature: results from a study of a drift fence and tunnel system at Henley-on-Thames, Buckinghamshire, England. Pages 145–152 in Amphibians and roads. Proceedings of the toad tunnel conference, Rendsburg, Federal Republic of Germany. T.E.S. Langton, editor. ACO Polymer Products, Shefford, UK.

MEA (Millennium Ecosystem Assessment). 2005. Ecosystems and human well-being: synthesis. Island Press, Washington, DC, USA.

Müller, S., and G. Berthoud. 1996. Fauna/Traffic Safety: manual for Civil Engineers. École Polytechnique Fédérale de Lausanne, Switzerland.

Podloucky, R. 1989. Protection of amphibians on roads— examples and experiences from Lower Saxony. Pages 15–28 in Amphibians and roads. Proceedings of the toad tunnel conference, Rendsburg, Federal Republic of Germany. T.E.S. Langton, editor. ACO Polymer Products, Shefford, UK.

Reck, H., B. Schulz, and C. Dolnik. 2011. Field guide of Holstein habitat corridors and the fauna passage Kiebitzholm. Holsteiner Lebensraum Korridore. Molfsee, Haselmaus, Hirsch, Molfsee, Germany. [In German.]

Regan, H. M., L. A. Hierl, J. Franklin, D. H. Deutschman, H. L. Schmalbach, C. S. Winchell, and B. S. Johnson. 2008. Species prioritization for monitoring and management in regional multiple species conservation plans. Diversity and Distribution 14:462–471.

Renard, M., A. A. Visser, W. F. de Boer, and S. E. van Wieren. 2008. The use of the "Woeste Hoeve" wildlife overpass by mammals. Lutra 51:5–16.

Roedenbeck, I. A., L. Fahrig, C. S. Findlay, J. E. Houlahan, J.A.G. Jaeger, N. Klar, S. Kramer-Schadt, and E. A. van der Grift. 2007. The Rauischholzhausen agenda for road ecology. Ecology and Society 12:11. http://www.ecologyandsociety .org/vol12/iss1/art11/. Accessed 26 April 2014.

Ryser, J., and K. Grossenbacher. 1989. A survey of amphibian preservation at roads in Switzerland. Pages 7–13 in Amphibians and roads. Proceedings of the toad tunnel conference, Rendsburg, Federal Republic of Germany. T.E.S. Langton, editor. ACO Polymer Products, Shefford, UK.

Savage, R. M. 1935. The influence of external factors on the spawning date and migration of the common frog, Rana temporaria. Proceedings of the Zoological Society of London 105:49–98.

Schmidt, B. R., and S. Z. Zumbach. 2008. Amphibian mortality and how to prevent it: a review. Pages 157–167 in Urban Herpetology. J. C. Mitchell, R. E. Jung Brown, and B. Bartholomew, editors. Herpetological Conservation No. 3. Society for the Study of Amphibians and Reptiles, Salt Lake City, Utah, USA.

Stolz, F. M., and R. Podloucky. 1983. Toad tunnels as a protective measure for migrating amphibians in Lower Saxony. State Administration Office, Fachbeh. Nature Conservation: Conservation Information Service 3, No. 1, Hannover, Germany. [In German.]

Terry, A., K. Ullrich, and U. Riecken, editors. 2006. The green belt of Europe. From vision to reality. International Union for Conservation of Nature and Natural Resources, Gland, Switzerland and Cambridge, UK.

Trocmé, M. 2006. The Swiss defragmentation program— reconnecting wildlife corridors between the Alps and Jura: an overview. Pages 144–149 in Proceedings of the 2005 international conference on ecology and transportation. C. L. Irwin, P. Garrett, and K. P. McDermott, editors. Center for Transportation and the Environment, North Carolina State University, Raleigh, North Carolina, USA.

USFWS (US Fish and Wildlife Service). 2012. Innovative approaches to wildlife/highway interactions. US Fish and Wildlife Service training video, Shepherdstown, West Virginia, USA. http://nctc.fws.gov. Accessed 25 April 2014.

van der Grift, E. A., F.G.W.A. Ottburg, and R.P.H. Snep. 2010. Monitoring wildlife overpass use by amphibians: do artificially maintained humid conditions enhance crossing rates? Pages 341–347 in Proceedings of the 2009 international conference on ecology and transportation. P. J. Wagner, D. Nelson, and E. Murray, editors. Center for Transportation and the Environment, North Carolina State University, Raleigh, North Carolina, USA.

van der Grift, E. A., R. van der Ree, L. Fahrig, S. Findlay, J. Houlahan, J.A.G. Jaeger, N. Klar, et al. 2013. Evaluating the effectiveness of road mitigation measures. Biodiversity and Conservation 22:425–448.

van der Ree, R., E. van der Grift, N. Gulle, K. Holland, C. Mata, and F. Suarez. 2007. Overcoming the barrier effect of roads—how effective are mitigation strategies? Pages 423–431 in Proceedings of the 2007 international conference on ecology and transportation. C. L. Irwin, D. Nelson, and K. P. McDermott, editors. Center for Transportation and the Environment, North Carolina State University, Raleigh, NC, USA.

2

Natural History and Physiological Characteristics of Small Animals in Relation to Roads

Scott D. Jackson, Tom A. Langen, David M. Marsh, and Kimberly M. Andrews

2.1. Natural History and Physiology of Movement

When thinking about animal movement, the great animal migrations often come to mind—caribou (*Rangifer tarandus*) across the tundra, neotropical birds from South to North America, bison (*Bison bison*) up and down western mountain slopes. However, movement is equally important for small animals like amphibians, reptiles, and rodents; their movements just occur at smaller scales. In fact, regular movement is critical to the survival of most small animals. Movements take animals to breeding sites, to winter sheltering sites, and to new habitats. To the extent that roads interrupt these movements, they can reduce the abundance of small animals, and over the long term, eliminate species from the landscape.

This chapter focuses on how the natural history and physiology of small animals affects their movement and vulnerability to roads. Small animals, including all amphibians and reptiles, have been noted to be particularly susceptible due to life history traits that predispose them to a suite of road effects (e.g., Rytwinski and Fahrig 2012). Caudates (salamanders) were found to exhibit a more negative response than anurans (frogs and toads), likely due to even lower reproductive rates. Anurans were found to be additionally vulnerable due to smaller body sizes and shorter periods of sexual maturity. Furthermore, life history traits of turtles and squamates (lizards and snakes), such as long lives, high adult survivorship, low rates of recruitment, and delayed sexual maturity, greatly increase their sus-

ceptibility to road effects. Here, we discuss how, why, and when different groups of small animals move, and how these movements may contribute to the relative vulnerability of different species to the effects of roads. We also consider the physiology of movement—that is, what kinds of conditions allow small animals to move efficiently and what conditions prevent movement from occurring. Although physiological considerations can vary widely from one species to the next, a solid understanding of physiological issues is necessary to design effective mitigation strategies that increase animal movement through passages or that reduce the tendency of animals to cross roads in the first place.

From an ecological standpoint, attention to spatial scale is critical to understanding the functions of animal movement. Ecologists typically think about animal movement at the scale of the individual animal, at the scale of the population, or at the scale of the metapopulation. Roads may affect movement at any one of these scales, or at all of them simultaneously. To understand the effects of roads on animals, it is important to first consider the ecological functions of animal movement across a range of scales.

2.1.1. Importance of Movement for Individuals

Movement is one of the more energetically expensive activities an animal can undertake. In addition, movement can expose animals to the risk of predation and physiological stress. Thus, animal movements must yield important benefits that outweigh these considerable costs. At the smallest scale, animals make daily movements within their home ranges. These movements may be associated with territory defense, with foraging, or with thermoregulation. Because small animals typically have small home ranges, these daily movements are unlikely to take individual animals onto roads. But small animals also do move long distances relative to their body size. Some of these longer movements are seasonal and can include migrations to breeding sites, migrations to sites with habitat needed to survive cold winters or hot summers, or migrations to seasonally shifting feeding sites. Other movements are associated not with seasons, but with specific life stages. For example, amphibians that metamorphose at aquatic breeding sites may carry out mass migrations to terrestrial habitats that are more suitable for adults. These are important movements (see also Section 2.4), because individuals that cannot get to breeding sites will fail to breed, individuals that cannot get to feeding sites may starve, and individuals that cannot get to overwintering shelter may freeze. Seasonal and stage-specific movements are thus essential to the survival and reproduction of the individual animals who undertake them.

2.1.2. Importance of Movement for Populations

Movement for animals, or the disruption of those movements, is not only important for individual animals but also has consequences at the population level. Movement within a population can affect every important aspect of population demography, including immigration and emigration rates, survival rates, and reproduction. Movement of individuals is also required to avoid genetic fragmentation within a population. Barriers to movement (even incomplete barriers) can result in demographic independence (e.g., birth and death rates, age structures, or altered sex ratios) and cause genetic differentiation between population segments separated by the barrier (e.g., Lowe and Allendorf 2010). The movement of individuals away from a population is important for density regulation and maintaining the population within the carrying capacity of the environment. The mountain pygmy-possum (*Burramys parvus*) of Australia is a small mammal with a life history that, under normal circumstances, involves the segregation of male and female home ranges except during mating. This separation helps maintain adequate resources within the females' home ranges to

raise the young. At one site where a newly constructed road prevented males from leaving the females' territories, reproductive success dropped significantly (Mansergh and Scotts 1989).

If movement-related mortality increases, population size will be expected to decrease. Similarly, if breeding success is reduced by restricted access to breeding sites, populations will be expected to decline. It should be noted that increased mortality or decreased breeding success does not always lead to population declines. If, for example, the size of a population is limited by the availability of breeding habitat, then increased adult mortality might simply open up breeding opportunities for other individuals. For this reason, effects of roads should optimally be evaluated both at the level of the individual (i.e., mortality rate) and of the population (i.e., population size), an issue considered in more detail in Chapters 8 and 12. However, when the links between individual and population effects are poorly understood (as they will be for most species), start with the assumption that increased mortality and decreased reproduction will tend to harm populations.

In addition to population declines, a host of other negative effects may be observed in populations where mortality increases or where movements are impeded by barriers (reviewed in Chapter 3). These effects include skewed sex ratios in species where one sex moves more than the other and skewed age distributions where one life stage is the primary disperser. Population consequences can also include sublethal effects such as reduced body condition or increased levels of stress hormones. Ultimately, these kinds of effects should lead to population declines. Sublethal effects can serve as early indicators of negative effects of reduced movement.

2.1.3. Importance of Movement for Metapopulations

Individual populations are typically linked to one another by occasional long-distance movements. Although long-distance movements may be infrequent, they can nevertheless be critical for the long-term persistence of a set of linked populations, or metapopulation (Levins 1969; Hanski 1999). Small populations of amphibians, reptiles, and small mammals may naturally exist where patches of habitat are themselves small. For example, many vernal pools are quite small and are unlikely to support the large number of individuals necessary to prevent a deterioration of genetic health due to genetic drift or inbreeding depression. As a general rule of thumb, population geneticists believe that the immigration of as few as one migrant per generation can maintain the genetic health of even small and isolated populations (Mills and Allendorf 1996).

Ecologists have long noted that the ultimate fate of any isolated population is extinction. Chance events like droughts or floods can eliminate a local population, or a population can disappear due to successional changes in its preferred habitat. However, when populations are linked together by occasional long-distance dispersal, suitable habitats will ultimately be colonized or recolonized. Thus, even though individual populations may go extinct, a metapopulation can persist through a balance between local extinction and colonization from nearby populations. When long-distance movements are a little more frequent, migrants can even potentially prevent populations from going extinct in the first place, a process known as the "rescue effect" (Brown and Kodric-Brown 1977).

One real-world implication of metapopulation dynamics is that long-distance movements can prevent extirpation (local extinction) over the long term even when these movements are infrequent. As a result, short-term studies may be unable to detect the kinds of movement that support metapopulation persistence. In fact, even thorough studies of animal movement often underestimate the frequency and extent of movements by the longest-distance dispersers—resident animals may regularly move outside the study area and immigrants may infrequently enter into populations. Although road mitigation strategies are often developed one road at a time, long-term persistence of animal populations requires consideration of metapopulation processes (i.e., long-distance movement) and the maintenance of movement corridors between populations. The potential effects of roads on metapopulations at these larger scales are considered in more depth in Chapter 3.

2.1.4. Case Study: Amphibian Use of Ephemeral, Isolated Ponds—Implications for Impacts of Roads and Highways

Scott D. Jackson and Cathryn H. Greenberg

Developing strategies for mitigating the effects of roads on amphibian populations is complicated by the temporal (seasonal or year-to-year variation) and spatial (variation based on location or landscape context) variability of those populations. Many amphibian species breed in ephemeral ponds that are separated by an intervening matrix of terrestrial habitat. An important objective of many amphibian road passage projects is to maintain ecological infrastructure by facilitating movement of animals between breeding and nonbreeding habitats. To be effective, these projects need to take into account the timing of migrations, distances traveled, and spatial arrangement of habitat patches for both adult and juvenile life stages and for multiple species. Another important aspect over the long-term is the occasional movement of individuals from one breeding population to another. These movements have the potential to knit local populations centered on individual breeding ponds into metapopulations and enhance population persistence over time (Marsh and Trenham 2001).

Capture data from a long-term (1994–2007) mark-recapture study of amphibians in longleaf pine-wiregrass sandhills in the Ocala National Forest, Florida, provide insight into the seasonal variability in movement for a number of amphibian species. Over a 14-year period of monitoring 8 small (0.1–0.37 ha), isolated, ephemeral sinkhole ponds, nearly 50,000 individuals of 22 amphibian species were captured. Pond use by adults as well as successful juvenile recruitment were both highly erratic for all species among both ponds and years (Greenberg and Tanner 2005a). Absence of a species from a particular pond for one or more years was not an indicator that the pond would not be used as a breeding site or serve as an important source of juvenile recruitment in later years.

Daily and seasonal variation in traffic volume on roads near breeding sites will influence mortality for amphibian species differently, depending on interspecific differences in temporal patterns of movement by adults and newly transformed juveniles (i.e., metamorphs). We found differences among species in the temporal patterns of both adult migrations to and from breeding sites and juvenile emigration from ponds after metamorphosis, with corresponding differences in vulnerability to road mortality (Greenberg 2001; Greenberg and Tanner 2004, 2005b, 2005c). The long-term data suggest that roads with high traffic volumes that are located between ephemeral breeding ponds and upland habitats are likely to adversely affect local amphibian populations, but effects may differ among species and years. For example, a large proportion of local eastern spadefoot (*Scaphiopus holbrookii*) populations could be affected by heavy road mortality in a single night during explosive breeding events or juvenile emigration periods.

The results of a long-term study of marbled salamanders (*Ambystoma opacum*) in western Massachusetts also provide information about dispersal movements among populations and how they may affect the fate of local populations. This landscape-level investigation monitored breeding populations of marbled salamanders in 14 ephemeral ponds over a 7-year period (Gamble et al. 2007). By tracking individual adult salamanders (via pattern recognition) and metamorphs by pond (via cohort-specific markings), scientists were able to calculate dispersal probabilities for first-time breeders and established breeders (adults documented breeding in at least one previous year). The study found that although marbled salamanders demonstrated a high degree of natal site fidelity (91.0%) and breeding site fidelity (96.4%), there was also some movement of individuals from one population to another (Figure 2.1). Among established breeders, a relatively small proportion (3.6%) dispersed to breeding ponds other than where they were first recorded as breeding adults. Dispersal distances ranged from 105–439 m and were best modeled with a normal distribution centered on the pond (distance = 0 m) and having a standard deviation of 331.5 m, meaning that about 68% of dispersal distances were predicted to fall between 0–331.5 m. Nine percent of first-time breeders dispersed to a pond other than their natal pond. Dispersal distances ranged from 142–1,297 m (standard deviation = 440.1 m) with 16% of those first-time breeders dispersing to ponds over 1,000 m from their natal ponds (Gamble et al. 2007).

All marbled salamander populations involved in this study were relatively small with effective population sizes where genetic drift and inbreeding depression (resulting in reduced genetic diversity) were potential concerns. However, rates of dispersal documented by this study were generally sufficient to counteract the genetic consequences of small population size. Presumably these dispersal movements also reinforce populations with low rates of native recruitment and provide opportunities to recolonize breeding ponds after local extinction events (Gamble et al. 2007).

Distance distributions for first-time breeders and emigrating juveniles were similar, suggesting that the first season after emergence may be a critical time for dispersing individuals (Gamble et al. 2006, 2007). Thus, the population-stabilizing benefits of dispersal are dependent to a large degree on the ability of small, juvenile salamanders to move, sometimes long distances, through the landscape. Strategies aimed at conserving amphibians that breed in ephemeral ponds not only have to take into account the breeding migrations of adults and emigration of metamorphs, but also the long-distance dispersal movements of juveniles.

Among the groups of vertebrate animals that are the focus of this book (amphibians, reptiles, and small mammals of North America), a handful of species can be considered to be fully aquatic at all life stages, meaning that they have a very limited capacity to move outside the medium of water. Among these are salamanders from families that are made up entirely of aquatic species: hellbenders (Cryptobranchidae), mudpuppies and waterdogs (Proteidae), sirens (Sirenidae), and amphiumas (Amphiumidae). Salamanders from other families can exhibit paedomorphism (i.e., mature adults retaining larval characters such as external gills) where individuals or populations can be considered fully aquatic. These include Cope's giant salamanders (*Dicamptodon copei*), Oklahoma salamanders (*Eurycea tynerensis*), Mexican axolotls (*Ambystoma mexicanum*), eastern tiger salamanders (*A. tigrinum*), and Pacific giant salamanders (*D. ensatus, D. aterrimus,* and *D. tenebrosus*).

Other amphibians are not fully aquatic but many do have aquatic life stages. For many amphibian species in North America, the larval life stage is aquatic. Little is known about how far larval amphibians travel. However, aquatic movement is probably important for meeting seasonal requirements (e.g., feeding, overwintering) and avoiding adverse habitat conditions (e.g., desiccation, thermal stress), particularly for some species such as spring salamanders (*Gyrinophilus porphyriticus*) and Pacific giant salamanders (*Dicamptodon* spp.) with extended larval periods of 18 months to 5 years.

There are many species of amphibians, reptiles, and small mammals that can be considered semi-aquatic, meaning that they readily use aquatic environments

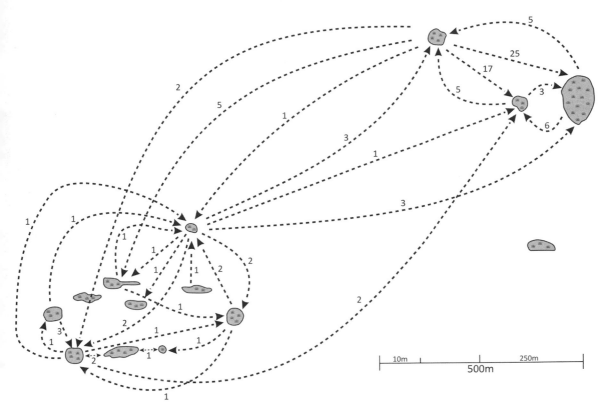

Figure 2.1. Dispersal of marbled salamanders (*Ambystoma opacum*) among breeding populations, displayed by origin and destination ponds over a six-year period. *Credit: © Victor Young Illustration, adapted from Gamble et al. 2007.*

but are capable of moving out of water. These species are generally found close to water and are likely to use water bodies and waterways as a primary means of moving through the landscape. Included in this group are stream salamanders and frogs, semi-aquatic snakes, many species of turtles, and a number of mammals including muskrat (*Ondatra zibethicus*), American mink (*Neovison vison*), water shrew (*Sorex palustris*), and star-nosed mole (*Condylura cristata*). Although these species are capable of moving overland, the best strategy for facilitating their movement across roads is to focus on aquatic crossing structures (Section 9.3).

More generally, river and stream networks are important for many species of small animals, including those that are not considered to be either aquatic or semi-aquatic. Rivers and streams and their associated wetlands represent important sources of water for amphibians, particularly in arid regions. Riparian zones in arid areas also may contain denser vegetation than the surrounding environment and therefore may support larger populations of reptiles and small mammals. In

more developed landscapes, river and stream corridors often represent the last remaining areas of contiguous, undeveloped habitat. Riparian areas can function as corridors of interconnected habitat important for maintaining continuous populations, and facilitating gene flow among populations. Road-stream crossings can be significant constriction points along these networks, and careful design of crossing structures and use of mitigation measures at these locations can benefit many species of small vertebrates.

2.1.5. Small Vertebrate Movement in Aquatic Environments

Little is known about the swimming abilities of small vertebrates, apart from fish. It is likely that most, if not all, of them are comparatively weak swimmers relative to fish. Likewise, with the exception of frogs, small aquatic or semi-aquatic vertebrates are not known for their leaping ability. The altered flow velocities and inlet or outlet drops that are common features of culverted stream crossings are likely to represent severe barriers to passage for most aquatic small vertebrates. Whereas semi-aquatic species might have some success crossing the road itself, for fully aquatic species, poorly designed and placed culverts are likely to fragment and isolate populations. For semi-aquatic species that are forced to cross the road, individuals are exposed to the risk of road mortality.

Many species of amphibians and reptiles and some small mammals, such as water shrews and star-nosed moles, are likely to be crawlers or short-burst swimmers in stream environments. They move along the substrate or bank edges among cover objects as they travel along stream channels. The movement of many aquatic species depends on the diversity of channel structure (e.g., woody debris, rocks, boulders, and bedrock outcrops) and hydrological features (e.g., scour pools, areas of variable velocity flow) typically found in natural streams. This diversity creates a variety of pathways within the channel and along the bank. Although some locations in natural channels may not be traversable (e.g., due to high velocity or high turbulence zones), other more appropriate pathways are generally available. For aquatic species or life stages, stream substrate (rocks, gravel, sand, or other materials that cover the stream bed) serves three principal functions with regard to movement. First, it provides cover for small animals that are vulnerable to predation when exposed in the water column (or in a culvert that lacks substrate). Second, substrate on the stream bottom and bank edges offers areas of lower water velocity where crawling and weak-swimming species can more easily move against the current. Lastly, substrate provides areas of quiet water (pools within stream channels) that animals can use between bouts of swimming, crawling, or breeding. Continuity of substrate (natural substrate extending the entire length of a culvert or bridge; see Chapter 9) is one of the most important features of road-stream crossings for aquatic small vertebrates.

Some movements of semi-aquatic species may occur in the water while other movements, as well as those of terrestrial species that utilize aquatic crossings, occur on the drier areas along the banks. Characteristics of the bank substrate again are likely to be important for facilitating the movement of some semi-aquatic small vertebrates. The absence of banks in the area where a stream passes through a culvert or beneath a bridge may inhibit crossing by these species. Some culverts

may appear passable, but when roadkill occurs above these culverts, it suggests that the current designs are inadequate to address species sensitivity to size, light, or openness within a structure (Chapter 9). This may be true for aquatic species and life stages as well as for semi-aquatic species.

2.2. Characteristics of Terrestrial Small Vertebrates

As used in this book, the term "terrestrial small vertebrate" refers to amphibians, reptiles, and small mammals that are not strongly associated with water. In practical terms, this means that movement patterns and pathways are not closely tied to water courses and wetlands. Although all amphibians are vulnerable to dehydration when exposed to dry conditions, many species including toads (Bufonidae) and spadefoots (Scaphiopodidae) are largely terrestrial and are not typically found in wetlands or water bodies (except for breeding). Many species of terrestrial salamanders in the family Plethodontidae are not dependent on wetlands for reproduction.

Although terrestrial species are generally not tied to aquatic features, riparian corridors may be the only travel pathways available in some landscapes (e.g., arid and highly developed areas). These corridors may be the only areas within the landscape where vegetation, moist soil conditions, or water are available, making them attractive pathways for terrestrial species. Where roads bisect wetlands or streams, mitigation measures such as ramps, elevated shelves (Figure 2.2), dry passage along banks, or benches may be necessary within crossing structures to facilitate passage by terrestrial species (Chapter 9).

2.2.1. Substrate

Many small vertebrates have preferences for particular microhabitats. The availability and composition of substrate, as one element of microhabitat, affects how these animals move through the landscape. For example, juvenile natterjack toads (*Epidalea calamita*) move more efficiently across sandy soil that is characteristic of their preferred habitat than across other substrates and cover types (Stevens et al. 2004). A study of eastern hog-nosed snakes (*Heterodon platirhinos*) found that they readily crossed unpaved sand roads

Figure 2.2. Structure for streamflow under a road in the Netherlands; the shelf inside the structure facilitates movement of terrestrial small animals. *Credit: Scott D. Jackson.*

but avoided crossing paved roads (Robson 2011). Moles (Talpidae) spend most of their time underground and have a specialized limb anatomy that makes it difficult to move on the surface. Paved surfaces or compacted surfaces of dirt roads are likely to present significant challenges for moles. Common wormsnakes (*Carphophis amoenus*), legless lizards (*Anniella* spp.), and amphisbaenians (worm lizards) are burrowing reptiles that often move underground. Other small mammals, such as pikas (*Ochotona* spp.) and some ground squirrels (Sciuridae), and some salamander species live in rocky substrates, such as talus, and move through spaces among rocks. These species may be reluctant to venture forth over flat ground surfaces because such behaviors expose them to predators. In Australia, endangered mountain pygmy-possums inhabit rocky talus slopes. To facilitate passage for these marsupials, stones were added to large culverts to simulate their preferred microhabitat (Mansergh and Scotts 1989).

Reptiles adapted to sandy or rocky substrates may be vulnerable to the interruption of natural substrate caused by roads and highways and may be reluctant to use wildlife passage structures that do not resemble conditions found in their preferred microhabitats. Specialists of Florida scrub habitats (dry, sandy communities characterized by evergreen shrubs and patches of bare sand), such as Florida sand skinks (*Plestiodon reynoldsi*) and Florida scrub lizards (*Sceloporus woodi*) as well as lizards that are adapted to rock face and boulder habitats may be particularly vulnerable to the fragmenting effects of roads. To accommodate these species, traditional crossing structure designs may be

inadequate, and creative modifications with these specialized needs in mind may be necessary.

2.2.2. Cover and Exposure to Predators

As a group, small vertebrates occupy a wide range of environments. Some are generalists and some are specialized to use particular microhabitats. Meadow voles (*Microtus pennsylvanicus*) and northern leopard frogs (*Lithobates pipiens*) may be most comfortable traveling through grasslands, while flying squirrels, wood frogs (*L. sylvaticus*), and timber rattlesnakes (*Crotalus horridus*) prefer forested environments. Thus, it is not surprising that woodland species such as white-footed mice (*Peromyscus leucopus*) and eastern chipmunks (*Tamias striatus*) are rarely found in or travel through open grassland environments (Wegner and Merriam 1979). Many wildlife species take the shortest distance possible when crossing a road, suggesting that they perceive the road as a potentially threatening environment. In an experiment investigating crossing propensities and speeds in snakes, no species deviated from a 90° angle when crossing (Andrews and Gibbons 2005). The propensity of animals to travel through less familiar microhabitats may be related to the mechanics of movement and the efficiency with which they can cross various substrates (Stevens et al. 2004; Zajitschek et al. 2012), or it may be related to a general reluctance to expose themselves to predation.

Although many small vertebrates are predators themselves they are also vulnerable to predation because of their size. Some small vertebrates are toxic or unpalatable to predators but most avoid exposure by relying on cryptic coloration and the use of cover. For some species, the composition and structure of vegetation may be a critical factor in avoiding predators. Cover objects (e.g., leaf litter, rocks, downed woody debris) not only provide some measure of protection from predators but also create favorable temperature and moisture conditions. Natural grassy environments offer excellent cover for green snakes (*Opheodrys* spp.), leopard frogs, and meadow voles that move through dense vegetation on the ground surface. Wood frogs and eastern chipmunks typically rely on leaf litter and downed woody debris for cover in forests. Roadside vegetation on the approaches to wildlife passage structures may be an important factor in determining which species are likely to cross a road, either over the road surface or through a passage structure. The presence of cover on the approaches, in the form of vegetation, rocks, and logs, may enhance passage use by a variety of small and mid-sized mammals (Hunt et al. 1987; Rodriguez et al. 1996; Rosell et al. 1997; Santolini et al. 1997; Clevenger and Waltho 1999).

Roads and most wildlife passage structures that lack vegetation or other cover serve as impediments to movement for many small vertebrates. Some amphibians, reptiles, and small mammals are reluctant to cross even relatively small and unpaved roads, presumably because of behavior adaptations designed to reduce predation risk (e.g., Case Study 9.6). For example, a narrow, seldom-used vehicle path was found to restrict movements of prairie voles (*Microtus ochrogaster*) and, to a lesser degree, hispid cotton rats (*Sigmodon hispidus*; Swihart and Slade 1984). Mader (1984) found that even forest roads closed to public traffic were significant barriers to two species of woodland mice (*Apodemus flavicollis* and *Chlethrionomys glareolus*). Data collected by Oxley et al. (1974) indicated that white-footed mice and eastern chipmunks rarely crossed roads, and when they did, they preferred roadsides with cleared areas of ≤30 m. They concluded that roadside clearance (rather than surface type or traffic volume) is the most important factor inhibiting the movement of forest mammals. More recently, Brehme et al. (2013) tracked four small mammal and two lizard species using florescent powder near a dirt road, a low-use paved road, and a rural two-lane highway in California. Two of the small mammals avoided crossing even the low-use paved road, and all species avoided crossing the rural highway.

Meadow voles, although abundant around the entrances to wildlife passage structures on Route 93 in Montana, United States, were not documented using the structures until small diameter pipes ("vole tubes") were installed (Foresman 2002, 2003). Presumably, these tubes simulated the tunnels that voles typically use to move about while limiting their exposure to predators. Some small animals avoid enclosed spaces that offer concealment cover for predators or areas that are frequently used by predators. Downes (2001) found that snake predator scent resulted in changes in behavior, mobility, and microhabitat use in garden skinks (*Lampropholis guichenoti*). There is some concern that predator use of wildlife passage structures may inhibit

prey species use, but there is not yet conclusive evidence that animals (small or large) actively avoid crossing structures frequented by predators.

2.2.3. Arboreal Connections

In Australia, roads and highways may represent significant barriers to movement for arboreal (tree-dwelling) mammals that rarely move across the ground (Andrews 1990). Numerous species of arboreal squirrels, salamanders, and treefrogs occur in North America. There are also tree-dwelling snakes, such as the rough greensnake (*Opheodrys aestivus*) and eastern ratsnake (*Pantherophis alleghaniensis*), as well as arboreal lizards that might be adversely affected by significant breaks in the tree canopy. However, it has not been documented whether any North American arboreal small vertebrates are deterred by gaps in tree or shrub canopy cover such as those caused by roads.

Many arboreal species move across the landscape by traveling from tree to tree through the canopy. Allowing tree branches to overhang roads may allow arboreal species to cross these roads without having to leave the trees. Utility companies and highway departments are increasingly clearing trees from roadsides and trimming branches to protect utility lines and increase sunlight exposure to reduce the buildup of snow and ice. These maintenance activities might be contributing to the fragmentation and isolation of tree-dwelling small vertebrate populations.

2.2.4. Moisture

In general, amphibians have permeable skin through which a significant amount of gas exchange occurs in place of, or in addition to, lungs or gills. In fact, the largest family of salamanders in the world, Plethodontidae, is entirely comprised of lungless salamanders. The Americas are the center of adaptive radiation for this family and many lungless salamander species occur in North America and play an important ecological role as abundant predator and prey species. The eastern redbacked salamander (*Plethodon cinereus*) accounts for more biomass in forested ecosystems than any other vertebrate (Burton and Likens 1975).

The trade-off for having permeable skin is that amphibians are more vulnerable than other vertebrates to dehydration when exposed to dry conditions. As a result, amphibians typically occur in moist environments, and those few species that dwell in drier habitats such as deserts demonstrate unusual and fascinating adaptations to avoid desiccation. When amphibians move through upland areas, they typically do so during or after rain events or during periods of high humidity.

Elongate amphibians, such as salamanders, have higher surface area to volume ratios and therefore tend to lose water more rapidly than more compact amphibians like toads. Small amphibians have more surface area relative to volume than larger amphibians. As a result, juvenile frogs and salamanders are more vulnerable to dehydration than adults of the same species. Although adult amphibians often breed during wet, rainy times of the year, newly metamorphosed juveniles often disperse from breeding areas during drier periods. The migrations of amphibians that take advantage of ephemeral water bodies for breeding are often triggered by seasonal or episodic rainy periods. For example, spadefoots typically migrate to highly ephemeral breeding sites during periods of unusually rainy weather or heavy downpours. Rainfall is an important stimulus for movement for these species. Rainy weather is also associated with increased nocturnal activity for small mammals (Doucet and Bider 1974; Vickery and Bider 1981).

Migrating amphibians risk increased water loss when they encounter developed areas that are unshaded and contain impervious surfaces (Mazerolle and Desrochers 2005). The lack of shade over the impervious surface of many roads means that these areas typically dry out more quickly than the surrounding landscape (Marsh and Beckman 2004). Dry conditions on and around roads present a significant risk of dehydration for migrating amphibians, especially for small species and juveniles. Maintenance of moist substrate conditions within wildlife crossing structures (Chapter 9) is likely to reduce stress and mortality risk for frogs and salamanders and provide appropriate conditions and stimuli for movement of amphibians and small mammals.

2.2.5. Temperature

All organisms have preferred temperature ranges within which they can optimally function (Brattstrom 1979). Temperatures outside those ranges will typically trigger behavioral or physiological stress responses, and animals may reduce their activity levels as they seek to avoid extreme temperatures or save energy. As temperatures move further out of those tolerable ranges, mortality risk increases.

Most mammals are endothermic, able to regulate their body temperatures physiologically, whereas amphibians and reptiles are ectothermic, relying on behavior to regulate body temperature using heat primarily from external sources. When their core temperature drops, ectotherms move to warmer locations to increase their body temperature. When conditions are too warm, they will seek shade, water, or the shelter of burrows or other cooler areas. Reptiles can warm themselves by basking in direct sunlight. Amphibians will bask while positioned in shallow water or on wet substrate, seek warmer areas of water bodies (shallows or surface water during sunny days), or spend time under rocks or other cover objects exposed to the sun. Small reptiles with a higher vulnerability to predation will also employ these indirect basking strategies.

For amphibians and reptiles, metabolism and activity levels are dependent on body temperature (Duellman and Trueb 1986; Zug 1993). If temperatures are cold, metabolism and activity levels slow down. Reduced metabolic activity can affect digestion, growth, and the speed at which animals can move (Huey and Stevenson 1979; Hertz et al. 1982; Rocha and Bergallo 1990). Amphibians and reptiles are not capable of moving the same distances during cold weather as they would be under more favorable temperatures. When temperatures are too hot, animals may avoid moving. These adaptations to extreme temperatures can reduce foraging time and may put individuals and populations at risk (e.g., Sinervo et al. 2010).

The sunny, exposed condition of roads and road shoulders may provide basking opportunities for amphibians and reptiles, especially for lizards and snakes. In temperate regions, forest edges provide important thermoregulatory habitat for reptiles (Blouin-Demers and Weatherhead 2002). Road shoulders may provide attractive basking habitat as well as nesting opportunities for reptiles but may serve as ecological traps (Chapter 4; Practical Example 1). On sunny days, road surfaces often heat up faster than surrounding areas, potentially offering opportunities for reptiles to warm themselves on the pavement or gravel (McClure 1951; Sullivan 1981; Ashley and Robinson 1996), though some may be attempting to cross and then halt in response to traffic (Andrews and Gibbons 2005; Chapter 4). These surfaces can also retain heat after sunset providing warm substrates that are both attractive to and used by amphibians and reptiles.

2.2.6. Light

Light coming from sunlight or celestial sources is a primary determinant of activity periods and levels for visually oriented animals. Small vertebrates may be diurnal (active primarily during the day), nocturnal (active at night), or crepuscular (active in the twilight hours of dusk and dawn). Light also serves as a navigation cue for some species. Based on a study of wood frogs and spring peepers (*Pseudacris crucifer*), Halverson et al. (2003) concluded that the amount of light available at the breeding pool (as thermal energy) is a strong predictor of distribution and reproductive success for amphibians.

It is not uncommon for the activity patterns of small nocturnal animals to track lunar cycles, with reduced activity during brighter moon phases, presumably to reduce risk of predation (Kotler et al. 2002). For example, aquatic snakes respond to lunar cycles by reducing activity during periods around the full moon (Madsen and OsterKamp 1982; Houston and Shine 1994). Effects of the moon phase have also been documented for a number of toads in the genus *Anaxyrus*, with some species being more active during the full moon and others more active during the new moon (Church 1960a; Church 1960b; Ferguson 1960; Church 1962; Fitzgerald and Bider 1974). Small mammals also exhibit activity patterns that are associated with phases of the moon (Jahoda 1973; Wolfe and Summerlin 1989; Daly et al. 1992).

One of this chapter's authors (S. D. Jackson) has observed that spotted salamanders (*A. maculatum*) are attracted to light during migrations to breeding sites. In the process of monitoring a pair of salamander tunnels in Amherst, Massachusetts, it was observed that providing artificial light (a flashlight beam directed into the tunnel) was sufficient to eliminate the salamanders' hesitation to use the tunnels (Jackson 1996). While tagging animals as part of this study, Jackson observed that salamanders could be diverted from their migratory path and made to follow a flashlight beam as it moved along the forest floor. Although spotted salamanders

are generally believed to shun artificial light, especially while in their breeding pools, it may be that light is one of the cues used by these salamanders to locate breeding pools in the late winter and early spring (a currently untested hypothesis).

Hatchling sea turtles use light as an orientation cue to direct them to the ocean (Carr and Ogren 1960). Studies have shown that artificial light from ocean-front hotels and streets misdirect hatchling turtles away from the ocean and into these developed areas, resulting in large numbers of turtle roadkills (McFarlane 1963; Witherington and Martin 1996; Tuxbury and Salmon 2005).

Little is known about how artificial lighting specifically associated with roads and highways may affect small vertebrates. It has been suggested that frogs lured by abundant of insect prey beneath street lights might be more susceptible to road mortality (Ferguson 1960; Baker 1990). Baker and Richardson (2006) found that artificial light caused green frogs (*L. clamitans*) to produce fewer advertisement calls and to move more often than under normal lighting conditions. Rapid increases in illumination such as that produced by the headlights of a passing vehicle slowed visual foraging in Cope's gray treefrogs (*Hyla chrysoscelis*; Buchanan 1993). Mazerolle et al. (2005) found that American toads, spring peepers, green frogs, and wood frogs were more likely to become immobile on the road when they encountered a vehicle-caused disturbance, possibly due to the effects of light (although noise may also have been involved). More research is needed on the effects of artificial light associated with roads and highways on small vertebrates and on how such light might be used to facilitate movement through wildlife passage structures.

2.2.7. Sound

Hearing is an important way for small vertebrates to perceive their environment, detect threats or potential prey; making and perceiving sound is also how they communicate with others of the same or different species. The use of sound is variably stronger or more weakly developed depending on the species. Snakes perceive sound primarily by detecting vibrations with their skin and jaw bones. Many of the species covered by this book are relatively mute, but others rely on sound—primarily vocalization—for communication, especially anurans (frogs and toads). Some species of turtles have recently been shown to vocalize at levels inaudible to humans (Ferrara et al. 2013).

Many frogs and toads have well-developed vocal abilities and produce advertising calls to attract mates or territorial calls to claim and defend territories. In some species, the calls combine into a loud, dense chorus. For other species, individuals are more dispersed and their calls can get lost among the loud choruses of other species. Some frogs such as pickerel frogs (*L. palustris*), northern red-legged frogs (e.g., *Rana aurora*), and Columbia spotted frogs (e.g., *R. luteiventris*) call underwater, which mutes the amplitude of the call and limits the distance outside of the water over which it can be detected. Thus, some anuran species may be more vulnerable than others to the negative effects of noise pollution.

Roads and highways are sources of noise that range from sporadic to loud and constant. Given the vulnerability of small animals to predation, any sudden and episodic noise such as the sound of a passing vehicle or vehicles hitting an expansion joint on a bridge could produce a startle response and affect the likelihood of an animal to cross over a road or under a bridge. Road vibrations also have the potential to interfere with sensory perception in snakes. One concern about open-top passage structures designed to let in moisture and light for amphibians is the potential for road noise transmitted into the structure to disrupt the movement of other taxa. If the traffic noise is loud or constant, it can cause hearing loss and an increase in levels of stress hormones, and can interfere with breeding communications (Dufour 1980; Forman and Alexander 1998). Road-generated noise can also disrupt cues necessary for amphibians to orient themselves and navigate to breeding sites. Lastly, vibrations can serve as a cue for some amphibians, such as spadefoot toads (*Scaphiopus* spp.), to emerge from dormancy; this action likely evolved as an alert to rainfall (Dimmitt and Ruibal 1980).

2.2.8. Navigation Cues

Small vertebrates use a variety of cues both to orient themselves and navigate in their environments. These include sound, olfactory and chemosensory cues, celestial cues, visible and polarized light, and magnetism

(for a review see Sinsch 1990). Although most species probably use multiple mechanisms to find their way around, roads and highways have the potential to disrupt many of these cues. These are all important considerations when designing wildlife passage structures.

2.3. Vagility and Movement Patterns of Small Vertebrates

Animal species vary in their vagility—that is, in how far, how fast, and how frequently they move from one location to another. To evaluate the likely impacts of a road on a population and to design an appropriate mitigation plan, it is important to understand what kinds of movements cause individuals of a particular species to encounter roads.

An animal's speed is, in part, a factor of its body size; larger animals have longer strides, while small vertebrates tend to be relatively slow. However, there is much variation among small vertebrates in the speed at which they move. At one extreme are salamanders, which move very slowly due to physical constraints on sustained rapid locomotion. At the other extreme are small carnivores, such as weasels, that are adapted for constant, rapid motion.

In general, small vertebrates that make their living by chasing down prey (pursuit predators) are faster than species that wait in ambush (sit-and-wait predators). Carnivorous predators are typically faster than species that feed on stationary food, like fruits or seeds. Individuals of ectothermic species are faster in warm conditions than cold. Endothermic small vertebrates are faster than ectothermic vertebrates when the air and ground temperature is cold, but this can be reversed in hot conditions. Not surprisingly, prey species that escape predators by running are faster than species that avoid predators by freezing and being cryptic, or deter predators via effective defensive body structures, such as porcupines (*Erethizon dorsatum*), skunks (Mephitidae), and armadillos (Dasypodidae).

In terms of road ecology, the speed at which an animal moves is important because it can affect how long an animal is in a roadway while crossing, and consequently its vulnerability to being struck by a vehicle. Among snakes, for example, timber rattlesnakes are substantially slower than North American racers

(*Coluber constrictor*) and are therefore at greater risk of mortality when crossing a road (Andrews and Gibbons 2005). Slow-moving porcupines are much more common as roadkill than would be expected from their abundance (Barthelmess and Brooks 2010). Speed also affects how long an animal is exposed to the microclimate and conditions of a roadside, or how long it takes to move through a crossing structure like a culvert. Slow-moving amphibians such as salamanders may be at significant risk of predation, thermal stress, and desiccation when moving, whether across roads, on sparsely vegetated road shoulders, or through crossing structures.

Small vertebrates vary dramatically not only in the speed at which they move, but also in the distances that they move overland on a daily, seasonal, and lifetime scale. Some highly aquatic turtles only travel tens of meters from the water to nest, whereas some terrestrial turtles may move tens of kilometers during a year (e.g., Steen et al. 2012). Some small vertebrates may only encounter roads once a season, when moving to and from a breeding site. In areas of largely intact ecological infrastructure, encounters with roads may be as seldom as once in a lifetime, as when dispersing from a birthplace to what will become their adult area of residence. Other species such as wide-ranging species of frogs, snakes, and turtles move far enough in a year that road encounters are relatively frequent in many developed parts of the world. Not surprisingly, wide-ranging species are more at risk of population declines due to road effects (e.g., roadkill and barrier effects of roads) than sedentary species (e.g., Bonnet et al. 1999). For example, leopard frogs are more vulnerable to road mortality than green frogs because leopard frogs range more widely from water into upland fields when foraging (Carr and Fahrig 2001). Semi-aquatic turtles and snakes that move widely among various wetlands and upland sites are more vulnerable to road effects than similar species that are more sedentary (Roe et al. 2006; Attum et al. 2008).

Although aquatic and many semi-aquatic species do not travel far from water, they are capable of making long-distance movements within rivers and streams and are vulnerable to the fragmentation effects of inadequate culverts at road-stream crossings. Movements of spiny softshell turtles (*Apalone spinifera*) are almost exclusively aquatic (except for nest-ing and basking). In Arkansas, average movements of 122–141 m per day were recorded with some individuals moving more than 900 m in a day. Average annual home range length for these animals is thought to be between 1,400–1,750 m (Plummer et al. 1997). These long-distance movements make it likely that individual turtles will encounter roads. Where opportunities for movement across these roads (through various road-stream passage structures) either do not exist or are poorly designed for turtle passage, their movements will be truncated and thus their amount of accessible habitat will be limited.

2.3.1. Types of Movements

To evaluate whether a road is likely to affect populations of a small vertebrate species, and to design an appropriate mitigation plan, it is essential to understand the circumstances in which species encounter roads. The types of movements that animals make can depend on their sex, age class, and species; therefore, the timing and patterns of encounters with roads and the effects of such encounters can be different for different groups of animals (Bonnet et al. 1999).

Breeding Migration and Mate Search

Many small vertebrates undergo annual migrations to suitable places to breed. Wood frogs and some salamander species move from upland forest habitat to vernal pools where breeding adults mate and lay eggs. Many species of frogs migrate to wetland sites appropriate for egg deposition; males gather at these sites and call to attract mates. Males of many small mammal and reptile species travel farther during a specific mating season in search of females. Female reptiles seek out localities that are appropriate for nesting. Many aquatic turtles and some snakes are attracted to roadsides for nesting, because they prefer sites that have warm, dry soil, as is typically the case on road shoulders (Chapter 4; Practical Example 1). Roads can severely affect small vertebrate populations where they bisect two habitats or cross migration corridors.

Natal Dispersal

A critical life-history event for many animals called natal dispersal is leaving the location of birth, hatching, or juvenile development in search of a new place to live. Small mammals that are sufficiently developed to live on their own are forced, or choose, to leave their parents and find an appropriate place to inhabit. Amphibians metamorphose from larval or tadpole forms into juveniles that set out in search of upland habitat in which to live or new wetlands to colonize. Hatchling aquatic turtles leave terrestrial nests and trek overland in search of water. Recently hatched lizards, snakes, and terrestrial turtles disperse from the nest in search of safe and food-rich habitat.

Roads are of greatest concern, in terms of their effects on natal dispersal, when a significant fraction of dispersers must cross them to access suitable habitat. Natal dispersal can be one of the longest movements that an animal will make in its life and thus is a period of high road mortality risk. Additionally, roads can also be barriers that prevent natal dispersers from accessing suitable habitat. Landscape configuration, vegetation cover, and topography can funnel natal dispersers toward particular road segments, which may then be the most suitable locations for targeted mitigation measures (Chapters 8 and 9).

Migration to and from Overwintering Sites

Where winters are cold, many amphibian and reptile species spend the season at sites that are suitable for dormancy without risk of freezing or dying from lack of oxygen. Snakes, for example, seek caves or deep rock fissures that allow them to move far enough underground to avoid freezing. Many snakes use traditional mass-denning sites where dozens to hundreds of snakes spend the winter each year. In northern latitudes, aquatic turtles and some frogs typically move to permanent water bodies that have a low risk of freezing completely and a low risk of winterkill (mass suffocation deaths of aquatic animals due to depleted dissolved oxygen under the ice). Streams, rivers, or other sites where water circulation is maintained even under ice are favored by some; these aquatic animals remain dormant, sometimes in mass numbers, at the bottom of water bodies.

To get to overwintering sites, amphibians and reptiles may move long distances overland. Severe impacts on local populations occur when roads lie between overwintering sites and habitat used for breeding or summer foraging. For example, in the La Rue-Pine Hills Research Natural Area of Shawnee National Forest in Illinois, a road is closed for two months in spring

and fall to provide safe passage for amphibians and reptiles making such movements. Prior to implementing the biannual road closures, the roadkill of snakes and other amphibians and reptiles was substantial and likely unsustainable (US Forest Service 2006). As a general rule, the closer a road is to a traditional overwintering or breeding site, the greater the chance that migrating animals will have to cross it and the greater the risk that it will affect the population of animals that use that site.

Movements Associated with Foraging for Food

The movements made most frequently by small vertebrates are those made while searching for food. There are two kinds of foraging movements: (1) those that involve relatively long-distance movements from one foraging location to another, and (2) short-distance movements while searching for food within a foraging site. In some species, individuals regularly move among multiple sites during a season or from one year to another. Blanding's turtles (*Emydoidea blandingii*), like a number of semi-aquatic amphibians and reptiles, can visit multiple wetlands both within a summer (Beaudry et al. 2008) and from year to year (Grgurovic and Sievert 2005). Individuals of such species encounter roads more frequently than those that remain in one foraging home range.

During daily foraging activities, animals with home ranges near a road may occasionally encounter it. There are some species that are attracted to roads and roadsides as foraging sites. Scavengers such as raccoons (*Procyon lotor*), Virginia opossums (*Didelphis virginiana*), and striped skunks (*Mephitis mephitis*) will patrol roadsides and enter roads to feed on discarded food, garbage, and roadkill. The use of roads as foraging habitat is reviewed in Chapter 4. In general, the effects of roads and road traffic on local foraging movements are probably minor for most species, as compared to the impacts of roads on other kinds of movements.

Thermoregulation

Snakes, turtles, and lizards may make short-distance movements to the open, managed vegetation of road shoulders to use them for sunny basking sites. After sundown, as the air temperature cools, nocturnal amphibians and reptiles may move to roads presumably to raise their body temperature using conductance from the still-warm road pavement and the open habitat of the road shoulder. It is suspected that many snakes that are killed on roads at night have been attracted to the road surface for thermoregulation (Dodd et al. 1989). However, these behaviors appear to be more typical of western snake species that are more adapted to open habitats and exhibit fewer avoidance behaviors of roads (e.g., Andrews et al. 2008 and references therein).

Movements Driven by Changes in Habitat Conditions

Mass movements of animals can be caused by dramatic changes in habitat quality. Flooding and fires may drive small mammals, snakes, and other small terrestrial vertebrates to move en masse. Mowing of hayfields or other forms of large-scale vegetation removal may also result in large movements of small vertebrates. Mass movements frequently occur in association with the drying of a water body, such as a marsh, swamp, or reservoir. Drying of water bodies due to drought or intentional drainage can force entire populations of amphibians, aquatic snakes, turtles, and aquatic mammals such as muskrats to disperse in search of new places. Mass road mortality events have been recorded where reservoirs or marshes adjacent to heavily trafficked roads have dried, and in some cases the number of aquatic animals entering a road can be high enough to create a hazard to motorists (Bernardino and Dalrymple 1992; Aresco 2005). At sites where major roads are adjacent to large water bodies that periodically dry, wildlife crossing structures and fencing can prevent roadkill and hazardous driving conditions, and allows animals to access alternate habitat and eventually recolonize the water body once it refills.

2.3.2. Timing and Seasonality of Movements

Time

The time within a 24-hour day that animals move is important because traffic volume can vary significantly. Traffic volume is often highest during morning and afternoon rush hours associated with motorists traveling to and from work, but at some locations may be consistent throughout the day. Traffic is usually lowest in the hours after midnight. Although road mortality is highest for animals entering a road when traffic is heaviest, even low traffic volumes may deter some animals from attempting to cross. For animals with slow

crossing speeds or immobilization responses, the risk of road mortality can still be high and they may be more likely to attempt to cross a road when traffic is light.

Many snakes, lizards, toads, terrestrial turtles, small mammals, and other small vertebrates are diurnal. They may be most likely to attempt crossing roads in the morning and late afternoon, because the pavement can be too hot for animals during the heat of the day. Turtles frequently attempt to nest along roads in the early morning or evening (Practical Example 1). Many small mammals, some snakes, and many amphibians are primarily nocturnal. For example, salamander breeding migrations to vernal pools or other water bodies occur primarily after dark.

Seasonality

Seasonality of roadkill for small animals has been well documented (e.g., Langen et al. 2007; Glista et al. 2008; Barthelmess and Brooks 2010). At northern latitudes, most species of amphibians migrate from overwintering to breeding sites in late winter and early spring. Other amphibians typically make their breeding migrations in late spring and early summer. However, there are exceptions, such as spadefoots that can breed any time during spring or summer and typically during periods of unusually heavy rain. Marbled salamander breeding migrations occur in late summer and fall. Movement to foraging habitat typically occurs after breeding, and some species may make movements that bring them in contact with roads throughout the summer (e.g., most frog species). Dispersal of newly metamorphosed amphibians occurs from late spring to early fall. Mass movements to overwintering sites occur in the fall.

Reptiles also encounter roads during migration, breeding, and foraging. Snakes encounter roads when moving to and from their overwintering sites in spring and early fall. The timing of snake movements associated with breeding depends on the species; for example, roadkill associated with mate search peaks in late summer for massasaugas (*Sistrurus catenatus*; Shepard et al. 2008). Turtles in northern latitudes nest along roadsides and in other areas in early summer, and hatchlings disperse from the nests in late summer to early fall, or early in the following spring. Terrestrial species such as box turtles (*Terrapene* spp.) and semi-terrestrial species such as Blanding's turtles make foraging movements that bring them in contact with roads throughout the summer. For many reptile species, there are small peaks in movements in spring and early fall associated with movements to and from overwintering sites.

The timing of mammal movements that result in peak encounters with roads is variable, but is often associated with the timing of natal dispersal. For example, young beavers (*Castor* spp.) leave their birth lodge in spring and muskrats disperse in late summer. For many other mammals, natal dispersal occurs in summer or early fall. Another peak in road encounters, especially for mature males, occurs during breeding periods; the timing depends on the natural history of the species.

Weather

Weather can dramatically affect the timing of movements for ectothermic small vertebrates (Glista et al. 2008). Most amphibians make their long-distance movements during mild, wet weather. Amphibians that breed early in the spring often engage in large, synchronous migrations to wetland spawning sites in response to weather cues. The first mild, rainy nights of spring often bring forth very large movements of amphibians; in some cases nearly all of a population's breeding adults will migrate on just a handful of nights. Spadefoots initiate their explosive breeding during heavy summer rains. Rain or even a prolonged mild drizzle after a long period of cold or dry weather can trigger mass movements of amphibians from spring through fall. Many terrestrial reptiles are most active in warm, dry weather, but aquatic species tend to move in association with warm, rainy conditions (e.g., Shepard et al. 2008). Warm days followed by cool nights may cause nocturnal reptiles (and some amphibians) to move to roads to increase their body temperature by lying on, or more commonly next to, the road surface.

2.3.3. Learning and Acclimation to Roads

Animals' behavior when encountering roads can be modified by individual learning—that is, learning based on the consequences of an individual's direct experience, as well as learning based on observing what other animals do (social learning) or detecting cues of what other animals have done (e.g., snake scent trails). In Yellowstone National Park, moose (*Alces alces*) have learned to calve near roads as a refuge from predators

that avoid roads (Berger 2007). In Banff National Park, there was a lag between the installation of crossing structures and widespread use by animals (Clevenger et al. 2009). Presumably, individual animals learned that such structures can be used to safely cross over or under a roadway and other animals followed, either as companions of crossers, by seeing animals crossing, or through attraction to scents and other cues left by crossing animals.

Individual and social learning also likely affect how some small vertebrates behave near roads. For mammals that have extended periods of parental care after offspring become mobile, such as small carnivores, there may be intergenerational transfer of learned behavior about roads and road crossing structures. By following family members or others, or by using scents or other cues to guide movements, animals may become attracted to roads over time, or learn to use particular locations and structures for crossing. Alternatively, animals may learn to avoid roads or road traffic.

Most amphibians, reptiles, and small mammals encounter roads as solitary individuals. For these animals, scent cues left by others may affect how individuals behave near roads. For example, male snakes are more likely to cross roads when a female has recently crossed; the likelihood of detecting a female's scent cues depends on the amount of time since her crossing, the weather, and the type of road surface (Shine et al. 2004). Scent trails may also be an important mechanism for young-of-the-year snakes to find communal hibernation sites (Heller and Halpern 1981; Brown and MacLean 1983; Costanzo 1989). It is plausible that mass migrating amphibians may move toward passage structures under roads by following others who already have learned the location or by following scent cues. There is a need for additional research on whether scent marks or other cues can be used to manipulate animal movements across roads.

Recent studies on animal behavior indicates that there can be strong genetic influences on "personality" traits such as a tendency to take or avoid risks (Stamps and Groothuis 2010). Many small vertebrates have short generation times and high reproductive potential; such species could potentially evolve rapidly through natural selection caused by the road environment. These adaptive behavioral changes might include changes in road avoidance or road attraction, or adaptive responses to the presence of traffic. For example, male frogs inhabiting wetlands adjacent to heavily trafficked roads adjust the fundamental frequency at which they call to avoid masking by traffic noise (Parris et al. 2009; Cunnington and Fahrig 2010).

It is unclear how frequently we can expect to see adaptive evolutionary changes in behavior related to roads and road traffic. If gene flow is substantial from populations distant from a road, evolutionary change may be slight despite substantial natural selection caused by the road environment. However, given that there are examples of genetic changes in populations caused by roads (Section 3.3.4), adaptive behavioral evolution seems quite possible for some small vertebrate species.

2.4. Vulnerability of Small Vertebrates on Roads

2.4.1. Hard to Detect and Avoid

2.4.1.1. Low Profile

Amphibians, reptiles, and small mammals, with the exception of adult American alligators (*Alligator mississippiensis*) and some large turtle species, have a low profile, making it more difficult for drivers to see them on the road simply because they are smaller and less visually apparent than larger animals. Additionally, their reduced size does not pose a threat to the safety of the driver or damage to the vehicle as with larger animals such as ungulates (e.g., white-tailed deer, *Odocoileus virginianus*), thus making drivers less inclined to avoid them. The fact that small vertebrates are low profile both physically and culturally represents an important challenge to better understanding the effects of roads on these species, and developing and implementing effective strategies for mitigating those impacts.

2.4.1.2. Threat Responses: "Freeze" versus Flee

An animal's response to an approaching vehicle is related to its natural predator defenses. For example, while some snakes exhibit flight as an initial defense (e.g., eastern racer) others remain stationary (immobilize) and rely on crypsis (e.g., timber rattlesnake; Andrews and Gibbons 2005). Immobilization behaviors in response to oncoming or passing vehicles would significantly increase the time spent crossing a road and

thereby the likelihood of mortality. Mazerolle et al. (2005) found that the strongest stimulus for immobilization behavior across six amphibian species was a combination of headlights and vibration. Andrews and Gibbons (2005) found high rates of immobilization in response to a passing vehicle in three snake species. Drivers are largely uneducated about wildlife behavior on roads and how animals respond to approaching vehicles (Sherman 1995).

2.4.2. Vulnerability to Collection and Persecution

2.4.2.1. Small Vertebrates as Targets

Many small vertebrates are furtive animals that are rarely encountered. This lack of steady interaction between small vertebrates and humans has led to a populace that is largely uneducated about these animals (Gibbons and Buhlmann 2001; Baskaran and Boominathan 2010), and this ignorance can transpire as a chronic fear or as a simple lack of awareness in watching out for small animals on roads. In particular, snakes are widely maligned and an irrational fear of these animals is ranked as the most prevalent phobia (ophidiophobia) of Americans, a condition that is more common in women than men and that decreases with increasing education levels (Mittermeier et al. 1992). Species of small vertebrates that are loathed (e.g., snakes) or held in low regard may be put at further risk by drivers who will not make any effort to avoid hitting these species.

While some drivers swerve to miss animals in the road, many people will swerve intentionally to hit them. Swerving to miss animals on roads is a common cause of single car accidents in the United States (Sherman 1995). Many species of snakes present a relatively large target as they crawl across roads, which may influence the frequency of intentional killing (Whitaker and Shine 2000). Langley et al. (1989) conducted a survey of college students and found that both males and females chose to intentionally run over a snake more than any other animal. Intentional killing on roads has been noted with turtles as well (Boarman et al. 1997; N. Weaver, Clemson University, unpublished data). Aside from intentional killing, much wildlife mortality on roads occurs from lack of driving awareness or training.

2.4.2.2. Threat of Collection

Small vertebrates such as turtles and snakes are collected by people for various reasons. Turtles and snakes—even venomous snakes—are kept as pets. Some rare turtles, such as bog turtles (*Glyptemys muhlenbergii*), command high prices on the black market. In some parts of North America, rattlesnake roundups are still a recreational activity. (Several have now been converted to festivals that do not involve killing wild animals and some even promote the education and conservation of species.) Roads increase human-wildlife interactions by facilitating human access to habitats and by making small vertebrates more visible to people as those animals cross roads. This access places local populations under pressure from human predation and collection (Dodd et al. 1989; Bennett 1991; Krivda 1993; Ballard 1994; McDougal 2000).

2.5. Conclusions

Movement is essential for maintaining healthy wildlife populations and metapopulations. These movements occur over various temporal and spatial scales, and species are quite diverse in their speed, distances traveled, and propensity for movement. Amphibians, reptiles, and small mammals possess a wide variety of physiological and natural history traits that make them more or less vulnerable to the negative effects of roads. These characteristics are also important considerations when designing wildlife passage structures or using other approaches for mitigating road impacts.

2.6. Key Points

- Regular movement is critical to the survival of small animals.
- Movement for animals, or the disruption of those movements, is not only important for those individual animals but also has consequences at the population level.
- Although long-distance movements may be infrequent, they can nevertheless be critical for the long-term persistence of a set of linked populations, or metapopulation.
- Even though individual populations may go extinct, a metapopulation can persist through a

balance between local extinction and colonization from nearby populations.
- To understand the effects of roads on animals, one must first consider the ecological functions of animal movement across a range of spatial scales.
- The natural history and physiology of small animals affect their movement and vulnerability to roads.
- Animal movement can be related to reproduction, natal dispersal, movement to and from overwintering sites, foraging, thermoregulation, changes in habitat conditions, time of day, seasonality, and weather.
- When designing wildlife crossing structures for small vertebrates, it is important to consider the physiological, behavioral, and life-history characteristics that may affect their movement patterns or propensity to use crossing structures.
- Movement or propensity to use crossing structures is affected by substrate, water depth and flow, bank edges, openness, cover and exposure to predators, arboreal connections, moisture, temperature, light, sound, and navigation cues.
- River and stream networks are important movement corridors for aquatic and semi-aquatic small vertebrates as well as for terrestrial species in arid or highly developed landscapes.
- The low profiles of most small vertebrates, both physically and culturally (i.e., many people don't know about or like them), contribute to their vulnerability to road mortality.

LITERATURE CITED

Andrews, A. 1990. Fragmentation of habitat by roads and utility corridors: a review. Australian Zoologist 26:130–141.

Andrews, K. M., and J. W. Gibbons. 2005. How do highways influence snake movement? Behavioral responses to roads and vehicles. Copeia 2005:772–782.

Andrews, K. M., J. W. Gibbons, and D. M. Jochimsen. 2008. Ecological effects of roads on amphibians and reptiles: a literature review. Pages 121–143 *in* Urban herpetology. J. C. Mitchell, R. E. Jung Brown, and B. Bartholomew, editors. Herpetological Conservation Vol. 3. Society for the Study of Amphibians and Reptiles, Salt Lake City, Utah, USA.

Aresco, M. J. 2005. The effect of sex-specific terrestrial movements and roads on the sex ratio of freshwater turtles. Biological Conservation 123:37–44.

Ashley, E. P., and J. T. Robinson. 1996. Road mortality of

amphibians, reptiles and other wildlife on the Long Point Causeway, Lake Erie, Ontario. Canadian Field Naturalist 110:403–412.

Attum, O., Y. M. Lee, J. H. Roe, and B. A. Kingsbury. 2008. Wetland complexes and upland-wetland linkages: landscape effects on the distribution of rare and common wetland reptiles. Journal of Zoology 275:245–251.

Baker, B. J., and J.M.L. Richardson. 2006. The effect of artificial light on male breeding-season behavior in green frogs, *Rana clamitans melanota*. Canadian Journal of Zoology 84:1528–1532.

Baker, J. 1990. Toad aggregations under street lamps. British Herpetological Society Bulletin 31:26–27.

Ballard, S. R. 1994. Status of the herpetofauna in the LaRue-Pine Hills / Otter Pond Research Natural Area in Union County, Illinois. Master's thesis, Southern Illinois University, Carbondale, Illinois, USA.

Barthelmess, E. L., and M. S. Brooks. 2010. The influence of body-size and diet on road-kill trends in mammals. Biodiversity and Conservation 19:1611–1629.

Baskaran, N., and D. Boominathan. 2010. Road kill of animals by highway traffic in the tropical forests of Mudumalai Tiger Reserve, southern India. JoTT Communication 2:753–759.

Beaudry, F., P. G. deMaynadier, and M. L. Hunter Jr. 2008. Identifying road mortality threat at multiple spatial scales for semi-aquatic turtles. Biological Conservation 141:2550–2563.

Bennett, A. F. 1991. Roads, roadsides and wildlife conservation: a review. Pages 99–118 *in* Nature conservation 2: The role of corridors. D. A. Saunders, and R. J. Hobbs, editors. Surrey Beatty and Sons, London, UK.

Berger, J. 2007. Fear, human shields and the redistribution of prey and predators in protected areas. Biological Letters 3:620–623.

Bernardino, F. S., Jr., and G. H. Dalrymple. 1992. Seasonal activity and road mortality of the snakes of the Pa-hay-okee wetlands of Everglades National Park, USA. Biological Conservation 62:71–75.

Blouin-Demers, G., and P. J. Weatherhead. 2002. Habitat-specific behavioural thermoregulation by black rat snakes (*Elaphe obsoleta obsoleta*). Oikos 97:59–68.

Boarman, W. I., M. Sazaki, and W. B. Jennings. 1997. The effects of roads, barrier fences, and culverts on desert tortoise populations in California, USA. Pages 54–58 *in* Proceedings of the conservation, restoration and management of tortoises and turtles—an international conference. J. Van Abbema, editor. New York Turtle and Tortoise Society, New York, New York, USA.

Bonnet, X., G. Naulleau, and R. Shine. 1999. The dangers of leaving home: dispersal and mortality in snakes. Biological Conservation 89:39–50.

Brattstrom, B. H. 1979. Amphibian temperature regulation studies in the field and laboratory. American Zoologist 19:345–356.

Brehme, C. S., J. A. Tracey, L. R. McClenaghan, and R. N. Fisher. 2013. Permeability of roads to the movement of scrubland lizards and small mammals. Conservation Biology 27:710–720.

Brown, J. H., and A. Kodric-Brown. 1977. Turnover rates in insular biogeography: effect of immigration on extinction. Ecology 58:445–449.

Brown, W. S., and F. M. MacLean. 1983. Conspecific scent-trailing by newborn timber rattlesnakes, *Crotalus horridus*. Herpetologica 39:430–436.

Buchanan, B. W. 1993. Effects of enhanced lighting on the behavior of nocturnal frogs. Animal Behaviour 45:893–899.

Burton, T. M., and G. E. Likens. 1975. Salamander populations and biomass in the Hubbard Brook Experimental Forest, New Hampshire. Copeia 1975:541–546.

Carr, A., and L. Ogren. 1960. The ecology and migration of sea turtles, 4: the green turtle in the Caribbean Sea. Bulletin of the American Museum of Natural History 121:1–121.

Carr, L. W., and L. Fahrig. 2001. Effect of road traffic on two amphibian species of differing vagility. Conservation Biology 15:1071–1078.

Church, G. 1960a. Annual and lunar periodicity in the sexual cycle of the Javanese toad *Bufo elanostictus* (Schneider). Zoologica 45:181–188.

Church, G. 1960b. The effects of seasonal and lunar changes on the breeding pattern of the edible Javanese frog, *Rana cancrivora* (Gravenhorst). Treubia 25:215–233.

Church, G. 1962. Seasonal and lunar variations in the numbers of mating toads in Bandung, Java. Herpetologica 17:122–126.

Clevenger, A. P., A. T. Ford, and M. A. Sawaya. 2009. Banff wildlife crossings project: integrating science and education in restoring population connectivity across transportation corridors. Final Report to Parks Canada Agency, Radium Hot Springs, British Columbia, Canada.

Clevenger, A. P., and N. Waltho. 1999. Dry drainage culvert use and design considerations for small and medium sized mammal movement across a major transportation corridor. Pages 263–277 *in* Proceedings of the third international conference on wildlife ecology and transportation. G. L. Evink, P. Garrett, and D. Zeigler, editors. FL-ER-73-99. Florida Department of Transportation, Tallahassee, Florida, USA.

Costanzo, J. P. 1989. Conspecific scent trailing by garter snakes (*Thamnophis sirtalis*) during autumn: further evidence for use of pheromones in den location. Journal of Chemical Ecology 15:2531–2538.

Cunnington, G. M., and L. Fahrig. 2010. Plasticity in the vocalizations of anurans in response to traffic noise. Acta Oecologica 36:463–470.

Daly, M., P. R. Behrends, M. I. Wilson, and L. F. Jacobs. 1992. Behavioral modulation of predation risk: moonlight avoidance and crepuscular compensation in a nocturnal desert rodent, *Dipodomys merriami*. Animal Behaviour 44:1–9.

Dimmitt, M. A., and R. Ruibal. 1980. Environmental correlates of emergence in spadefoot toads (*Scaphiopus*). Journal of Herpetology 14:21–29.

Dodd, C. K., Jr., K. M. Enge, and J. N. Stuart. 1989. Reptiles on highways in north-central Alabama, USA. Journal of Herpetology 23:197–200.

Doucet, G. J., and J. R. Bider. 1974. The effects of weather on the activity of the masked shrew. Journal of Mammalogy 55:348–363.

Downes, S. 2001. Trading heat and food for safety: costs of predator avoidance in a lizard. Ecology 82:2870–2881.

Duellman W. E., and L. Trueb. 1986. Biology of amphibians. McGraw-Hill, New York, St. Louis, San Francisco, USA.

Dufour, P. A. 1980. Effects of noise on wildlife and other animals: review of research since 1971. Report No. 550/9-80-100. US Environmental Protection Agency, Office of Noise Abatement and Control, Washington, DC, USA.

Ferguson, D. E. 1960. Movements and behaviour of *Bufo fowleri* in residential areas. Herpetologica 16:112–114.

Ferrara, C. R., R. C. Vogt, and R. S. Sousa-Lima. 2013. Turtle vocalizations as the first evidence of posthatching parental care in Chelonians. Journal of Comparative Psychology 127:24–32.

Fitzgerald, G. J., and J. R. Bider. 1974. Influence of moon phase and weather factors on locomotory activity in *Bufo americanus*. Oikos 25:338–340.

Foresman, K. R. 2002. Small mammal use of modified culverts on the Lolo South Project of Western Montana. Pages 581–582 in Proceedings of the 2001 international conference on ecology and transportation. Center for Transportation and the Environment, North Carolina State University Raleigh, North Carolina, USA.

Foresman, K. R. 2003. Small mammal use of modified culverts on the Lolo South Project of Western Montana—an update. Pages 342–343 in Proceedings of the 2003 international conference on ecology and transportation. C. Leroy Irwin, P. Garrett, and K. P. McDermott, editors. Center for Transportation and the Environment, North Carolina State University, Raleigh, North Carolina, USA.

Forman, R.T.T., and L. E. Alexander. 1998. Roads and their major ecological effects. Annual Review of Ecology and Systematics 29:207–231.

Gamble, L. R., K. McGarigal, and B. W. Compton. 2007. Fidelity and dispersal in the pond-breeding amphibian, *Ambystoma opacum*: implications for spatio-temporal population dynamics and conservation. Biological Conservation 139:247–257.

Gamble, L. R., K. McGarigal, C. L. Jenkins, and B. C. Timm. 2006. Limitations of regulated "buffer zones" for the conservation of marbled salamanders. Wetlands 26:298–306.

Gibbons, J. W., and K. A. Buhlmann. 2001. Reptiles and amphibians. Pages 372–390 in Wildlife of southern forests: Habitat and management. J. G. Dickson, editor. Hancock House Publishers, Surrey, British Columbia, Canada.

Glista, D. J., T. L. DeVault, and J. A. DeWoody. 2008. Vertebrate road mortality predominantly impacts amphibians. Herpetological Conservation and Biology 3:77–87.

Greenberg, C. H. 2001. Spatio-temporal dynamics of pond use and recruitment in Florida gopher frogs (*Rana capito aesopus*). Journal of Herpetology 35:74–85.

Greenberg, C. H., and G. W. Tanner. 2004. Breeding pond selection and movement patterns by eastern spadefoot toads (*Scaphiopus holbrookii*) in relation to weather and edaphic conditions. Journal of Herpetology 38:569–577.

Greenberg, C. H., and G. W. Tanner. 2005a. Chaos and continuity: the role of isolated, ephemeral wetlands on amphibian populations in xeric sandhills. Pages 79–90 in Amphibians and reptiles: Status and conservation in Florida. W. E. Meshaka, and K. J. Babbitt, editors. Krieger, Malabar, Florida, USA.

Greenberg, C. H., and G. W. Tanner. 2005b. Spatial and temporal ecology of eastern spadefoot toads on a Florida landscape. Herpetologica 61:20–28.

Greenberg, C. H., and G. W. Tanner. 2005c. Spatial and temporal ecology of oak toads (*Bufo quercicus*) on a Florida landscape. Herpetologica 61:422–434.

Grgurovic, M., and P. R. Sievert. 2005. Movement patterns of Blanding's turtles (*Emydoidea blandingii*) in the suburban landscape of eastern Massachusetts. Urban Ecosystems 8:203–213.

Halverson, M. A., D. K. Skelly, J. M. Kiesecker, and L. K. Freidenburg. 2003. Forest mediated light regime linked to amphibian distribution and performance. Acta Oecologica 234:360–364.

Hanski, I. 1999. Metapopulation ecology. Oxford University Press, Oxford, UK.

Heller, S., and M. Halpern. 1981. Laboratory observations on conspecific and congeneric scent trailing in garter snakes (Thamnophis). Behavioral and Neural Biology 33:372–377.

Hertz, P. E., R. B. Huey, and E. Nevo. 1982. Fight versus flight—body-temperature influences defensive responses of lizards. Animal Behaviour 30:676–679.

Houston, D., and R. Shine. 1994. Movements and activity patterns of arafura filesnakes (Serpentes: Acrochordidae) in tropical Australia. Herpetologica 50:349–357.

Huey, R. B., and R. D. Stevenson. 1979. Integrating thermal physiology and ecology of ectotherms—discussion of approaches. American Zoologist 19:357–366.

Hunt, A. H., J. Dickens, and R. J. Whelan. 1987. Movement of mammals through tunnels under railway lines. Australian Zoologist 24:89–93.

Jackson, S. D. 1996. Underpass systems for amphibians. Pages 224–227 in Trends in addressing transportation related wildlife mortality, Proceedings of the transportation-related wildlife mortality seminar. G. L. Evink, P. Garrett, D. Zeigler, and J. Berry, editors. FL-ER-58-96. Florida Department of Transportation, Tallahassee, Florida, USA.

Jahoda, J. C. 1973. The effect of the lunar cycle on the activity pattern of *Onychomys leucogaster breviauritus*. Journal of Mammalogy 54:544–549.

Kotler, B. P., J. S. Brown, S.R.X. Dall, S. Gresser, D. Ganey, and A. Bouskila. 2002. Foraging games between gerbils and their predators: temporal dynamics of resource depletion and apprehension in gerbils. Evolutionary Ecology Research 4:495–518.

Krivda, W. 1993. Road kills of migrating garter snakes at The Pas, Manitoba. Blue Jay 51:197–198.

Langen, T. A., A. Machniak, E. K. Crowe, C. Mangan, D. F. Marker, N. Liddle, and B. Roden. 2007. Methodologies for surveying herpetofauna mortality on rural highways. Journal of Wildlife Management 71:1361–1368.

Langley, W. M., H. W. Lipps, and J. F. Theis. 1989. Responses of Kansas motorists to snake models on a rural highway. Transactions of the Kansas Academy of Science 92: 43–48.

Levins, R. 1969. Some demographic and genetic consequences of environmental heterogeneity for biological control. Bulletin of the Entomological Society of America 15:237–240.

Lowe, W. H., and F. W. Allendorf. 2010. What can genetics tell us about population connectivity? Molecular Ecology 19:3038–3051.

Mader, H. J. 1984. Animal habitat isolation by roads and agricultural fields. Biological Conservation 29:81–96.

Madsen, T., and M. Osterkamp. 1982. Notes on the biology of the fish-eating snake Lycodonomorphus bicolor in Lake Tanganyika. Journal of Herpetology 16:185–188.

Mansergh, I. M., and D. J. Scotts. 1989. Habitat continuity and social organization of the mountain pygmy-possum restored by tunnel. Journal of Wildlife Management 53:701–707.

Marsh, D. M., and N. G. Beckman. 2004. Effects of forest roads on the abundance and activity of terrestrial salamanders. Ecological Applications 14:1882–1891.

Marsh, D. M., and P. C. Trenham. 2001. Metapopulation dynamics and amphibian conservation. Conservation Biology 15:40–49.

Mazerolle, M. J., and A. Desrochers. 2005. Landscape resistance to frog movements. Canadian Journal of Zoology 83:455–464.

Mazerolle, M. J., M. Huot, and M. Gravel. 2005. Behavior of amphibians on the road in response to car traffic. Herpetologica 62:380–388.

McClure, H. E. 1951. An analysis of animal victims on Nebraska's highways. Journal of Wildlife Management 15:410–420.

McDougal, J. 2000. Conservation of tortoises and terrestrial turtles. Pages 180–206 in Turtle conservation. M. W. Klemens, editor. Smithsonian Institution Press, Washington, DC, USA.

McFarlane, R. W. 1963. Disorientation of loggerhead hatchlings by artificial road lighting. Copeia 1963:153.

Mills, L. S., and F. W. Allendorf. 1996. The one-migrant-per-generation rule in conservation and management. Conservation Biology 10:1509–1518.

Mittermeier, R. A., J. L. Carr, I. R. Swingland, T. B. Werner, and R. B. Mast. 1992. Conservation of amphibians and reptiles. Pages 59–80 in Herpetology: Current research on the biology of amphibians and reptiles. K. Adler, editor. Society for the Study of Amphibians and Reptiles, St. Louis, Missouri, USA.

Oxley, D. J., M. B. Fenton, and G. R. Carmody. 1974. The effects of roads on populations of small mammals. Journal of Applied Ecology 11:51–59.

Parris, K. M., M. Velik-Lord, and J.M.A. North. 2009. Frogs call at a higher pitch in traffic noise. Ecology and Society 14: 25. http://www.ecologyandsociety.org/vol14/iss1/art25/. Accessed 26 April 2014.

Plummer, M. V., N. E. Mills, and S. L. Allen. 1997. Activity, habitat, and movement patterns of softshell turtles (Trionyx spiniferus) in a small stream. Chelonian Conservation and Biology 2:514–520.

Robson, L. E. 2011. The spatial ecology of Eastern Hognose Snakes (Heterodon platirhinos): habitat selection, home range size, and the effect of roads on movement. Master's thesis, Ottawa-Carleton Institute of Biology, Ottawa, Ontario, Canada.

Rocha, C.F.D., and H. G. Bergallo. 1990. Thermal biology and flight distance of Tropidurus oreadicus (Sauria, Iguanidae) in an area of Amazonian Brazil. Ethology, Ecology, and Evolution 2:263–268.

Rodriguez, A., G. Crema, and M. Delibes. 1996. Use of non-wildlife passages across a high speed railway by terrestrial vertebrates. Journal of Applied Ecology 33:1527–1540.

Roe, J. H., J. Gibson, and B. A. Kingsbury. 2006. Beyond the wetland border: estimating the impact of roads on two species of water snakes. Biological Conservation 130:161–168.

Rosell, C., J. Parpal, R. Campeny, S. Jove, A. Pasquina, and J. M. Velasco. 1997. Mitigation of barrier effect on linear infrastructures on wildlife. Pages 367–372 in Habitat fragmentation and infrastructure. K. Canters, editor. Ministry of Transportation, Public Works and Water Management, Delft, The Netherlands.

Rytwinski, T., and L. Fahrig. 2012. Do species life history traits explain population responses to roads? A meta-analysis. Biological Conservation 147:87–98.

Santolini, R., G. Sauli, S. Malcevschi, and F. Perco. 1997. The relationship between infrastructure and wildlife: problems, possible solutions and finished works in Italy. Pages 202–212 in Habitat fragmentation and infrastructure, proceedings of the international conference on habitat fragmentation, infrastructure and the role of ecological engineering. K. Canters, editor. Ministry of Transport, Public Works and Water Management, Delft, The Netherlands.

Shepard, D. B., M. J. Dreslik, B. C. Jellen, and C. A. Phillips. 2008. Reptile road mortality around an oasis in the Illinois corn desert with emphasis on the endangered eastern massasauga. Copeia 2008:350–359.

Shepard D. B., A. R. Kuhns, M. J. Dreslik, and C. A. Phillips. 2008. Roads as barriers to animal movement in fragmented landscapes. Animal Conservation 11:288–296.

Sherman, J. 1995. Roadkill: body counts. Audubon July/ August:21–22.

Shine, R., M. Lemaster, M. Wall, T. Langkilde, and R. Mason. 2004. Why did the snake cross the road? Effects of roads on movement and location of mates by garter snakes (*Thamnophis sirtalis parietalis*). Ecology and Society 9:9. http://www.ecologyandsociety.org/vol9/iss1/art9/. Accessed 26 April 2014.

Sinervo, B., F. Mendez-de-la-Cruz, D. B. Miles, B. Heulin, E. Bastiaans, M. Villagran-Santa Cruz, R. Lara-Resendiz, et al. 2010. Erosion of lizard diversity by climate change and altered thermal niches. Science 328:894–899.

Sinsch, U. 1990. Migration and orientation in anuran amphibians. Ethology, Ecology and Evolution 2:65–79.

Stamps, J., and T.G.G. Groothuis. 2010. The development of animal personality: relevance, concepts and perspectives. Biological Reviews 85:301–325.

Steen, D. A., J. P. Gibbs, K. A. Buhlmann, J. L. Carr, B. W. Compton, J. D. Congdon, J. S. Doody, et al. 2012. Terrestrial habitat requirements of nesting freshwater turtles. Biological Conservation 150:121–128.

Stevens, V. M., E. Polus, R. A. Wesselingh, N. Schtickzelle, and M. Baquette. 2004. Quantifying functional connectivity: experimental evidence for patch-specific resistance in the Natterjack toad (*Bufo calamita*). Landscape Ecology 19:829–842.

Sullivan, B. K. 1981. Observed differences in body temperature and associated behavior of four snake species. Journal of Herpetology 15:245–246.

Swihart, R. K., and N. A. Slade. 1984. Road crossing in *Sigmodon hispidus* and *Microtus ochrogaster*. Journal of Mammalogy 65:357–360.

Tuxbury, S. M., and M. Salmon. 2005. Competitive interactions between artificial lighting and natural cues during seafinding by hatchling marine turtles. Biological Conservation 121:311–316.

US Forest Service. 2006. Snake migration LaRue-Pine Hills. Shawnee National Forest, Mississippi Bluffs Ranger District flier. US National Forest Service. http://www.fs.usda.gov/Internet/FSE_DOCUMENTS/stelprdb5106391.pdf. Accessed 26 April 2014.

Vickery, W. L., and J. R. Bider. 1981. The influence of weather on rodent activity. Journal of Mammalogy 62:140–145.

Wegner, J. F., and G. Merriam. 1979. Movements by birds and small mammals between a wood and adjoining farmland habitats. Journal of Applied Ecology 16:349–357.

Whitaker, P. B., and R. Shine. 2000. Sources of mortality of large elapid snakes in an agricultural landscape. Journal of Herpetology 34:121–128.

Witherington, B. E., and R. E. Martin. 1996. Understanding, assessing, and resolving light-pollution problems on sea turtle nesting beaches. Florida Marine Research Institute Technical Report TR-2. Florida Marine Research Institute, St. Petersburg, Florida, USA.

Wolfe, J. L., and T. Summerlin. 1989. The influence of lunar light on nocturnal activity of the old-field mouse. Animal Behaviour 37:410–414.

Zajitschek, S.R.K., F. Zajitschek, and J. Clobert. 2012. The importance of habitat resistance for movement decisions in the common lizard, *Lacerta vivipara*. BMC Ecology 12:1–9. http://www.biomedcentral.com/1472-6785/12/13. Accessed 26 April 2014.

Zug, G. R. 1993. Herpetology: An introductory biology of amphibians and reptiles. Academic Press, San Diego, California, USA.

3

Direct Effects of Roads on Small Animal Populations

David M. Marsh and
Jochen A. G. Jaeger

3.1. Introduction

When people think about the effects of roads on animals, they typically think of roadkill. In fact, road mortality may be high and may place populations of some species at risk of extinction, but it is not the only way that roads affect animal populations. Beyond increasing mortality of individuals, roads can reduce animal movement, diminish habitat quality, alter sex ratios and age structure, and even change the genetic structure of adjacent animal populations. All of these can be considered direct effects of roads on animals.

In evaluating the direct effects of roads on animal populations, it can be useful to distinguish effects that occur at the local scale (i.e., in the areas immediately adjacent to roads) from effects at the landscape scale (i.e., where multiple populations are linked by dispersal into a larger metapopulation). Figure 3.1 presents a graphical scheme that summarizes these various direct effects of roads. Local-scale effects include roadkill, but also include road avoidance and other effects of roads on behavior, such as decreased chorusing in frogs or altered song characteristics in birds. Landscape-scale effects occur outside the immediate vicinity of a road and influence the processes that promote the long-term persistence of metapopulations. Landscape effects include reduction in movement across the landscape and disruption of migration and colonization at large scales. In addition, effects of roads on genetic diversity and population fitness are typically seen at the landscape scale. Landscape-scale effects are often harder to detect than local effects, and they may

be manifested over longer time periods, making landscape effects challenging to monitor. However, from an ecological perspective, landscape effects are usually considered to be at least as, if not more, important than local effects in terms of influences on population viability.

3.2. Local Effects

3.2.1. Roadkill

Small animals may be at a high risk of dying while crossing a road. They are rarely seen by drivers, and may be unable to perceive vehicles until it is too late. Crossing a road can be a long journey for some small animals. Whereas large mammals may be able to cross a road in seconds, small animals could require minutes or even hours to cross a road. Some small animals may simply avoid roads, or they may cross them only at night when there is little traffic (Chapter 2). Other animals may be lucky enough to pass underneath vehicles without being harmed. So how frequent is road mortality for small animals?

A sample of estimated mortality rates from research studies (Table 3.1) shows that mortality rates are highly variable. Much of this variation is among species. For example, common toad (*Bufo bufo*) populations in Poland showed an annual road mortality >20%, whereas for fire-bellied toads (*Bombina bombina*) at the same sites, road mortality was more than 5 times lower (Brzeziński et al. 2012). Within a species, mortality risk can also vary widely from one road to another. For spotted salamanders (*Ambystoma maculatum*) in

Table 3.1. Sample roadkill mortality rates for small vertebrates, ordered from lowest to highest mortality rate. Rates are either measured as mortality per road crossing attempt or per year for the population

Species	Data type	Mortality	Location	Citation
Black ratsnake	per crossing	2.6%	Secondary road, Ontario, CA	Row et al. (2007)
European hare	per crossing	<2%	Rural road in Denmark	Hels and Buchwald (2001)
Hedgehog	per crossing	<4%	Rural road in Denmark	Hels and Buchwald (2001)
5 spp. of amphibians	per crossing	34–61%	Rural road in Denmark	Hels and Buchwald (2001)
5 spp. of amphibians	per crossing	89–98%	Highway in Denmark	Hels and Buchwald (2001)
Turtles (6 spp.)	per crossing	98%	Highway in FL	Aresco (2003)
Mustelid spp.	per year	1.2%	Sweden (nationwide)	Seiler et al. (2004)
Fire-bellied toads	per year	<4%	3 ponds near roads, Poland	Brzezinski et al. (2012)
Common newts	per year	<4%	3 ponds near roads, Poland	Brzezinski et al. (2012)
Small pond turtles	per year	<5%	Across the United States	Gibbs and Shriver (2002)
European hare	per year	9.1%	Sweden (nationwide)	Seiler et al. (2004)
European badger	per year	11%	Sweden (nationwide)	Seiler et al. (2004)
Land turtles	per year	5–25%	Across the United States	Gibbs and Shriver (2002)
Common toads	per year	~20%	3 ponds near roads, Poland	Brzezinski et al. (2012)
Common frogs	per year	~20%	3 ponds near roads, Poland	Brzezinski et al. (2012)
Spotted salamanders	per year	17–37%	Central, western MA	Gibbs and Shriver (2005)

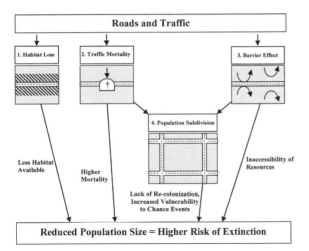

Figure 3.1. The four main effects of transportation infrastructure on wildlife populations. Both traffic mortality and the barrier effect contribute to population subdivision and isolation, and all four effects contribute to a higher risk of population extinction. *Credit: Jochen A. G. Jaeger, adapted from Jaeger, Bowman, et al. (2005).*

Massachusetts, mortality per attempted road crossing ranged from 7 to 37% with traffic volumes that varied from 4.2 to 15.6 vehicles per hour (Gibbs and Shriver 2005). Because of this variability, it can be difficult to extend inferences between species or sites, so site-specific data are always preferable for predicting and mitigating the impacts of roads.

Nevertheless, some generalizations can be made with respect to road mortality. First, slow-moving taxa will be at a particularly high risk of dying while crossing roads. Slow-moving groups include turtles, amphibians, and some snakes, both of which can have mortality risks as high as 98% per crossing attempt (Hels and Buchwald 2001; Andrews and Gibbons 2005; Aresco 2005a). Road-crossing risks for small mammals tend be lower (Table 3.1), although animals that spend time in roadside habitats (e.g., voles and squirrels) may be exceptions. Second, mortality will tend to increase with traffic volume. Relationships between traffic volume and mortality have been found for amphibians (Fahrig et al. 1995), turtles (Gibbs and Shriver 2002), and armadillos (Inbar and Mayer 1999). In some cases, traffic volume may reach a threshold above which animals of a given species will almost never be able to cross successfully (Hels and Buchwald 2001). As mortality tends to depend on traffic volume, data on traffic can be used to predict animal mortality (see Box 3.1 for a simple method of making these calculations).

The third generalization about road mortality is that it is not randomly distributed, but rather is spatially and temporally concentrated in hotspots (e.g., Langen et al. 2009, 2012) and critical time periods of peak movements (e.g., "hot moments" in Crawford, Maerz, Nibbelink, Buhlmann, Norton, and Albeke 2014; Practical Example 1). Roadkill hotspots usually occur in areas

Box 3.1. A Simple Model to Predict Animal Mortality from Road Traffic

Hels and Buchwald (2001) designed a model to predict traffic mortality. They applied their model to amphibians, but it is applicable to any small vertebrate. Hels and Buchwald derive the probability of successfully crossing the road as:

$$p_{successful\_crossing} = e^{(-Na\,/\,v\,/\,\cos(\alpha))}$$

where N = traffic volume (number of vehicles per time unit), a = effective width of the vehicle (the width of the vehicle's tires), v = movement speed of the animal, and α = angle of the movement direction ($\alpha = 0$ for movements perpendicular to the road). In this model, the probability of successfully crossing a road does not decrease linearly with increase in traffic volume, but rather exponentially. In addition, probability of crossing increases exponentially with the movement speed, v, of the animal crossing the road. When the crossing angle $\alpha = 0$ (the animals go straight across the road), the probability of crossing successfully is highest. In this model, the number of animals killed does not decrease when traffic volume is high because it assumes no road avoidance (i.e., when traffic volumes are particularly high, animals may stop trying to cross as often). Because this is a model for small animals, Hels and Buchwald (2001) use the width of the vehicle's tires as the "effective width," plus 5% to account for animals that are killed by air pressure waves (from the tires) alone (Hummel 2001). For larger animals that do not pass between the tires or are killed by air pressure wave under the vehicle (Hummel 2001), the entire width of the vehicle would be used.

This model can be easily applied given data for traffic density, tire width, and animal movement speed. Hels and Buchwald (2001) studied a road with a traffic density of 3,200 vehicles per day, and they assumed car tires to have a width of 0.22 m and truck tires to have a width of 0.38/ 0.64 m (single/double tires). A movement rate of 0.5 m/min for newts (*Triturus* spp.) led to a mortality risk of nearly 98%, whereas a movement rate of 2.0 m/min for common frogs (*Rana temporaria*) led to a mortality risk closer to 50%.

One caution in applying this model is that traffic volume can vary widely throughout the day. Since most animals move only during certain times of the day (or night), measures of daily traffic volume can present a misleading picture of road mortality. Ideally, traffic volume estimates should be based on the activity periods of any animal species of interest.

---

with particularly high densities of animals, or in areas that represent common animal migration routes. For amphibians and reptiles, sites where wetlands are located close to roads are particularly likely to be roadkill hotspots (e.g., Langen et al. 2009, 2012). Where roadkill hotspots are easily discernable, mitigation of road effects becomes much more tractable because mitigation measures can be focused on these specific areas.

Although roadkilled animals are frequently seen, actual rates of mortality can be very difficult to estimate accurately (e.g., Seiler et al. 2004). Roadkill can be quickly removed by scavengers, and in areas with high traffic, carcasses may be rapidly destroyed by vehicles (Santos et al. 2011; Guinard et al. 2012; K. M. Andrews, University of Georgia, personal communication). In addition, animals may be knocked off the road by vehicles or be injured and die in surrounding habitat. As a result, counts of dead animals on roads will almost always underestimate mortality (Zimmermann Teixeira et al. 2013). In most cases, relative mortality (rather than absolute mortality) is frequently used to determine where hotspots occur and whether mitigation methods are effective. Where absolute mortality rates are needed, a mark-recapture approach that allows estimation of the number of animals killed on the road that disappear before they are observed in surveys can be employed (Hels and Buchwald 2001; Section 12.4.3.2).

3.2.1.1. Effects of Roadkill on Populations
High levels of road mortality will typically reduce the abundance and persistence of populations (e.g., for adult

female turtles, see Aresco 2005*b*; Crawford, Maerz, Nibbelink, Buhlmann, and Norton 2014; Practical Example 1). Studies on a variety of animal groups have found a relationship between road mortality and the abundance of animals in populations near roads (e.g., Fahrig et al. 1995; Eigenbrod et al. 2008; Fahrig and Rytwinski 2009). For example, traffic mortality in otter (*Lutra lutra*) populations in Mecklenburg-Vorpommern (eastern Germany) was found to be 59% of all mortality (Binner et al. 1999). In fact, traffic has generally become the most important source of mortality for otters in Germany and is considered a serious threat to the persistence of this protected population (Hauer et al. 2002). Similarly, traffic has been demonstrated to be the most important source of mortality for badgers (*Meles meles*) in the United Kingdom (Clarke et al. 1998), and one of the most significant sources of mortality for the Iberian lynx (*Lynx pardinus*) in southwest Spain (Ferreras et al. 1992).

However, road mortality will not always lead to population declines. If populations are limited by other resources (e.g., the availability of mates, habitat, or food), moderate levels of mortality might simply increase opportunities for other animals in the population, thus keeping it viable. This principle is the basis by which fish and other wildlife species are harvested sustainably (at least in theory) and hunting quotas are determined. In fact, several studies of small animals have noted no population declines in the face of moderate road mortality (e.g., Adams and Geis 1983; Mazerolle 2004).

To determine if road mortality is leading to population decline, mortality can be explicitly connected to population effects through population viability analysis (PVA) or related modeling approaches (Beissinger and McCullough 2002). The basic idea of PVA is to construct a demographic model for a species based on information about its growth, reproduction, and survival. Estimates of road mortality are used to simulate the effects of reduced survival on population dynamics over time. These simulations can estimate the probability of a population going extinct or dropping below a target abundance within a specified time horizon (usually decades to a few hundred years). Further details on using PVA to estimate the effects of road mortality on small animal populations, along with an example, are given in Box 3.2.

The time scales of PVAs (i.e., decades to hundreds of years) illustrate an important point—road impacts can take a long time to show up in measures of population size (e.g., Findlay and Bourdages 2000). For example, Row et al. (2007) observed only three deaths of radio-tracked eastern ratsnakes (*Pantherophis alleghaniensis*) during an eight-year study. Yet, extrapolating this added mortality to long-term population dynamics led to an estimated 90–99% local extinction probability over 500 years. Without this road mortality, the probability of extinction for this population was less than 10%.

The potential for long-term effects of roads leads to two important observations. First, the ultimate effects of roads on populations do not appear immediately, so newer roads may cause substantial mortality before population-level effects are observed. Second, the apparent absence of population effects should be interpreted cautiously; in some cases, it may take decades before the effects of road mortality on population persistence can be determined.

3.2.1.2. Not Just Death: Sex Ratios, Age Structure, and Selection

Increased mortality risk can result in reduced population persistence, but it can also have less obvious consequences for populations. The consequences of mortality are not evenly distributed among individuals within a population. For example, female land turtles and pond turtles migrate in order to nest, so road mortality affects females more than males (e.g., Steen et al. 2006). As a result, turtle sex ratios may be male biased, which has been observed for painted turtles (*Chrysemys picta*) and snapping turtles (*Chelydra serpentina*, Steen and Gibbs 2004), diamond-backed terrapins (*Malaclemys terrapin*, Grosse et al. 2011), and several other species of North American freshwater turtles (Aresco 2005*b*; Gibbs and Steen 2005). In general, loss of females is more of a concern than loss of males, because one male can mate with many females. Because turtles can live for decades, it is not yet known whether the death of females in pond turtle populations will ultimately lead to population declines and extinctions. At a minimum, biased sex ratios can be considered an early indicator that road mortality is occurring at a significant rate.

In addition to being sex-biased, road mortality can also be age or stage-biased. In spotted salamander

Box 3.2. Road Mortality and Population Viability Analysis (PVA)

The Problem

Road mortality occurs in most animal populations living near roads. However, simply documenting mortality explains very little about the long-term effects of that mortality on populations. Many populations may be limited by factors other than adult mortality, so that a small amount of additional adult mortality can be easily absorbed by the population. In other cases, reproductive females could be extremely important to maintaining the population, so that road mortality has population effects much larger than expected from the roadkill numbers alone. Similarly, mitigation that reduces roadkill might have very large or very small effects on the future survival of a population.

PVA refers to a set of modeling tools used to link environmental impacts from roads or the effectiveness of mitigation structures to the future status of a population of interest. The basic idea of a PVA in a road context is to construct a model for survival and reproduction in a population and then add the effects of road mortality. Typically, data are compiled from published sources, so PVAs are most commonly used on species that have been well studied.

An Example

Gibbs and Shriver (2005) used a variety of data on spotted salamanders in Massachusetts to forecast the effects of road mortality on population survival. First, they calculated the probability of a salamander successfully crossing a road based on their movement speed, road width, and traffic volume (see Box 3.1). Then, using maps of actual road networks in Massachusetts, they converted this value into a yearly probability of road mortality.

Additionally, they constructed a basic population model for spotted salamanders from data on adult survival, adult fecundity, larval survival, and juvenile survival. Importantly, their model included a larval carrying capacity—that is, they incorporated the typical characteristic of amphibian populations often being limited by resources for larvae in their aquatic habitats. They then created simulations that combined the road mortality model with the population model to reflect increased mortality from roadkill. By simulating population trajectories multiple times under different sets of assumptions, they were able to calculate probabilities of population extinction for populations suffering different levels of road mortality.

Outcome

The model of Gibbs and Shriver (2005) showed that an annual road mortality rate of less than 10% could typically be absorbed by a population without increasing the risk of extirpation. Populations suffering from road mortality greater than 20% were predicted to go extinct. Gibbs and Shriver (2005) suggest that the critical level of road mortality for spotted salamander populations will be somewhere between 10 and 20% per year. Note that none of this information about thresholds was obvious from the different types of raw data.

populations in the Adirondacks, egg masses near roads are smaller than egg masses further from roads (Karraker and Gibbs 2011). This effect likely occurs because high adult mortality has shifted these populations toward smaller, younger individuals.

Finally, road mortality may be selective in subtle ways that could affect the fitness of individuals in a population. In many populations, some individuals migrate while others remain as residents in the areas where they were born. If only more fit individuals migrate (Bélichon et al. 1996), and migration exposes animals to road mortality, then selection could reduce the mean individual fitness of the population. Although this issue has not been studied, it once again illustrates that the effects of road mortality can go beyond removing individuals from the population.

These selective aspects of road mortality raise the prospect that populations could, over time, adapt in ways that reduce their risk of road mortality. Given that exploratory and migratory behaviors are often heritable, it is possible that populations could evolve more efficient road-crossing behaviors or increased road avoidance over many generations. Indeed, a study in Massachusetts reported that spotted salamander eggs from roadside pools had higher rates of hatching in these habitats than eggs transferred from forest populations distant from roads, suggesting some local adaptation to roadside water quality (Brady 2012, Chapter 4). Nevertheless, the hatching success of eggs from populations in these roadside pools was still much lower than that of eggs left in forest pools. Whether evolutionary adaptation to roads occurs is certainly an interesting issue, but it is unlikely to go very far toward compensating for the negative effects of roads.

3.2.2. Barrier Effects

The second major class of road effects is barrier effects. That is, roads may act as barriers to movement even if they don't kill individuals. Because many animals move between habitats to breed, forage, or find shelter, barrier effects can be as or more important than road mortality in terms of how much they affect populations. If animals cannot reach the habitats they require for survival and reproduction, they can be threatened by roads even in the absence of road mortality.

Roads can be barriers to animal movement for several distinct reasons. First, animals may avoid going near moving vehicles, either because of noise or because of some visual cue (Chapter 4). Second, animals accustomed to natural substrates may be deterred from crossing road surfaces (e.g., Case Study 9.6), and small mammals, such as mice and chipmunks, may be particularly likely to avoid pavement (McGregor et al. 2008). Third, animals may not avoid roads per se, but instead avoid some related features, such as steep roadside verges (e.g., Marsh et al. 2005) or predators that inhabit road edges. In most cases, biologists and transportation planners will not know precisely what cues animals are avoiding, only that roads appear to be barriers to movement.

Barrier effects may be most detrimental to animals that make regular migrations between different habitats (Gibbs 1998). These movements are characteristic of many amphibians, which migrate from terrestrial habitats to wetlands (usually in the spring) in order to breed, and then migrate back to terrestrial habitats for summer or winter shelter or foraging. Movements between habitats are also common for turtles, which move between foraging and mating habitat and nesting habitat. Barrier effects may also be critical for wide-ranging species, such as snakes that follow pheromone trails for long distances in order to find mates (Shepard et al. 2008). Not all species make regular interhabitat movements, however. Many small mammals are territorial, and once they establish a territory, they make few long-distance movements. Lizards and terrestrial salamanders may also be highly territorial, and thus do not require movements between habitats in order to complete their life cycle.

For a number of animal taxa, there is good evidence for the barrier effects of roads. Several studies have shown that salamanders avoid crossing roads (Gibbs 1998; deMaynadier and Hunter 2000), as do frogs (Mazerolle et al. 2005), turtles, snakes (Andrews and Gibbons 2005; Shepard et al. 2008), and small mammals (McGregor et al. 2008). In some cases, roads can be almost complete barriers to movement. Frank et al. (2002) suggested an estimated average threshold of 15 vehicles per minute at which roads become an absolute barrier to otters (for a crossing duration of 20 seconds). Baur and Baur (1990) found that only 1 of 169 recaptured marked land snails (*Arianta arbustorum*) in Sweden had successfully crossed an 8 m wide paved road (with only 500 vehicles/day, or about 21/hour) after 3 months. Even more daunting, Wirth et al. (1999) found that none of 560 marked land snails (*Helicella itala*) crossed roads of 6 m and 9 m widths, which would indicate that these roads are probably complete barriers for snails. Although road planning will rarely be focused on snails, these animals share many attributes with amphibians, such as moving slowly and desiccating quickly.

More commonly, though, roads are partial barriers to movement that reduce but do not eliminate movement across roads. The degree to which these barrier effects will reduce populations varies, but at least one study showed that the fragmentation of habitats (in this case frog breeding and foraging habitats) was associated with reduced species richness (Becker et al. 2007). Roads may be particularly important in this regard since rural roads often follow streams, rivers, or lakeshores, thus separating aquatic habitats from terrestrial habitats over long distances.

3.2.3. Road Mortality versus Barrier Effects

Roadkill effects and barrier effects are not independent from one another. In many cases, both effects will occur together; roads with heavy traffic may be avoided by animals, and animals that do attempt to cross them may be at high risk of death or injury. However, there is also an important trade-off between road mortality and barrier effects. This trade-off occurs because species that avoid roads get killed less often but are more subject to barrier effects. The relationship between mortality and avoidance is particularly relevant to mitigating the effects of roads, because some approaches that reduce mortality (e.g., erecting fences or other barriers) may also reduce movement across roads. Installation of barriers may be justified when barrier effects are much less detrimental to a population than road mortality caused by attempted road crossings. More generally, approaches that decrease mortality but increase barrier effects can be effective when road mortality is very high (i.e., animals cannot make it across the road) but tend to be less useful when road mortality is low (Jaeger and Fahrig 2004). In most cases, fences along roads should be combined with a sufficient density of wildlife passages to make the fenced road permeable for animals.

3.2.4. Habitat Loss and the Road-Effect Zone

New road construction often eliminates valuable wildlife habitats. Some area is lost to the road itself, and additional habitat may be lost to road shoulders and verges. Still, most roads are narrow and the area of habitat lost to roads tends to be small relative to other forms of development. Thus, the greater concern with respect to roads and habitat loss is typically not the roads themselves, but the degradation of habitats adjacent to roads. The specific effects of roads on roadside habitats are dealt with in detail in Chapter 4.

The road-effect zone refers to the area over which the influence of a road on adjacent ecosystems is detectable (Forman and Alexander 1998). The road-effect zone can consist of habitat that is too degraded to support viable populations, and it can also be an area that is avoided by animals. Typically, the road-effect zone is

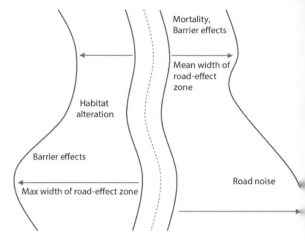

Figure 3.2. Example of a road-effect zone. Road-effect zones are typically asymmetrical over the length of a road—that is, narrow in some places but wider in others, depending on the landscape and the nature of the surrounding habitats. In addition, different kinds of road effects may extend over different spatial scales. For example, the effects of road mortality may be more localized than the effects of road noise. Barrier effects can occur at different scales, depending on the dispersal ability of the species and whether or not multiple habitats are required for the life history of an individual animal. *Credit: David M. Marsh.*

measured in the context of the abundance or presence of some target species (or group of species). The width of a road-effect zone will depend on factors such as the size of the road, traffic volume, local topography, and sensitivity of the habitats and species in roadside areas. A road-effect zone is not expected to be uniform; road effects will usually be strongest close to a road and then decline with distance from the road (Figure 3.2). When increasing distance from the road is no longer correlated with ecological changes from the road, the limit of the road-effect zone has been reached.

Several studies have attempted to quantify road-effect zones in terms of their influence on small animals. Forman and Deblinger (2000) examined a variety of ecological factors that affected birds and amphibians along a four-lane highway in Massachusetts. All factors affected adjacent habitat over distances greater than 100 m from roads. Some effects extended only a little beyond this (e.g., increased plant invasion, altered wetland drainage, presence of road deicing salts) whereas other effects were still detectable at greater than 1000 m from roads (e.g., road noise, avoidance by birds). Forman and Deblinger (2000) also found that the road-effect zone was highly asymmetrical. In many areas of the road, the road-effect zone was relatively narrow, but other areas had dramatically wider road-effect zones due to topography and streamflow patterns.

Other studies demonstrate that road-effect zones are species- and site-dependent. In the Mojave Desert of California, Boarman and Sazaki (2006) found that Mohave desert tortoise (*Gopherus agassizii*) abundance was reduced for 400–800 m from a highway, presumably as a result of road mortality. At the upper extreme, road-effect zones have been estimated to be as high as 2800 m for birds and 17 km for mammals (Benítez-López et al. 2010). Meanwhile, smaller roads are predicted to have road-effect zones more narrow than those found for highways. For gravel-surfaced, one- to two-lane roads through eastern forests, Marsh and Beckman (2004) found road-effect zones for woodland salamanders more on the order of 40–80 m, depending on the weather and time of year.

3.3. Landscape Effects

3.3.1. Introducing the Concept of Landscape-Scale Effects

Local effects of roads, such as increased mortality and decreased ability to move between habitats, can negatively affect animal populations in areas adjacent to roads. In some cases, even a single road running through an otherwise undeveloped area can have major impacts on ecological infrastructure. For example, the 2,594 km transcontinental Inter-Oceanic Highway in Brazil and Peru, completed in 2011, bisects the Amazonian rain forest and will almost certainly lead to increased logging, burning, and hunting in areas near the road (Brandon et al. 2005). But more commonly, a single road will have less dramatic effects than a road network. In the latter situation, the critical feature for animal populations may be the total density and configuration of roads in the landscape. One reason why road density matters is that the local effects of all the individual roads will accumulate across the landscape. Members of mobile species may encounter many roads, and their risk of dying as roadkill increases every time they try to cross another road. With enough roads, mortality may be too high for a population to sustain, and

population declines will result (Robinson et al. 2010). The collective local effects of roads at the landscape scale are one class of landscape effects.

However, another application of the term "landscape effects" refers to ecological effects that only appear at larger spatial scales. These effects occur when processes that are unimportant to local populations affect the persistence of populations at the landscape scale. An example of this can be seen with groups of animal populations (e.g., frogs) that are linked by occasional dispersal into a metapopulation (Levins 1969; Compton et al. 2007). Most individual frogs may stay in their natal pond, but typically a small percentage will disperse and find a different pond. Over the span of decades, each breeding population has some real probability of local extinction. The pond may dry out, too many predators may get in, or a disease might pass through the population. Typically, with occasional dispersal, the pond will eventually be recolonized, and the population can begin to grow again (see also Section 8.3). With many ponds and populations in the landscape, the metapopulation can persist when random (natural or nonanthropogenically driven) local extinctions are balanced by recolonization and re-establishment of new populations.

Now consider what may happen when roads intervene and reduce dispersal between ponds. If roads are barriers to movement, animals trying to disperse will be forced to remain at their natal ponds. No local effects would be seen here; animals would still have access to their pond and to adjacent upland habitat. However, should that population go extinct, the barrier effects of roads can prevent recolonization. Later, another population might go extinct, then another, and another. Ultimately, reduced dispersal at the landscape scale can slowly drive a metapopulation to extinction. Simulations of population dynamics based on dispersal and survival data suggest that this scenario is a real possibility (e.g., Hels and Nachmann 2002), and it has been observed in the field (Means 1999).

Metapopulation dynamics are just one example of a landscape-scale process. Other kinds of landscape processes include predator-prey dynamics, the balancing of population density between spatially connected areas, and source-sink dynamics. All of these processes can present challenges for researchers and conservation planners because the landscape-scale movements that underlie each are notoriously difficult to study. Furthermore, studying populations across multiple sites requires greater investment and longer time scales than does a study at a single location (Chapter 12). Typically, mitigation for the effects of roads on wildlife is done on a road-by-road basis. However, even when practical considerations require a local approach, it is important to remember that landscape considerations are quite real and can have a large impact on the ultimate success of any conservation strategy.

3.3.2. Road Density

At the landscape scale, road density may be a key factor affecting the viability of animal populations (Figure 3.3). Road density can be measured in terms of the total length of roads within a specific area, in terms of the total road surface area (which takes into account the length of roads and their width), or in terms of the area of roads plus adjacent road verges. When road mortality (as opposed to, say, barrier effects) is the concern, traffic volume across the landscape—that is, average number of cars per unit time traversing roads within the region—should be considered.

However road density is measured, evidence from a range of different animal species has shown that road density can have a strong influence on where animal populations do and do not persist. In Ontario wetlands, road density within a few hundred meters of ponds is negatively related to species richness of both amphibians and reptiles (Findlay and Houlahan 1997). In Switzerland, European treefrog (*Hyla arborea*) presence is negatively associated with road surface area over distances from 100 m to 1 km (Pellet et al. 2004). Moreover, this effect on treefrogs was even stronger when the researchers also considered traffic intensity on the roads, suggesting that cars were directly responsible for the missing frogs.

Not all species will be influenced by road density within the landscape. The same study that found effects of road density on amphibians and reptiles in Ontario did not find effects on the species richness of small mammals (Findlay and Houlahan 1997). A more recent study in Ontario comparing forest habitats surrounded by either high or low road densities found no effects of road density on the abundance of white-footed mice (*Peromyscus leucopus*, Rytwinski and Fahrig 2007). Burrow densities of black-tailed prairie

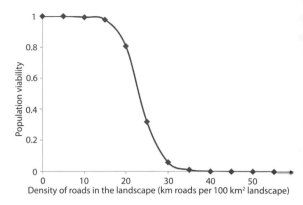

Figure 3.3. The road density threshold. The viability of wildlife populations decreases when the density of roads in the landscape increases. The specific value of the threshold depends on the particular species, traffic volumes, the amount and quality of habitat remaining, and other human impacts present in the landscape. Once the threshold is passed, it would be very difficult to rescue a declining population. *Credit: Jochen A. G. Jaeger, adapted from Jaeger and Holderegger (2005).*

dogs (*Cynomys ludovicianus*) in remnant grasslands in Colorado were positively correlated with road density, possibly because areas with many roads provided refuges for prairie dogs from their predators (Johnson and Collinge 2004). While it is too soon to make generalizations about the effects of road density on small mammals, it may be that some mammals are successful enough at avoiding cars that the cumulative effects of road mortality are lower for small mammals as compared to slower animals like amphibians and turtles.

In addressing the effects of road density, it is important to realize that the effects observed for individual roads at the local scale need not be mirrored in the effects of road density at the landscape scale. For example, if the primary effects of roads are barrier effects, there may be no apparent effects of roads on animal abundance in adjacent habitats (as compared to habitats far from the road), but a large negative effect of road density on abundance could be occurring at the landscape scale. Similarly, if animals are attracted to roads for feeding or thermoregulation but suffer mortality as a result, a positive effect of individual roads on abundance could occur on a local scale with a negative effect of road density at the landscape scale (i.e., fewer individuals are alive, but those that remain are found near roads). Although it is difficult to predict in advance how road density will affect a given animal population, it is feasible to determine this empirically. By measuring animal abundance or species richness at sites with higher and lower road densities, it can be apparent whether road density is likely to be an important factor for the persistence of an animal population (Roedenbeck et al. 2007). Given that the home ranges of wide-ranging animals often overlap with a number of roads and that road networks subdivide wildlife populations into a patchwork of subpopulations, future studies in road ecology should directly address ecological effects at the landscape scale (Roedenbeck et al. 2007; van der Ree et al. 2011).

3.3.3. Habitat Fragmentation

Roads are an important cause of habitat fragmentation, but they are certainly not the only cause. Other types of intervening land use that often co-occur with road density, including urban and suburban development, row-crop agriculture, and railways, can affect animal populations as much as roads do (e.g., van der Grift 1999; Andrews et al. 2008). The direct effects of roads—population isolation, increased mortality of dispersing animals, decreased habitat quality due to edge effects—are also frequently seen with other types of land use. Ultimately, population persistence in fragmented landscapes will be influenced not just by roads, but by all land uses that contribute to fragmentation. Thus, in addition to roads, landscape conservation planning should take into account the combined effects of all major land uses on animal survival and movement.

A clear example of the detrimental effect of landscape fragmentation by roads is the continued decline of brown hare (*Lepus europaeus*) populations in Switzerland as a consequence of the increased density of major roads in combination with intensive agricultural practices. These landscape alterations have made the hare populations much more vulnerable to unfavorable weather conditions, and hare densities are currently low despite the implementation of various habitat improvement measures (Zellweger-Fischer et al. 2011). The hare is currently listed as endangered, and its extinction may be impossible to prevent. To conserve the brown hare in Switzerland, large roadless areas of high ecological quality would need to be created and maintained (Roedenbeck and Voser 2008), which is unlikely because it would require the removal of existing roads.

Another, more encouraging, example is the story of the European badger (*Meles meles*) in the Netherlands. The observed decline of the badger populations in the 1970s was addressed by a national defragmentation program established in 1984 (van der Grift 2005). It included the construction of numerous culverts as badger pipes, among other measures. The first badger pipe was constructed in 1974 and populations have since increased (Dekker and Bekker 2010). This example, in concert with the example of the brown hare, emphasizes that wildlife passages can only be effective if there is still enough ecological infrastructure in place for the populations in the landscape to persist (Fahrig 2001, 2002).

Because landscape fragmentation contributes to the destruction of established ecological connections between adjoining areas of the landscape (Jaeger, Grau, and Haber 2005), it affects not only individual habitats but also entire communities and ecosystems. Landscape fragmentation also lowers the resilience of animal populations (i.e., their ability to recover after disturbances, Thomas and Jones 1993). Thus, landscape fragmentation increases the risk of populations becoming extinct, because isolated populations are more sensitive to natural stress factors such as natural disturbances (e.g., weather conditions, fires, diseases). Roads also enhance human access to wildlife habitats, facilitate the spread of invasive species, and alter habitat conditions for wildlife populations (Chapter 4).

Entire books have been written on the subject of habitat fragmentation (e.g., Lindenmayer and Fischer 2006); here, we just summarize a few important principles that have emerged from fragmentation research. First, maintaining ecological infrastructure (e.g., corridors of intact habitat that connect larger habitat blocks) can help mitigate some of the negative effects of landscape-scale fragmentation. Such corridors can allow movement of a wide range of species, so establishment of corridors need not be done on a single-species basis (Tewksbury et al. 2002). Second, when habitats are fragmented, threshold levels of habitat loss are likely to exist beyond which the probability of a population persisting declines rapidly, from nearly certain to persist to nearly certain to disappear (Figure 3.4). Such thresholds will be at different levels of habitat loss for different species, depending on how fragmentation affects survival and movement ability (Fahrig 2001).

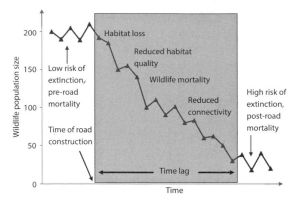

Figure 3.4. Four ecological impacts of roads on animal populations and the time lag for the cumulative effect. After the time lag (often on the order of decades), populations are smaller, they exhibit greater relative fluctuations over time, and they are more vulnerable to extinction. *Credit: Jochen A. G. Jaeger, adapted from Forman et al. (2003).*

Generalization is difficult, but typical thresholds may be observed with loss of anywhere from 25–80% of intact habitat (Fahrig 2001). Third, recently fragmented landscapes may incur an "extinction debt" (see Tilman 1994) in which these landscapes contain more species than are able to persist over the long term. As local extinction is a process that may require tens to hundreds of years, a recently fragmented landscape will likely contain many species that will eventually disappear. Because of the extinction debt, fragmentation effects (genetics, Jackson and Fahrig 2011) may continue to build over time, even in the absence of any additional development.

3.3.4. Population and Landscape Genetic Effects

Increasingly, genetic data are being used to understand the barrier effects of roads and to map animal movement across a fragmented landscape (e.g., Balkenhol and Waits 2009). An advantage of genetic studies is that they effectively average movement over long periods of time. Because long-distance dispersal occurs infrequently and can be difficult to observe directly, genetic studies can be much more efficient for examining the barrier effects of roads. Basic population genetic techniques have become much more accessible and inexpensive over the past 10 years, putting these kinds of studies well within the reach of smaller agencies and environmental consultants (e.g., Simmons et al. 2010).

Most genetic studies examine differentiation between populations separated by roads as compared to populations not separated by roads (or those separated by other landscape features). The assumption is that reduced movement will, over time, lead to genetic differentiation among populations. The magnitude of genetic differentiation can be used to infer the extent of reductions in movement due to roads or other landscape features. In spite of the fact that most modern roads are only a few decades old, researchers have demonstrated the feasibility of this approach. For example, population genetic analysis showed that an interstate highway in Virginia contributed to differentiation in eastern red-backed salamander populations (*Plethodon cinereus*), whereas smaller roads did not produce such effects (Marsh et al. 2008). Similarly, bank voles (*Myodes glareolus*) in Europe showed genetic subdivision related to a highway but not to a smaller road (Gerlach and Musolf 2000). Lastly, populations of agile frogs (*Rana dalmatina*) located ≤180 m from a 20-year-old major highway showed significantly higher rates of population subdivision than control populations located farther from the road (≥550 m; Lesbarrères et al. 2006).

In addition to being a tool to study movement across roads, genetics may be important to population persistence in fragmented environments. If populations are isolated by roads, they can lose genetic diversity over time. Reduced diversity can lead to decreased fitness, decreased ability to respond to future environmental changes, and an increased likelihood of extinction. Recent studies have indeed found reduced genetic diversity in some populations isolated by roads. For example, three lizard species and one bird species all showed concordant losses of genetic variation associated with isolation by highways and development near Los Angeles, California (Delaney et al. 2010). Similarly, timber rattlesnake (*Crotalus horridus*) populations in New York isolated by roads had reduced genetic diversity and increased differentiation from other populations (Clark et al. 2010). The ramifications of genetic diversity for population persistence will take a long time to become apparent—much longer than the time scale of most mitigation or monitoring projects (Chapter 12). However, if mitigation measures are successful in allowing movement across roads, these genetic issues are likely to take care of themselves. The amount

Table 3.2. Summary of the direct effects of roads on animal populations and communities. Effects are categorized at the scale of local versus landscape and by the time scales over which they would typically be observed

	Short term (a few years)	Medium term (decades)	Long term (decades to centuries)
Local Effects	• Road mortality • Behavioral changes • Decreased dispersal between habitats • Reduced population abundance • Altered population age structure or sex ratios	• Altered population age structure or sex ratios • Reduced population abundance • Changes in community structure • Extinction risk to populations	• Population extinction • Decreased species richness • Changes in community structure and food webs
Landscape Effects	• Reduced abundance in areas with high road densities • Impaired dispersal and migration in areas with high road densities	• Decreased recolonization rates • Loss of genetic diversity • Extinction debt in areas with high road densities	• Metapopulation extinction • Changes in communities • Decreased species richness • Reduced fitness and evolutionary capacity to respond to environmental change

of movement needed to sustain genetic diversity is on the order of 1–10 migrants per generation (Mills and Allendorf 1996), assuming these migrants actually reproduce. This is much lower than the amount of movement that will be needed to keep populations viable when dispersal or migration is part of the typical life cycle.

3.4. Key Points

• Roads have direct effects on populations at both local and landscape scales (Table 3.2). At a local scale, these effects may be due to road mortality, but barrier effects and habitat loss can also be important. At the landscape scale, roads can disrupt the dynamics of metapopulations and the processes that contribute to long-term persistence.

• The time scales of these different kinds of effects will vary considerably (Table 3.2). Roadkill effects on mortality rates may be seen very quickly, particularly in short-lived animals. Barrier effects may also have rapid consequences for populations if roads disrupt the ecological infrastructure needed for animals to complete their life cycle.

• Over intermediate time scales (e.g., decades), these processes may lead to effects such as genetic changes in populations and substantial decreases in population abundance. Other processes, though no less important, may take even longer to become apparent. Effects of roads on population extinction

and patterns of biodiversity may require several decades or even a few hundred years to be observed.

• Even though mitigation measures may predominantly address the local and short-term effects of roads, landscape-scale and long-term effects also need to be considered in any mitigation framework.

• The ecological impacts of road networks will increase tremendously in the coming years given the ambitious road construction plans in many countries, and many ecological effects of fragmentation will only manifest themselves over time.

• Remaining unfragmented areas are under enormous pressure, and these areas are critical to conserving biodiversity (e.g., Selva et al. 2011). Accordingly, the recent European report on landscape fragmentation (EEA and FOEN 2011) identified the following three highest priority measures: (1) immediate protection of large unfragmented areas, ecologically significant areas, and wildlife corridors; (2) monitoring of landscape fragmentation; and (3) application of fragmentation analysis as a tool in transportation planning and regional planning. These measures should also be highly relevant to road planning and mitigation in North America.

LITERATURE CITED

Adams, L. W., and A. D. Geis. 1983. Effects of roads on small mammals. Journal of Applied Ecology 20:403–415.

Andrews, K. M., and J. W. Gibbons. 2005. How do highways influence snake movement? Behavioral responses to roads and vehicles. Copeia 2005:772–782.

Andrews, K. M., J. W. Gibbons, and D. M. Jochimsen. 2008. Ecological effects of roads on amphibians and reptiles: a literature review. Pages 121–143 *in* Urban herpetology. J. C. Mitchell, R. E. Jung Brown, and B. Bartholomew, editors. Herpetological Conservation Vol. 3. Society for the Study of Amphibians and Reptiles, Salt Lake City, Utah, USA.

Aresco, M. J. 2003. Highway mortality of turtles and other herpetofauna at Lake Jackson, Florida, USA, and the efficacy of a temporary fence/culvert system to reduce roadkills. Pages 433–449 *in* Proceedings of the 2003 international conference on ecology and transportation. C. L. Irwin, P. Garrett, and K. P. McDermott, editors. Center for Transportation and the Environment, Raleigh, North Carolina, USA.

Aresco, M. J. 2005*a*. Mitigation measures to reduce highway mortality of turtles and other herpetofauna at a north Florida lake. Journal of Wildlife Management 69:549–560.

Aresco, M. J. 2005*b*. The effect of sex-specific terrestrial movements and roads on the sex ratio of freshwater turtles. Biological Conservation 123:37–44.

Balkenhol, N., and L. P. Waits. 2009. Molecular road ecology: exploring the potential of genetics for investigating transportation impacts on wildlife. Molecular Ecology 18:4151–4164.

Baur, A., and B. Baur. 1990. Are roads barriers to dispersal in the land snail *Arianta arbustorum*? Canadian Journal of Zoology 68:613–617.

Becker, C. G., C. R. Fonseca, C.F.B. Haddad, R. F. Batista, and P. I. Prado. 2007. Habitat split and the global decline of amphibians. Science 318:1775–1777.

Bélichon, S., J. Clobert, and M. Massot. 1996. Are there differences in fitness components between philopatric and dispersing individuals? Acta Oecologica 17:503–507.

Beissinger S. R., and D. R. McCullough. 2002. Population viability analysis. University of Chicago Press, Chicago, Illinois, USA.

Benítez-López, A., R. Alkemade, and P. A. Verweij. 2010. The impacts of roads and other infrastructure on mammal and bird populations: a meta-analysis. Biological Conservation 143:1307–1316.

Binner, U., A. Hagenguth, R. Klenke, and A. Waterstraat. 1999. Auswirkungen und Funktion unzerschnittener störungsarmer Landschaftsräume auf Wirbeltierarten mit großen Raumansprüchen. Analyse des Einflusses von Zerschneidungen und Störungen auf die Population des Fischotters (*Lutra lutra*) in Mecklenburg-Vorpommern. Endbericht zum Teilprojekt 3.2 im BMBF-Verbundprojekt Gesellschaft für Naturschutz und Landschaftsökologie e.V., Kratzeburg, Germany. [In German.]

Boarman, W. I., and M. Sazaki. 2006. A highway's road-effect zone for desert tortoises (*Gopherus agassizii*). Journal of Arid Environments 65:94–101.

Brady, S. P. 2012. Road to evolution? Local adaptation to road adjacency in an amphibian (*Ambystoma maculatum*). Scientific Reports 2:235–239.

Brandon, K., G.A.B. Da Fonseca, A. B. Rylands, and J.M.C. Da Silva. 2005. Introduction to special section: Brazilian conservation: challenges and opportunities. Conservation Biology 19:595–600.

Brzeziński, M., E., George, and M. Żmihorski. 2012. Road mortality of pond-breeding amphibians during spring migrations in the Mazurian Lakeland, NE Poland. European Journal of Wildlife Research 58:685–693.

Clark, R. W., W. S. Brown, R. Stechert, and K. R. Zamudio. 2010. Roads, interrupted dispersal, and genetic diversity in Timber Rattlesnakes. Conservation Biology 24:1059–1069.

Clarke, G. P., P.C.L. White, and S. Harris. 1998. Effects of roads on badger *Meles* populations in south-west England. Biological Conservation 86:117–124.

Compton, B. W., K. McGarigal, S. A. Cushman, and L. R. Gamble. 2007. A resistant-kernel model of connectivity for amphibians that breed in vernal pools. Conservation Biology 21:788–799.

Crawford, B. A., J. C. Maerz, N. P. Nibbelink, K. A. Buhlmann, T. M. Norton, and S. E. Albeke. 2014. Hot spots and hot moments of diamondback terrapin road-crossing activity. Journal of Applied Ecology 51:367–375.

Crawford, B. A., J. C. Maerz, N. P. Nibbelink, K. A. Buhlmann, and T. M. Norton. 2014. Estimating the consequences of multiple threats and management strategies for semi-aquatic turtles. Journal of Applied Ecology 51:359–366.

Dekker, J.J.A., and H. Bekker. 2010. Badger (*Meles meles*) road mortality in the Netherlands: the characteristics of victims and the effects of mitigation measures. Lutra 53:81–92.

Delaney, K. S., S.P.D. Riley, and R. N. Fisher. 2010. A rapid, strong, and convergent genetic response to urban habitat fragmentation in four divergent and widespread vertebrates. PLoS ONE 5(9):e12767.

deMaynadier, P. G., and M. L. Hunter, Jr. 2000. Road effects on amphibian movements in a forested landscape. Natural Areas Journal 20:56–65.

Eigenbrod, F., S. J. Hecnar, and L. Fahrig. 2008. The relative effects of road traffic and forest cover on anuran populations. Biological Conservation 141:35–46.

EEA and FOEN (European Environment Agency and Swiss Federal Office for the Environment). 2011. Landscape fragmentation in Europe. EEA Report No 2/2011. Publications Office of the EU, Luxembourg. http://www.eea.europa.eu /publications/landscape-fragmentation-in-europe/. Accessed 26 April 2014.

Fahrig, L. 2001. How much habitat is enough? Biological Conservation 100:65–74.

Fahrig, L. 2002. Effect of habitat fragmentation on the extinction threshold: a synthesis. Ecological Applications 12:346–353.

Fahrig, L., J. H. Pedlar, S. E. Pope, P. D. Taylor, and J. F. Wegner. 1995. Effect of road traffic on amphibian density. Biological Conservation 73:177–182.

Fahrig, L., and T. Rytwinski. 2009. Effects of roads on animal abundance: an empirical review and synthesis. Ecology and Society 14:21.

Ferreras, R., J. J. Aldama, J. F. Beltran, and M. Delibes. 1992. Rates and causes of mortality in a fragmented population of Iberian lynx *Felis pardini* Temminck, 1824. Biological Conservation 61:197–202.

Findlay, C. S., and J. Bourdages. 2000. Response time of wetland biodiversity to road construction on adjacent lands. Conservation Biology 14:86–94.

Findlay C. S., and J. Houlahan. 1997. Anthropogenic correlates of species richness in southeastern Ontario wetlands. Conservation Biology 11:1000–1009.

Forman, R.T.T., and L. E. Alexander. 1998. Roads and their major ecological effects. Annual Review of Ecology and Systematics 29:207–231.

Forman, R.T.T., and R. D. Deblinger. 2000. The ecological road-effect zone of a Massachusetts (USA) suburban highway. Conservation Biology 14:36–46.

Forman, R.T.T., K. Heanue, J. Jones, F. Swanson, T. Turrentine, T. C. Winter, D. Sperling, et al. 2003. Road ecology: Science and solutions. Island Press, Washington, DC, USA.

Frank, K., P. Eulberg, K. Hertweck, and K. Henle. 2002. A simulation model for assessing otter mortality due to traffic. IUCN Otter Specialist Group Bulletin (Proceedings VIIth international otter colloquium) 19A:64–68.

Gerlach, G., and K. Musolf. 2000. Fragmentation of landscape as a cause for genetic subdivision in bank voles. Conservation Biology 14:1066–1074.

Gibbs, J. P. 1998. Distribution of woodland amphibians along a forest fragmentation gradient. Landscape Ecology 13:263–268.

Gibbs, J. P., and W. G. Shriver. 2002. Estimating the effects of road mortality on turtle populations. Conservation Biology 16:1647–1652.

Gibbs, J. P., and W. G. Shriver. 2005. Can road mortality limit populations of pool-breeding amphibians? Wetlands Ecology and Management 13:281–289.

Gibbs, J. P., and D. A. Steen. 2005. Trends in sex ratios of turtles in the United States: implications of road mortality. Conservation Biology 19:552–556.

Grosse, A. M., J. C. Maerz, J. Hepinstall-Cymerman, and M. E. Dorcas. 2011. Effects of roads and crabbing pressures on diamondback terrapin populations in coastal Georgia. Journal of Wildlife Management 75:762–770.

Guinard, É., R. Julliard, and C. Barbraud. 2012. Motorways and bird traffic casualties: carcasses surveys and scavenging bias. Biological Conservation 147:40–51.

Hauer, S., H. Ansorge, and O. Zinke. 2002. Mortality patterns of otters (*Lutra lutra*) from Eastern Germany. Journal of Zoology 256:361–368.

Hels, T., and E. Buchwald. 2001. The effect of road kills on amphibian populations. Biological Conservation 99:331–340.

Hels, T., and G. Nachman. 2002. Simulating viability of a spadefoot toad *Pelobates fuscus* metapopulation in a landscape fragmented by a road. Ecography 25:730–744.

Hummel, D. 2001. Protection of amphibians through speed limits—an aerodynamic study. Natur und Landschaft 76:530–533. [In German.]

Inbar, M., and R. T. Mayer. 1999. Spatio-temporal trends in armadillo diurnal activity and road-kills in central Florida. Wildlife Society Bulletin 27:865–872.

Jackson, N. D., and L. Fahrig. 2011. Relative effects of road mortality and decreased connectivity on population genetic diversity. Biological Conservation 144:3143–3148.

Jaeger, J.A.G., J. Bowman, J. Brennan, L. Fahrig, D. Bert, J. Bouchard, N. Charbonneau, et al. 2005. Predicting when animal populations are at risk from roads: an interactive model of road avoidance behavior. Ecological Modelling 185:329–348.

Jaeger J.A.G., and L. Fahrig. 2004. Effects of road fencing on population persistence. Conservation Biology 18:1651–1657.

Jaeger, J., S. Grau, and W. Haber. 2005. Introduction: landscape fragmentation and the consequences. GAIA 14: 98–100. [In German.]

Jaeger, J., and R. Holderegger. 2005. Thresholds of landscape fragmentation. GAIA 14:113–118. [In German.]

Johnson, W. C., and S. K. Collinge. 2004. Landscape effects on black-tailed prairie dog colonies. Biological Conservation 115:487–497.

Karraker, N. E., and J. P. Gibbs. 2011. Contrasting road effect signals in reproduction of long- versus short-lived amphibians. Hydrobiologia 66:213–218.

Langen T. A., K. Gunson, C. Scheiner, and J. Boulerice. 2012. Road mortality in freshwater turtles: identifying causes of spatial patterns to optimize road planning and mitigation. Biodiversity and Conservation 21:3017–3034.

Langen, T. A., K. M. Ogden, and L. L. Schwarting. 2009. Predicting hot spots of herpetofauna road mortality along highway networks. Journal of Wildlife Management 73:104–114.

Lesbarrères, D., C. R. Primmer, T. Lodé, and M. Merilä. 2006. The effects of 20 years of highway presence on the genetic structure of Rana dalmatina populations. Ecoscience 13:531–538.

Levins, R. 1969. Some demographic and genetic consequences of environmental heterogeneity for biological control. Bulletin of the Entomological Society of America 15:237–240.

Lindenmayer, D. B., and J. Fischer. 2006. Habitat fragmentation and landscape change: An ecological and conservation synthesis. Island Press, Washington, DC, USA.

Marsh D. M., and N. G. Beckman. 2004. Effects of forest roads on the abundance and activity of terrestrial salamanders. Ecological Applications 14:1882–1891.

Marsh D. M., G. S. Milam, N. A. Gorham, and N. G. Beckman. 2005. Forest roads are partial barriers to salamander movement. Conservation Biological Conservation 19:2004–2008.

Marsh, D., R. Page, T. Hanlon, R. Corritone, E. Little, D. Seifert, and P. Cabe. 2008. Effects of roads on patterns of genetic differentiation in red-backed salamanders, *Plethodon cinereus*. Conservation Genetics 9:603–613.

Mazerolle, M. J. 2004. Amphibian road mortality in response to nightly variations in traffic intensity. Herpetologica 60:45–53.

Mazerolle, M. J., M. Huot, and M. Gravel. 2005. Behavior of amphibians on the road in response to car traffic. Herpetologica 62:380–388.

McGregor, R. L., D. J. Bender, and L. Fahrig. 2008. Do small mammals avoid roads because of the traffic? Journal of Applied Ecology 45:117–123.

Means, D. B. 1999. The effects of highway mortality on four species of amphibians at a small, temporary pond in northern Florida. Pages 125–128 *in* Proceedings of the third international conference on wildlife ecology and transportation. G. Evink, P. Garrett, and D. Zeigler, editors. Center for Transportation and the Environment, North Carolina State University, Raleigh, North Carolina, USA.

Mills, L. S., and F. W. Allendorf. 1996. The one-migrant-per-generation rule in conservation and management. Conservation Biology 10:1509–1518.

Pellet, J., A. Guisan, and N. Perrin. 2004. A concentric analysis of the impact of urbanization on the threatened European tree frog in an agricultural landscape. Conservation Biology 18:1599–1606.

Robinson, C., P. N. Duinker, K. F. Beazley. 2010. A conceptual framework for understanding, assessing, and mitigating ecological effects of forest roads. Environmental Reviews 18:61–86.

Roedenbeck, I. A., L. Fahrig, C. S. Findlay, J. E. Houlahan, J.A.G. Jaeger, N. Klar, S. Kramer-Schadt, E. A. van der Grift. 2007. The Rauischholzhausen agenda for road ecology. Ecology and Society 12:11. http://www.ecologyandsociety.org/vol12/iss1/art11/. Accessed 26 April 2014.

Roedenbeck, I. A., and P. Voser. 2008. Effects of roads on spatial distribution, abundance and mortality of brown hare (*Lepus europaeus*) in Switzerland. European Journal of Wildlife Research 54:425–437.

Row, J. R., G. Blouin-Demers, and P. J. Weatherhead. 2007. Demographic effects of road mortality in black ratsnakes (*Elaphe obsoleta*). Biological Conservation 137:117–124.

Rytwinski, T., and L. Fahrig. 2007. Effect of road density on abundance of white-footed mice. Landscape Ecology 22:1501–1512.

Santos, S. M., F. Carvalho, and A. Mira. 2011. How long do the dead survive on the road? Carcass persistence probability and implications for road-kill monitoring surveys. PLoS ONE 6:e25383.

Seiler, A., J. O. Helldin, and C. Seiler. 2004. Road mortality in Swedish mammals: results of a drivers' questionnaire. Wildlife Biology 10:225–233.

Selva, N., S. Kreft, V. Kati, M. Schluck, B-G. Jonsson, B. Mihok, H. Okarma, and P. L. Ibisch. 2011. Roadless and low-traffic areas as conservation targets in Europe. Environmental Management 48:865–877.

Shepard, D. B., A. R. Kuhns, M. J. Dreslik, and C.A. Phillips. 2008. Roads as barriers to animal movement in fragmented landscapes. Animal Conservation 11:288–296.

Simmons, J. M., P. Sunnucks, A. C. Taylor, and R. van der Ree. 2010. Beyond roadkill, radiotracking, recapture and Fst—a review of some genetic methods to improve understanding of the influence of roads on wildlife. Ecology and Society 15:9. http://www.ecologyandsociety.org/vol15/iss1/. Accessed 17 September 2014.

Steen, D. A., M. J. Aresco, S. G. Beilke, B. W. Compton, C. K. Dodd Jr., H. Forrester, J. W. Gibbons, et al. 2006. Relative vulnerability of female turtles to road mortality. Animal Conservation 9:279–273.

Steen, D. A., and J. P. Gibbs. 2004. Effects of roads on the structure of freshwater turtle populations. Conservation Biology 18:1143–1148.

Tewksbury, J. J., D. J. Levey, N. M. Haddad, S. Sargent, J. L. Orrock, A. Weldon, B. J. Danielson, et al. 2002. Corridors affect plants, animals, and their interactions in fragmented landscapes. Proceedings of the National Academy of Sciences of the United States of America 99:12923–12926.

Thomas, C. D., and T. M. Jones. 1993. Partial recovery of a skipper butterfly (*Hesperia comma*) from population refuges: lessons for conservation in a fragmented landscape. Journal of Animal Ecology 62:474–481.

Tilman, D., R. M. May, C. L. Lehman, and M. A. Nowak. 1994. Habitat destruction and the extinction debt. Nature 371:65–66.

van der Grift, E. A. 1999. Mammals and railroads: impacts and management implications. Lutra 42:77–98.

van der Grift, E. A. 2005. Defragmentation in the Netherlands: a success story? GAIA 14:144–147.

van der Ree, R., J.A.G. Jaeger, E. A. van der Grift, and A. P. Clevenger. 2011. Effects of roads and traffic on wildlife populations and landscape function: road ecology is moving towards larger scales. Ecology and Society 16:48. http://www.ecologyandsociety.org/vol16/iss1/art48/. Accessed 26 April 2014.

Wirth, T., P. Oggier, and B. Baur. 1999. Effect of road width on dispersal and genetic population structure in the land snail *Helicella itala*. Zeitschrift für Ökologie und Naturschutz 8:23–29. [In German.]

Zellweger-Fischer, J., M. Kéry, and G. Pasinelli. 2011. Population trends of brown hares in Switzerland: the role of land-use and ecological compensation areas. Biological Conservation 144:1364–1373.

Zimmermann Teixeira F., A. V. Pfeifer Coelho, I. Beraldi Esperandio, and A. Kindel. 2013. Vertebrate road mortality estimates: effects of sampling methods and carcass removal. Biological Conservation 157:317–323.

4 — Road Effects on Habitat Quality for Small Animals

Tom A. Langen,
Kimberly M. Andrews,
Steven P. Brady,
Nancy E. Karraker,
and Daniel J. Smith

4.1. Introduction

As reviewed in Chapter 3, roads and road traffic directly affect small vertebrate animals by elevating mortality risk, due to roadkill, and by interfering with habitat and population connectivity between the two sides of a road. Connectivity can be reduced or completely blocked because animals are unsuccessful at crossing roads, or else because they avoid crossing the road when they are deterred by vehicle traffic or by the environmental conditions on and adjacent to the road. What are these environmental conditions that deter some species from approaching roads? And what conditions along roads are attractive to other small animals? Road corridors, due to the thermal properties of pavement and the properties of road right-of-way (ROW) vegetation and soil, typically have dramatically different microclimates from the surrounding landscape. Loud noises and strong vibrations in the soil occur as vehicles pass. Roadside berms often consist of a broad strip of mowed, herbicide-treated vegetation that has had most microstructure (coarse woody debris, rocks) removed (Figure 4.1). Roadside soil is highly compacted, low in nutrients, and contaminated with metals, hydrocarbons, and, in snowy regions, salt. Dust and road salt reduce the health of roadside plants, as do the soil conditions. Roadside plants can be lower in nutrients and higher in toxic metals than the same species growing distant from roads. Ditches and runoff collection basins near roads result in temporary water bodies whose water quality is strongly affected by the chemical contaminants and sediment from the road.

Roadsides are often areas of colonization and invasion for exotic invasive plants and animals. Humans access the surrounding landscape from roads and, in areas where disposal of refuse is inconvenient or expensive, dump garbage along the roadside or in waterways crossed by the road. In arid regions, fires caused by discarded cigarettes or arson typically start along roadsides.

Depending on the type of environmental impact, the "road-effect zone" may be limited to less than 50 m from the road for the most severe impacts on microclimate, soil structure, and vegetation cover. The zone extends beyond 100 m for wind-blown dust, salt, and vehicle exhaust contaminants. It can further extend more than 1,000 m for invasive species colonization, human trespass, and contamination of streams and other water bodies. Some impacts of roads on the surrounding landscape take years to manifest; one study in Ontario, Canada, concluded that loss of wetland-associated species of plants, birds, amphibians, and reptiles due to the direct and indirect effects of roads continued to result in species loss decades after the roads were built (Findlay and Bourdages 2000).

Synergies and feedbacks among the various categories of road impacts present daunting challenges to understanding the totality of effects of roads on small animal populations and can greatly complicate the job of designing mitigation measures. Synergies are cases where two (or more) environmental stressors, when present together, have a greater effect than would be predicted from merely adding the effects of each alone. A positive feedback occurs when two (or more)

Figure 4.1. Browned vegetation resulting from herbicide application in the right-of-way around guardrails and road signs. To the left and just out of view of this image lies a river, which likely receives herbicide runoff from this management practice. *Credit: Steven P. Brady.*

Roads and road traffic affect habitat quality in various ways, making roads a hostile zone for many small vertebrates, and a favorable habitat for others. Understanding how environmental conditions are altered by roads and road traffic can help us to understand why some species are attracted to roads, why some are repelled, and why roadside habitat conditions may result in impaired health for some animals. To design effective mitigation for small vertebrates, one must anticipate the effects of roads on habitat quality. More general information on the effect of roads and road traffic on environmental quality are reviewed in Spellerberg (1998), Jones (2000), Trombulak and Frissell (2000), Forman et al. (2003), and Jaeger (2012); see also Table 1.

4.2. Contamination

Runoff and aerial spray from roads transport a variety of contaminants into adjacent habitats (Maltby et al. 1995). In general, the concentration of contaminants is strongly dependent upon the volume of traffic, but even in relatively undeveloped landscapes where traffic volumes are low, toxic levels can be found along roads. Heavy metals—such as copper, cadmium, and zinc—originate from vehicle wear, and hydrocarbons are spread from fuel combustion exhaust and leaks, lubricants, and tires. In areas with cold winters, deicing chemical agents are a significant source of various ions, especially chloride. An asphalt-paved road surface itself is a major source of polycyclic aromatic compounds. Unpaved roads can generate prodigious amounts of particulate dust and therefore are often treated with substances that may be harmful to organisms near the road. Contaminants associated with roads, singly or interacting in various complicated and poorly understood ways, have a significant effect on the organisms in the roadside environment through direct toxicity to small animals and through degradation of the environment affecting populations.

stressors interact in such a way that makes each worse over time. For example, dust, heat, low humidity, short vegetation, compacted soil, soil vibrations, intense sunlight or street lights, or contaminated water may act synergistically to deter amphibians from approaching a road, or else make these animals prone to thermal stress and desiccation should they move into the vicinity of the roadside. This environmental barrier to amphibians may be desirable if a management objective is to discourage the animals from approaching a road, but if the objective is to encourage amphibians to use an underpass to maintain connectivity between roadsides, roadside habitat quality will need to be improved.

Feedbacks among environmental stressors that affect roadside habitat quality are common, and time lags in environmental responses to feedbacks can cause the full extent of impacts to take years to manifest. For example, microclimate and human-caused roadside fires can select for grasses and forbs, and the spread of these plants puts the zone at greater risk of subsequent fires, expanding the width of the fire-prone zone. As another example, in the first years that road salt is used, it may appear to only have minor, fleeting negative effects on the roadside environment. After many years of use as a deicer, however, the roadside soil's chemical and physical structure can become profoundly altered, the roadside becomes dominated by salt-tolerant plants, and trees bordering the road develop signs of physiological stress and necrosis.

It is not uncommon for contaminants to reach toxic levels near roads, causing direct mortality to small vertebrates. For example, populations of amphibians and fish that are exposed to runoff from a road have lower survivorship than populations inhabiting similar habitat that does not receive such runoff (Turtle 2000; Feist et al. 2011). Exposure to contaminants may also affect physical development, health, and behavior such that animals are at greater risk of mortality from predators, pathogens, or other factors (reviewed by Trombulak and Frissell 2000). Effects such as increased mortality, decreased reproduction, decreased rates of metamorphosis, increased rates of malformation, and decreased swimming performance have been documented among fish, amphibians, and small mammals (e.g., Leblond et al. 2007; Karraker et al. 2008; Denoël et al. 2010; Meland et al. 2010; Brady 2012). Among small vertebrates, amphibians appear to have the most heightened sensitivity to the chemical effects of roads because of high skin permeability. Parasite burdens have been observed to increase (Patz et al. 2000), as has the prevalence of malformations (Reeves et al. 2008). Some of the heavy metal and hydrocarbon contaminants found in runoff are known mutagens (Handa et al. 1984; Hankinson 1995).

An additional indirect way that environmental contaminants affect small vertebrates is through their diet (Petranka and Doyle 2010). Some contaminants may affect the abundance and quality of phytoplankton, zooplankton, and periphyton in aquatic communities, or soil microbial organisms and plants in terrestrial communities. Small vertebrates that are exposed to contaminant-laden diets can incur high burdens of contaminants through bioaccumulation and biomagnification (Chen et al. 2000).

The United States Environmental Protection Agency (EPA) maintains a list of contaminants and specifies the concentrations at which they have been documented to be toxic to particular species. These concentrations are estimated by conducting experiments designed to reflect acute (i.e., short-term) and chronic (i.e., long-term) exposures to contaminants. Toxicity studies evaluate a very small number of species (for aquatic vertebrates, typically 1 species of fish) for a short time (acute = 24 hours, chronic = 7 days), usually only quantify lethality, and rarely investigate synergies with other contaminants to which an animal may be exposed at the same time. The actual toxicity for a given contaminant in the natural environment is likely to be higher due to synergies with other stressors, such as pathogens, predator-induced stress, and additional contaminants, and because eggs, larvae, and adults are typically exposed for far longer durations than occurs in the typical laboratory toxicity test

(Relyea and Mills 2001). Sublethal effects of contaminants on development and behavior may appear to be of minor consequence in laboratory studies of toxicity, but in fact can have major impacts on predation risk, competitive ability, reproductive success, and other traits for animals exposed in the wild (Relyea and Hoverman 2006; Boone et al. 2007).

4.2.1. Nonchloride Chemicals
4.2.1.1. Hydrocarbons

A diverse assortment of hydrocarbons is deposited into the roadside environment (Maltby et al. 1995; Miguel et al. 1998; Harrison et al. 2003), and they have been found to readily disperse from the roadside (Zakaria et al. 2002). Hydrocarbon contamination results from leakage and combustion, and originates from automobile fuel and lubricants, tires, and road surface materials such as asphalt (Miguel et al. 1998). Hydrocarbons have been documented to impose a broad array of negative biological effects including those that are toxic, carcinogenic (Grimmer et al. 1984), and mutagenic (Perera et al. 1992). For example, tire debris, which has been shown to contribute approximately 33% of the particulate in runoff (Breault and Granato 2000), is known to have lethal effects on trout and minnows (Day et al. 1993). Most of our understanding of these consequences is based on human health research, especially related to industrial production (Perera et al. 1988) and paving applications (Boffetta et al. 1997), and more recently to dust and particulates in heavily urbanized settings (Nielsen et al. 1996). Evidence from studies on the impacts of oil spills on wild populations also points to severe negative consequences from hydrocarbons, such as inhibited tadpole growth or failure to metamorphose in green treefrogs (*Hyla cinerea*) due to petroleum contamination (Mahaney 1994).

Over 100 different types of polycyclic aromatic hydrocarbons (PAHs) found in fuels, tires, and road surfaces have been shown to be emitted by automobiles (Rogge et al. 1993). Many PAHs have been classified as carcinogenic (see Boffetta et al. 1997). Effects on Pacific herring (*Clupea pallasii*) embryos exposed to PAH concentrations of just 0.4 to 7.6 parts per billion (roughly equivalent to 1 or 2 drops of PAH dissolved in 500 barrels of water) increased mortality and malformations and reduced swimming ability after hatching (Carls et al. 1999). While little is known about the consequences of hydrocarbon contamination on the roadside environment, what is known about the health effects of PAHs and other hydrocarbons contaminants warrants concern.

4.2.1.2. Metals

Roads are a significant source of contamination from a diverse assortment of metals resulting from automobile wear (on tires, catalytic converters, brake pads, and welded parts; Day et al. 1993; Long et al. 1995), fuel and lubricant leakage, and incomplete fuel combustion (Nriagu and Pacyna 1988; Trombulak and Frissell 2000). Metals from fuel and emission products include chromium, lead, zinc, vanadium, tin, and several others that accumulate in roadside soils (Sezgin et al. 2004; Zielinska et al. 2004). By mass, tire debris contains about 1% zinc (Davis et al. 2001), a metal that harms some aquatic animals such as wood frog (*Lithobates sylvaticus*) larvae, causing decreased rates of development to metamorphosis (Camponelli et al. 2009).

As with hydrocarbons, metal contamination can have direct and indirect impacts on the health and welfare of small vertebrates occupying habitat near roads. In fish, ecologically relevant levels of copper have been found to impair olfaction in salmon, posing a risk to their ability to navigate and to avoid predators (Sandahl et al. 2007). Brown trout (*Salmo trutto*) have shown increases in malformed red cells and elevated levels of white blood cells in response to metals (Meland et al. 2010). Gene expression has been shown to be altered in lizards when their eggs are exposed to soil containing cadmium (Trinchella et al. 2010). Scanlon (1979) and Getz et al. (1977) documented high concentrations of heavy metals in small mammals living alongside roads. Fill used during road construction in Great Smoky Mountain National Park leached toxic metals that entered streams via runoff (Kucken et al. 1994), thereby eliminating a significant proportion of the populations of 2 salamander species (black-bellied salamander, *Desmognathus quadramaculatus*; Blue Ridge two-lined salamander, *Eurycea wilderae*) and reducing the populations of an additional 2 species by 50%. Rowe et al. (1998) reported that sediments containing high levels of arsenic, barium, cadmium, chromium, and selenium caused oral deformities in American bullfrog (*Lithobates catesbeianus*) tadpoles. Metals can also interact in complex ways with other environmental factors,

and the toxicity of a metal to aquatic organisms is also influenced by pH, temperature, and dissolved oxygen, all of which affect the chemical speciation—and thus toxicity—of a given metal (reviewed by Wang 1987).

Wetland plants serve important roles in the uptake of heavy metals (Weis and Weis 2004) and also absorb nutrients including nitrogen and phosphorus (Tanner 1996). Plants may remove large quantities of contaminants from the water column, but many of these contaminants, and especially metals, may continue to cycle through the wetland ecosystem as plants die and decompose or when herbivorous animals feed on plant tissue and assimilate their contamination burdens. Contaminant loads will increase in wetlands over the long term, although phytoremediation, which includes harvest and removal of contaminant-accumulating plant structures, can be used to remove metals or other contaminants (Salt et al. 1995).

4.2.1.3. Nutrients

The term "nutrients" typically refers to nitrogen and phosphorous (or their various forms such as nitrate or phosphate), but may also include sulfur, potassium, calcium, and magnesium. The principal road-associated nutrient contaminant is nitrogen, which is a component of vehicle emissions (Vaze and Chiew 2002). Phosphorus, calcium, and magnesium are components of chemical deicers used on some roads. The principal effect of nutrient contamination is to alter ecosystem dynamics by affecting the productivity and composition of microbial and plant communities. This, in turn, affects the availability and quality of food for small vertebrates, sometimes positively. Plants higher in nitrogen and other nutrients are more nutritious to herbivores. However, road-associated nutrient enrichment can favor aggressively invasive exotic plants (and some natives, such as *Typha* cattails), degrading habitat conditions for some small vertebrates while potentially improving habitat for others. An example of nitrogen enrichment caused by vehicle exhaust that alters a roadside plant community is described in Lee et al. (2012).

4.2.1.4. Herbicides

Vegetation along road verges is often kept low by mowing or herbicide spraying for motorist safety and to prevent the growth of unwanted vegetation (see Figure 4.1), which can crack and damage road surfaces, block signs from view, interfere with power lines, attract large ungulates, and obstruct viewsheds. The widths of these "roadside clear zones" vary based on traffic volume and slope but, in the United States, typically range from 4–10 m (American Association of State Highway and Transportation Officials 2006).

Much research has been directed toward understanding the ecological effects of commonly applied herbicides (Solomon et al. 1996; Giesy et al. 2000). Because herbicides are so widely distributed in agricultural applications, the consequences of herbicides on wildlife—especially aquatic organisms—have been widely studied. Nevertheless, there are many information gaps that are critical for evaluating the safety of herbicides used on roadsides (e.g., the toxicity of sulfometuron-methyl, a popular herbicide along roads, to amphibians; Klotzbach and Durban 2005).

In addition to intended and unintended impacts in the roadside environment, roadside herbicide applications may compromise water quality adjacent to and distant from a road (Wood 2001). Kohl et al. (1994) found that 2–33% of herbicides applied to road shoulders can leach out into runoff in the first storm following application. Leaching rates can attain levels from 10–73% of the applied herbicide depending on the compound (Ramwell et al. 2002).

Some herbicides contain a suite of compounds capable of affecting hormone regulation. These contaminants, referred to as endocrine disrupting chemicals, or EDCs, have been associated in fish and amphibians with reproductive malformations and reduced sperm counts (Fort et al. 2004), altered sex ratios (Orton et al. 2006), and even animals developing intersexual reproductive tissues (Hayes et al. 2002; Rohr et al. 2006; Kinnberg et al. 2007). Further impacts on amphibians include direct mortality, reductions in growth and development, increases in malformations, and changes in behavior (Relyea 2004, 2005; Svartz et al. 2012). Egg-laying reptiles such as turtles and crocodilians have been found to experience changes in reproductive functioning, including smaller clutch sizes with lower egg viability and developmental disruption of reproductive organs, as a result of exposure to herbicides (Crain and Guillette 1998; Sparling et al. 2006). Thus, while little is directly known about the ecological consequences of roadside herbicide application, there

is reason to be concerned about its impact, especially near wetlands.

4.2.2. Chloride Chemicals

While there are many chemical deicers used for winter road management (reviewed in Langen et al. 2006), sodium chloride road salt remains the agent of choice. Chloride ions from salts are relatively persistent in the environment and are especially concentrated during winter and spring. The use of road salt as a deicer has been on the rise since 1940 (Jackson and Jobbágy 2005), and chloride ions have been increasing in aquatic environments as a result (Siver et al. 1996; Kaushal et al. 2005). Beginning in 2005, application of road salt for deicing exceeded all other uses of salt (Mullaney et al. 2009). A recent study of 168 sites in northern US metropolitan areas found that in 55% of streams surveyed, chloride concentrations exceeded EPA water quality criteria for chronic exposure (Corsi et al. 2010). Similarly, by one estimate, if application at current rates was to continue for the next century, and assuming conservatively that the amount of impervious surface remains the same, chloride concentrations in many streams—even in rural settings—will exceed 250 mg/L, becoming harmful to sensitive freshwater species and nonpotable for human consumption (Kaushal et al. 2005).

At ecologically relevant levels, road salts have been associated with a suite of negative environmental effects, including direct toxicity to aquatic animals, and habitat degradation caused by the harmful effects on plants and soil conditions (reviewed in Environment Canada and Health Canada 1999; Langen et al. 2006; Karraker 2008). For example, adult amphibians suffer increased mortality when experimentally exposed to water contaminated by road salt at concentrations typical of roadside conditions (Sanzo and Hecnar 2006; Karraker et al. 2008; Collins and Russell 2009; Alexander et al. 2012), as do eggs and larvae when developing in roadside ponds (Turtle 2000; Brady 2012). Road salt contamination can alter animal behavior by, for example, reducing swim speed and escape movement distance in developing tadpoles of the common frog (*Rana temporaria*; Denoël et al. 2010). The consequences for freshwater fish are similarly negative (Meland et al. 2010). For some terrestrial mammals, road salt acts as an attractant, much like a salt lick, and increases the risk of vehicle collisions (Leblond et al. 2007). The effects on plant composition are dramatic along roadsides where road salt is regularly applied; for example, an ongoing roadside invasion of *Phragmites* (great reed) in North America is being facilitated by salt contamination (Lelong et al. 2007).

The presence of road salt can also dramatically influence the toxicity of metals and in some cases appears to reduce toxicity by limiting the uptake of metals by an organism. For example, sodium chloride may reduce the toxicity of copper in two species of amphibians (Brown et al. 2012). However, other studies have shown that sodium chloride from deicing salt can mobilize chromium, lead, nickel, iron, cadmium, and copper in roadside soils, increasing the risk of exposure to animals (Amrhein et al. 1992).

Efforts have been made to reduce the use of sodium chloride, the most widely applied and least expensive form of deicing salt, by substituting it with other chloride-based compounds, such as magnesium and calcium chloride or several acetate-based compounds; this last includes calcium magnesium acetate and potassium acetate. While these have been marketed as environmentally friendly alternatives to sodium chloride, several are detrimental to amphibians (Harless et al. 2011) and their effects on small mammals and reptiles are unknown. Chloride is the element considered to be harmful to freshwater aquatic life. Thus, simply replacing one chloride-based compound with another does not solve the problem for sensitive species (Harless et al. 2011).

More recently developed deicers that contain a form of sugar (sugar beet juice or molasses as the principal ingredient) show promise. While the effects of sugar-based deicers on the environment are not well studied, it is likely that these compounds will be consumed rapidly by microbes in wetlands and streams. The current high cost of these new products, relative to sodium chloride, has limited their use, but future demand may help reduce prices.

Significant reductions in use of chemical deicers can be achieved with improved technology (e.g., use of brine or prewetting salt before application), better information on road conditions (e.g., sensors and roadside weather stations), improved winter maintenance training, and changes in motorist behavior (using better tires, reducing travel speed during winter weather events). A detailed review of such alternatives is provided in Langen et al. (2006). Given the amount of chemical deicers applied on roads in regions with severe winters, even proportionally small reductions are likely to have significant benefits to the environment in terms of reduced contamination from the deicer and associated contaminants.

4.2.2.1. Case Study: Road Salt Contamination and Vernal Pool-Breeding Amphibians

Nancy E. Karraker and Steven P. Brady

Vernal pools, also known as ephemeral or temporary pools, are a unique type of wetland differentiated primarily by hydroperiod (the length of time it holds water), and their hydrologic isolation from other wetlands (Colburn 2004; Calhoun and deMaynadier 2008). Vernal pools have neither inlet nor outlet streams; they fill by direct precipitation, overland flow, or from groundwater. Wetlands of this type occur globally, but their characteristics can vary greatly by region. In temperate regions, vernal pools generally hold water for at least two months and dry up every year. This annual drying shapes a vernal pool biota that is devoid of important predators requiring permanent water, such as fish, making these pools optimal breeding habitats for many amphibians. Vernal pools are of great conservation concern because they are relatively unprotected legally, in contrast to other kinds of wetlands, and their ecological value is often unrecognized (Calhoun and deMaynadier 2008).

Deicing salt, most commonly sodium chloride, accumulates in plowed snow piles adjacent to roads, dissolves in runoff, dissociates to chloride and sodium ions, and is mobilized during the spring thaw. Chloride is highly soluble and mobile and is not taken up in large quantities by plants or animals. Thus, vernal pools located near roads may receive runoff with high chloride concentrations. For example, in the Adirondack region of New York, deicing salts can be transported over 170 m from a state highway into the adjacent forest, resulting in elevated chloride concentrations in vernal pools, especially within 50 m of the road (Karraker et al. 2008). In addition to traveling in runoff, chloride can move from roads in aerosol spray with transport distances related to the amount of traffic and its speed, and the presence and composition of roadside vegetation (Viskari and Karenlampi 2000). Chloride ions have been transported in aerosol spray up to 40 m from roads (McBean and Al-Nassri 1987; Blomqvist and Johansson 1999), and if they are deposited onto snow they may be transported farther as runoff when the snow melts. Once present, chloride ions do not biodegrade or adsorb to mineral surfaces in appreciable amounts (Bowen and Hinton 1998), and with no outflow of water from the pool, chloride accumulates. Chloride concentrations in contaminated vernal pools and small ponds near roads have been documented to range from approximately 80 to 1,000 mg/L (Sanzo and Hecnar 2006; Collins and Russell 2009), while uncontaminated pools generally contain 1 mg/L chloride (Karraker et al. 2008).

Vernal pools represent important habitat for a specific set of amphibians, namely wetland-breeding species with relatively short embryonic and larval stages that are largely intolerant of predatory fish. Nearly every region of the United States has amphibian species that breed largely or exclusively in vernal pools, including spotted salamanders (Ambystoma maculatum) and wood frogs in the eastern United States and western spadefoots (Spea hammondii) and California tiger salamanders (Ambystoma californiense) in the west. Vernal pool-associated amphibians are often categorized as explosive breeders such that migration from terrestrial habitats is triggered by an increase in temperature and precipitation, and reproduction is completed in two to four weeks.

Amphibians that reproduce in vernal pools may be particularly sensitive to multiple impacts associated with road deicing salt. In temperate regions where deicing salt is applied, amphibians begin their migrations to breeding pools with the first spring rains—the same rains that transport chloride from nearby roads. This peak in the concentration of chloride in pools near roads is one of two that coincides with the times of greatest vulnerability of amphibians to contaminants (Karraker et al. 2008). This first peak occurs during egg laying and development of embryos. Amphibian eggs include an embryo surrounded by several membranes and jelly layers and these may be embedded within a larger jelly mass, depending upon the species. When eggs are deposited in vernal pools, they immediately begin absorbing water into the jelly layers, increasing their volume by as much as 10 times. However, amphibian eggs are unable to take up water at a natural rate, or at all, when deposited in an environment of highly concentrated solutes including chloride. Egg masses that have taken up water in uncontaminated environments will subsequently lose water if those environments receive a flush of chloride with spring snowmelt and may not be able to resume uptake when concentrations become diluted by spring rains (Karraker and Gibbs 2011). Uptake of water is critical to embryonic development because it provides adequate spacing between eggs for embryonic growth, reducing the risk of malformations associated with cramped conditions. Malformations are more frequent in amphibian embryos exposed to road salt, so high chloride concentrations during the embryonic period can be particularly detrimental to reproductive effort in a given year (e.g., Brady 2013).

The second peak in chloride concentration occurs as vernal pools begin to dry up, which triggers tadpoles to undergo metamorphosis—the transition from a tadpole to a subadult frog or salamander. At metamorphosis, larval amphibians in vernal pools transform from gilled, swimming aquatic animals to terrestrial, hopping or crawling animals with lungs (Duellman and Trueb 1986). The move from water to land, during the few days to a week over which metamorphosis occurs, can present significant physiological challenges for amphibians, whose skin must always remain moist and cool to facilitate respiration; the impacts of elevated salinity in drying vernal pools on the water balance in metamorphosing amphibians may result in significant physiological stress.

Wetlands contaminated with road salt may exclude some amphibian species, which may contribute to local extirpations in these ponds. For example, egg mass densities of wood frogs and spotted salamanders were twice as high in uncontaminated pools (Karraker et al. 2008), and both species were absent from ponds in which mean chloride concentrations exceeded 150 mg/L (Collins and Russell 2009), even though breeding-age adults inhabited the area. This suggests that local populations of these salt-intolerant species may have already been affected at these sites due to low reproductive success in these ponds. When salt-intolerant amphibians breed in contaminated pools, the cumulative impacts of road salt can exact a toll on survival of those whose aquatic stage can last from approximately three months, as in wood frogs, to up to six months, as in spotted salamanders (Bishop 1941). Karraker et al. (2008) estimated that average embryonic survival rates were <5% for spotted salamanders and 40% for wood frogs in highly contaminated pools versus survival rates of both species (between 80 and 90%) in uncontaminated pools.

Larval amphibians appear to be a bit more tolerant of road salt than embryos but are also affected. Effects of road salt on amphibian survival vary regionally, but several studies have shown that approximately 50% of wood frog and spotted salamander larvae die in waters with chloride concentrations of 1,000–1,500 mg/L versus a greater than 80% survivorship in uncontaminated water (Sanzo and Hecnar 2006; Collins and Russell 2009). In some places mortality is as high as 80–90% at concentrations of about 950 mg/L chloride (Karraker et al. 2008). In addition to mortality, wood frog and spotted salamander larvae exhibit an increased frequency of malformations and behavioral abnormalities at higher chloride concentrations (Sanzo and Hecnar 2006; Collins and Russell 2009).

While some amphibians may be excluded from highly contaminated pools, others that are breeding for the first time, those that lack uncontaminated pools within their home ranges, or those that may have some tolerance for some level of chloride concentration may choose to breed in contaminated pools (Collins and Russell 2009). We still have little understanding as to what this means for amphibian populations or local amphibian biodiversity, but one study on spotted salamanders suggested that populations are at risk of local extirpation in moderately contaminated vernal pools (Karraker et al. 2008). Efforts should be made to reduce the application or runoff of road deicing salt in areas where conservation of amphibians is a priority. Such efforts could be coupled with road salt management plans aimed at reducing contamination of public drinking water supplies so that both human health and amphibian populations are better protected.

4.2.3. Litter and Dumping

The disposal of rubbish into roadside habitat can elevate the concentrations of contaminants carried through precipitation runoff, as well as introduce a new suite of contaminants. These may include, for example, flame retardants from upholstered furniture and coolants from refrigerators (Tunnel 2008). In regions where items classified as hazardous waste (e.g., paint, solvents, batteries, pesticides) are difficult or expensive to safely dispose of, clandestine disposal along roads may be a frequent occurrence (US EPA 1998). Animals occupying or moving through areas near roads encounter these items directly and in some cases may even be attracted to them, especially if they possess food residues or offer refugia, which can further increase the risk of collision with vehicles or exposure to harmful substances in the refuse. While road runoff normally delivers comparatively dilute concentrations of contaminants with each runoff event, dumped refuse may act as a local point source of much higher levels of contaminations (Tunnel 2008). For example, an animal that investigates a dumped refrigerator or car battery may be exposed to very high concentrations of coolant or battery acid. A particularly hazardous activity is dumping directly into waterways where roads cross or pass close to water bodies (e.g., bridges and causeways). Unfortunately, clandestine dumping into wetlands is not uncommon in some rural areas and is presumably done to reduce the risk of detection. Refuse in waterways can provide a very severe source of contamination, especially when hazardous waste is disposed of in this manner (US EPA 1998).

In the 1960s, US First Lady Bird Johnson famously said, "Where flowers bloom, so does hope," in reference to addressing the wretched state of US highway roadsides. She contended that if we worked to make highways more attractive by reducing sign clutter and planting native wildflowers, people would be less inclined to discard refuse along them. Despite significant US legislative efforts in the 1960s and 1970s to beautify highways, littering and dumping of garbage along roads remain a problem in many parts of the United States, as it is elsewhere in the world (US EPA 1998). More recently, enforcement of regulations on littering and the establishment of volunteer-based highway cleanup programs have reduced the quantities of refuse along roadsides. One simple approach, the placing of garbage cans at stops along highways, has been shown to reduce litter by nearly 30% (Finnie 1973). However, these efforts have not entirely solved the problem. Community involvement seems to have had the greatest impact on reducing littering and dumping along roads (US EPA

1998). When individuals and organizations are involved with highway cleanup activities, others in the community will be more likely to treat these areas with greater respect. In rural areas where curbside waste pickup is unavailable, providing affordable and accessible collection stations for waste, including hazardous waste, may reduce clandestine dumping.

4.2.4. Transport of Contaminants

Contaminants can be classified into one of four classes depending on the timing and frequency of spread into the surrounding environment: temporary, chronic, seasonal, or accidental (terminology following Forman et al. 2003). The timing and frequency of spread is important, because they affect the severity and form of the impacts on small vertebrates, as well as the management options that may be most effective. Temporary spread of pollutants, including petroleum-based organic compounds, is the result of road construction and maintenance projects. Chronic spread of chemicals such as PAHs and heavy metals occurs at low levels as a result of road traffic and inadequate vehicle maintenance. Heavy loads of contaminants may be spread seasonally, as in the case of deicing salts and sand abrasives, or may be spread accidentally when vehicle fluids or hazardous liquid materials are spilled during collisions. Chronic and seasonal inputs are the most detrimental to the environment because of the temporal and spatial scales over which they occur (Forman et al. 2003). These inputs are transported away from the road as runoff, stream flow, and aerial spray; the mode of transport will influence the magnitude of the impact on the environment. It is not always possible to determine the means by which a pollutant has been delivered to the environment adjacent to a road, but most contaminants probably spread by multiple modes.

4.2.4.1. Runoff

In storm events, 20–80% of rainfall travels as runoff, and the remainder infiltrates the ground surface (van Bohemen and van de Laak 2003). The balance between runoff and infiltration is determined in part by soil saturation, the amount of impervious surface, and plant cover. Contaminants from roads traveling as runoff are deposited onto adjacent soils or are collected in roadside ditches and transported elsewhere.

Runoff transports dissolved contaminants, known as the dissolved load, and suspended contaminants, termed the suspended load (Forman et al. 2003). The distances that road-associated contaminants move depend greatly upon the type of contaminant, its weight, solubility, biodegradability, capacity for adsorption, mineral binding potential, and uptake by plants. In addition to characteristics of the contaminants, transport distances are affected by features of the area immediately adjacent to the road, including soil texture, vegetative structure, and topography (e.g., Turer and Maynard 2003; Mangani et al. 2005).

Pollutants accumulate on roads during dry periods, and deposited particles become smaller in size over time (Vaze and Chiew 2002). Only a small proportion of the contaminant load is removed with each precipitation event; pollutant deposition on roads, especially those that carry heavy traffic volumes, far exceeds the capacity of rainfall in an average storm to completely wash it from the road surface (Vaze and Chiew 2002). In cold regions, spring thaws and rains result in the transport of shock loads of pollutants, which can coincide with important life history events of animals including pool-breeding amphibians (Baker et al. 1996; Lepori and Ormerod 2005; see also Case Study 4.2.2.1).

Heavy metals, PAHs (van Bohemen and van de Laak 2003), nutrients such as nitrogen and phosphorus (Vaze and Chiew 2002), and petroleum products from leaking vehicles (Lefcort et al. 1997), are generally adsorbed onto sediments. The distances that they travel are related to the sizes of particles to which they adhere. Nickel and chromium deposited by road runoff have been detected up to 100 m from a highway (Zehetner et al. 2009), and zinc, copper, and cadmium occur at elevated levels up to 200 m away from the road (Turer et al. 2001; van Bohemen and van de Laak 2003). Other substances, such as tire particles (reviewed in Wik and Dave 2009), tend not to adhere to sediments and are not soluble but nevertheless are carried in road runoff. Highly soluble pollutants, such as road deicing salt, can be transported long distances with surface water flow and groundwater (Environment Canada and Health Canada 1999; Langen et al. 2006; Zehetner et al. 2009). Sodium chloride has been shown to have the additional effect of leaching heavy metals including lead (Vandenabeele and Wood 1972), copper, zinc, and nickel from soils, thereby remobilizing them in runoff (Zehetner et al. 2009).

REDIRECTING AND TREATING RUNOFF. Runoff that is collected into roadside ditches or channels is generally diverted into streams, rivers, or stormwater detention or retention basins, depending upon the surrounding landscape. Such infrastructure allows road-associated pollutants to travel much farther than they would otherwise and often contributes to the contamination of wetlands and natural water courses. In some regions at present, runoff is directed toward streams or natural wetlands as was the common practice in the past. Such legacy designs of stormwater management should be pinpointed within road networks and reconfigured so that runoff is redirected away from sensitive water bodies. Culverts installed to direct runoff downslope and away from wetlands and streams would help to reduce contamination of sensitive aquatic habitat.

A variety of methods have been developed for containing and treating road runoff. The most commonly used approach entails the redirection of runoff to stormwater detention basins or constructed wetlands. These two types of water bodies generally have sediment bottoms and wetland-emergent plants lining the edges. The objective is to contain stormwater runoff for a period of time sufficient to allow for reductions of contaminant loads via natural processes. Both the bottom sediment and wetland plants in a constructed wetland are important for sequestering heavy metals, including cadmium, copper, lead, and zinc (Mungur et al. 1995; Weis and Weis 2004) and nutrients (Tanner 1996).

Stormwater retention basins and constructed wetlands can be effective at minimizing the impacts of road runoff in natural habitat, but careful consideration must be given to their potential influence on animal populations. Small mammals, amphibians, and reptiles using these water bodies as foraging or breeding habitat may be exposed to harmful levels of heavy metals or contaminants in their food and through skin and respiratory passage absorption; their eggs and offspring may also be exposed. While constructed wetlands may appear to provide wildlife habitat (e.g., Lacki et al. 1992; Knight 1997), those with high contaminant burdens may be an "ecological trap" attracting animals to habi-

tat that is a population sink, meaning that reproductive output is not sufficient to maintain a population. Sink populations persist only because additional animals migrate into an ecological trap from elsewhere. One cannot assume that a constructed wetland is suitable habitat for a species simply because individuals are present in the wetland; only careful population monitoring can establish whether the wetland can sustain a population. If a constructed wetland is suspected to be an ecological trap, a mitigation measure may be to create a barrier that prevents colonization by animals susceptible to the trap.

In urban and suburban areas, stormwater runoff may pass through a filtration or infiltration system before being redirected into a detention or retention basin, constructed wetland, or stream. These systems vary in complexity and effectiveness, and the quantity of contaminants in runoff or the horizontal space necessary for implementation of a particular system dictates which type is used. Filtration systems may be as simple as a roadside swale in which grass is seeded and maintained. This swale should slow the flow of contaminants, facilitate soil sequestration of heavy metals, and enhance plant uptake of nutrients (Bäckström 2003). Sand filters are used for capturing particulate matter, nutrients, and dissolved metals from stormwater (Urbonas 1999). Infiltration trenches and basins are used to treat stormwater runoff by relying on passive adsorption of metals and other contaminants onto soils or sediments in the trench or basin (Birch et al. 2005). Redirecting runoff to a constructed wetland or through an infiltration system can result in a reduction in contaminant concentrations prior to runoff waters entering a natural habitat or groundwater.

4.2.4.2. Stream Flow
Contaminants from roads make their way to streams directly at bridge or culvert crossings or indirectly through other conduits such as roadside ditches. The concentration of pollutants entering streams is influenced by whether transport is direct or indirect, and if indirect, how far the pollutants travel before they enter the stream. In rural areas, there is often no diversion of runoff away from streams at bridges or other stream crossings. The amount of runoff that makes its way to a roadside ditch during a precipitation event depends upon the distance of the ditch from the road and the

structure of the intervening vegetation (i.e., whether there are trees or shrubs). In some cases, as little as 0.2–0.5% of runoff may be transported to a ditch (van Bohemen and van de Laak 2003). Whether traveling as overland flow or via a roadside channel, surges of pollutants into streams will be greater during major storm events because of the amount of water discharged and the capacity of heavy rainfall to dislodge pollutants from the road surface (Krein and Schorer 2000).

When streams receive direct runoff of contaminants, a broader array of pollutants may be transported because of the higher velocity of stream flow as compared to that of runoff or aerial spray. Streams readily transport lightweight pollutants, such as nitrogen and phosphorus, but they may also carry heavy metals that cannot be transported great distances by other means (Wilber and Hunter 1977). These metals are most often bound to organic matter and sediments and eventually settle out into streambeds (Paul and Meyer 2001). Stream size and discharge will affect the concentrations of contaminants transported. Larger streams will rapidly dilute entering pollutants, thereby reducing impacts on biota. By comparison, smaller, shallower streams may accumulate high concentrations of contaminants. For example, sediment from forest roads has been shown to be detrimental to stream-dwelling animals in streams that exhibit discharge of <10 liters per second. However, in larger, turbulent streams with discharge of >500 liters per second, sediment is diluted sufficiently to render it benign to stream fauna (Sheridan and Noske 2007). The degree to which the toxicity of pollutants is reduced by dilution will also depend strongly upon the type of pollutant and its physical and biochemical properties.

The mobility of pollutants and their propensity to bind to sediment also affect the extent of the downstream zone of a waterway that is affected. In one study, zinc, cadmium, and lead concentrations were elevated in streams at highway crossings but declined to background levels within 200 m downstream (Mudre and Ney 1986). Heavy metals readily bind to sediment; concentrations in sediment near roads will be chronically high and attenuate at short distances downstream. Metals slowly leach from sediment due to chemical processes, including those associated with acidic water conditions. Physical processes, such as significant discharge from storm events, can also result in the releases of metals from sediment. Other road-associated contaminants do not bind to sediment and may be transported long distances downstream of road crossings. In one of the longest distances reported for stream transport of a pollutant, chlorides associated with road deicing salt remain elevated above background levels over 20 km downstream of major highway crossings (Hoffman et al. 1981).

4.2.4.3. Aerial Spray, Dust, and Particulates

Pollutants from roads are transported away from a road via aerial dispersal during storm events, or when roads are wet, by two mechanisms—spray from rapidly moving vehicles and (in winter) scattering from the act of plowing snow from road surfaces (Figure 4.2). It has been estimated that up to 70–90% of deposited pollutants are transported from roads through these means (van Bohemen and van de Laak 2003). For road deicing salt, up to 45% of the salt deposited within 100 m of a road is transported as aerial spray (Lundmark and Olofsson 2007). Deposition patterns of aerial-transported contaminants are variable and are affected by traffic volume, number of road lanes, types of vehicles that use the road, and the density and structure of roadside vegetation. The quantity of zinc and lead transported increases with traffic volume (van Bohemen and van de Laak 2003). Although some contaminants may be transported >200 m from the road by wind and aerial spray (PAHs; Viskari et al. 1997; van Bohemen and van de Laak 2003), most contaminants, such as zinc, lead, and cadmium, are deposited within 30–60 m of the road (Motto et al. 1970; Rodriguez-Flores and Rodriguez-Castellon 1982; Viskari et al. 1997) or even within 15 m (Yang et al. 1991). Dispersion of pollutants is greater in quantity and farther in distance on the downwind side of the highway (Viskari et al. 1997). In terms of vegetation type and structure, greater contamination occurred, and at farther distances, in open fields than at forested sites for both PAHs (Hautala et al. 1995) and chlorides associated with road deicing salt (Lundmark and Olofsson 2007).

Dust and particulate matter contribute to the contamination of the roadside environment. For the most part, the same contaminants found in runoff or aerial spray are present in dust and dry particulate form. Dust abatement practices for unpaved roads occasionally include spraying calcium chloride. One study has

Figure 4.2. In spring, this roadside wetland (pictured here in late winter) will host thousands of breeding frogs and salamanders. The accumulated snow in the foreground is the result of plowing. Levels of contaminants in these wetlands spike during this time of year when piles of snow like these melt and release road pollutants. *Credit: Steven P. Brady.*

documented mass deaths of blue-spotted salamanders (*Ambystoma laterale*) along a US Forest Service road due to desiccation from exposure to calcium chloride (deMaynadier and Hunter 1995). Contaminants in the form of dust and particulate can be especially hazardous because of the potential for their inhalation, which may have cardiopulmonary consequences for animals inhabiting areas near roads, as it does for humans (e.g., Morgenstern et al. 2007). Dust settling onto the leaves of plants blocks stomata and blocks light, reducing photosynthesis, transpiration of water, and respiration (Farmer 1993). Coatings of dust can be very detrimental to animals that depend on respiration through the skin, such as amphibians (deMaynadier and Hunter 1995). In northern climates, sand used as an abrasive on snowy winter roads is a significant source of particulates (Environment Canada and Health Canada 1999; Langen et al. 2006). Improved street sweepers remove substantial amounts of dust and particulate. However, street sweeping is normally restricted to municipalities, and therefore does little to reduce particulate exposure to animals outside of urban areas.

BENEFITS OF ROADSIDE VEGETATION TO REDUCE CONTAMINANT SPREAD. Vegetation can serve as an effective barrier for reducing transport of contami-

nants to adjacent landscapes; roadside vegetation intercepts runoff and aerial spray from roads. For example, a roadside Norway spruce forest (*Picea abies*) greatly restricted the transport of inorganic anions associated with deicing salt and PAHs from a highway when compared with transport of pollutants into an adjacent open field (Hautala et al. 1995). Scots pine (*Pinus sylvestris*) near roads trapped a significant proportion of aerial spray of deicing salt (Viskari and Karenlampi 2000). While salt damages coniferous tree needles, trees near the road present an effective barrier to its transport, thereby protecting more interior trees and forest soil.

A broad array of plants potentially could serve as effective barriers for contaminants, but for mitigation it is crucial to determine whether a candidate plant species is tolerant of the contaminants it will be exposed to. Tolerance of plants to road-associated contaminants has been researched most extensively with regard to road deicing salt. Only a small subset of trees and shrubs are both salt tolerant and substantial enough to serve as a partial barrier to the aerial spray of deicing salt (Bryson and Barker 2002; Ruter 2003; Woodsen 2003).

Most pollutant deposition occurs within 30 m of roads, so establishing appropriate 30 m wide vegetated strips along roads would help to contain pollution near roads and reduce impacts to the surrounding landscape. This is particularly true in locations where native vegetation has been cleared and replaced with grass. Using native vegetation to manage road-related pollutants should reduce contamination of soils, wetlands, and streams, reduce risk of colonization and spread of exotic plants (Gelbard and Belnap 2003), and provide habitat to some birds (Laursen 1981), butterflies (Munguira and Thomas 1992), bees (Hopwood 2008), and small mammals (Bellamy et al. 2000). Restoration or maintenance of native trees and shrubs along roadsides could be made a priority along road segments bordering conservation areas, sensitive wetlands, or other important ecological areas. Signs warning motorists to reduce speed and drive with care along these vegetated verge zones, and signage explaining the purpose of the vegetation cover, may both reduce risk and educate the public about the importance of plants for managing contamination from roads.

4.3. Environmental Conditions

4.3.1. Noise

At 15 m from its source, the average automobile passing down a road has a noise level of 68 dBA (A-weighted decibels, a measure of sound pressure level commonly used for environmental measurement), which is within the range of sound intensity that can cause negative physiological effects including hearing loss and stress. At 1 km from its source, noise from an automobile is estimated to be louder than the low-frequency background sound (Barber et al. 2010), though this depends on the intervening landscape. For example, an open plain may have dramatically different effects on the distance noise travels from the road than a heavily forested landscape. Yet even in large conservation reserves, traffic noise can be intense well beyond the area adjacent to roads. For example, in Glacier National Park, daytime sound levels are estimated to be 41.8 dBA as far as 500 m from the road, despite a relatively low average traffic volume of 3,700 vehicles per day (Barber et al. 2011).

Animals are keenly aware of their acoustical environment and depend on their sense of hearing to inform them about their surroundings. For small animals, their soundscape is composed of both environmental sounds, (including informative sounds such as those produced by stalking predators or scurrying prey) and intentional communication regarding behaviors and functions such as mating or predator avoidance. Though few studies have directly assessed the effects of road noise on small animals, it is clear that noise can have a substantial negative influence on animals in habitat adjacent to roads. Traffic from roads is thought to be the most pervasive source of anthropogenic noise (Nega et al. 2012), and it impinges upon the ability of animals to perceive natural sound. This effect, referred to as "masking," is known to alter animal behavior and influence spatial distributions (Barber et al. 2011). There is a burgeoning literature on the behavioral responses of songbirds to masking by road noise (e.g., Parris and Schneider 2008). Parris et al. (2009) found that the southern brown treefrog (*Litoria ewingii*) has increased the pitch of its call in response to traffic noise, a strategy that appears to be adaptive for countering masking effects. Even when noise from roads

does not mask meaningful sounds, it can distract animals from critical activities such as foraging, vigilance for predators, or sleep. Loud noises in particular can be a source of increased stress, which can render animals more susceptible to disease or weaken their ability to perform routine functions.

Bat activity and abundance is reduced up to 1.6 km from the road, possibly due to the interference caused by road noise on echolocation (Berthinussen and Altringham 2011). Exposure to loud noise (120 db) induced immobility in northern leopard frogs (*Lithobates pipiens*; Nash et al. 1970). Noise from motorcycles and dune buggies (≥100 db) decreased receptiveness of lizards to acoustical cues, and exposures of eight minutes in duration induced hearing loss (Bondello and Brattstrom 1979). Background noise often resulted in modification of calling behavior in male treefrogs (*Hylidae* spp.), which may impair the ability of females to discriminate among call types to discern locations of males during breeding migrations (Schwartz and Wells 1983; Schwartz et al. 2001).

Sound barriers can be effective tools to reduce the amount of road noise entering habitat adjacent to a road. However, such barriers can severely limit animal passage, a consequence that is rarely considered. As one solution, barriers can be constructed with options for passage included. Alternatively, natural vegetation management can be used to limit noise, since this is more conducive to animal passage. Culverts and bridges can be equipped with materials to insulate vibrations and dampen noises (Chapters 9 and 10).

4.3.1.1. Vibrations through Soil

In addition to altering the acoustic environment, motor vehicles generate substantial vibrations that travel from the road surface into the soil and adjacent environment. Such vibrations are known to cause damage to surrounding transportation infrastructure and have been cited as sources of stress for human populations living adjacent to roads. Very little is known about the effect of road vibrations on animal populations. However, we do know that many animals perceive vibrations through the ground as means of detecting prey or avoiding predators; some animals even communicate through vibrations (Bradbury and Vehrencamp 2011). Thus, it is likely that road-induced vibrations that influence the ability of animals to detect natural vibrations might also act as sources of physiological stress. Subterranean animals, such as moles and animals that inhabit burrows, may be especially prone to disturbance by ground vibrations. Some subterranean amphibians such as spadefoot toads (*Scaphiopus* spp.) emerge to breed when they detect low frequency vibrations that signal heavy rainfall; there is anecdotal evidence that vibrations caused by passing vehicles may deceive toads into emerging.

4.3.1.2. Sound Environment Under Bridges and Culverts

The undersides of bridges and the interior of culverts can function as echo chambers, intensifying road noises and vibrations. As a result, animals moving through or residing in these structures can experience some of the most severe levels of noise and vibration. This may present a problem, particularly since such structures are used by animals as passages to safely cross to the other side of roads. Currently, it is unknown whether sound and vibrations deter small vertebrates that might otherwise use bridges and culverts as passageways, and thus whether there is a need to design and install technologies that would dampen noise and vibrations (Chapters 9 and 10).

4.3.2. Light

Like sound, light is critically important for how most animals perceive and interact with their environment. The consequences of artificial lighting on animals are diverse and substantial (reviewed by Rich and Longcore 2006). Many small animals are nocturnal and well adapted for being active in low light environments. The addition of artificial lighting can have a variety of effects on behavior, including foraging, predator avoidance, and dispersal patterns. Changes in the light environment can also influence the circadian rhythm of animals living near a road (Morin 2013).

The open space above a road increases the amount of incident light in a forested area and alters the spectral quality of light, which influences the type of visual environment animals perceive (Endler 1993). Similarly, when light is reflected, it can become linearly polarized as it passes through the atmosphere, reflecting off of atmospheric gases and water droplets. Many animals have evolved sensory adaptations to detect polarized

light; these are used in many types of behavior, including orientation and navigation. However, the process of polarization also occurs when light is reflected off the road surface, buildings, and other human structures. Polarized light from these sources has been shown to induce negative behavioral responses in animals (reviewed by Horvath et al. 2009). For example, many aquatic insects are drawn to polarized light that is reflected off cars or roads, and will choose these sites to lay their eggs.

4.3.2.1. Vehicle Lights and Reflection

Headlights are a predominant source of artificial lighting along roads, especially where streetlights are absent. In addition to creating an altered light environment on and adjacent to the road, headlights appear to influence the mobility of animals on the road, in some cases causing them to remain stationary despite an approaching vehicle (Mazerolle et al. 2005). Exposure to artificial light can cause nocturnal frogs to suspend normal foraging and reproductive behavior and remain motionless long after the light has been removed (Buchanan 1993). Thus, headlights may further increase the risk of small animal roadkill. Headlights could also become a source of partially polarized light after reflecting off the road surface or surrounding structures and vegetation, thereby altering the light environment for small animals.

4.3.2.2. Streetlights and Structural Lighting

Streetlights have resulted in the loss of naturally dark environments throughout the world (Hoelker et al. 2010) and have dramatically altered the nighttime light environment for animals in roadside habitats. Many invertebrates are attracted to sources of light, which can in turn attract small animals such as toads to forage on or near roadways. Yet the effect of street lighting is not limited to the nighttime environment. Streetlights have recently been shown to change the community composition of ground-dwelling invertebrates such as ants and beetles, which can in turn influence food-web dynamics (Davies et al. 2012). Herpetofauna are susceptible to alterations in foraging, reproductive, and defensive behaviors in response to artificial lighting along roads and urban areas (e.g., frogs and toads, Buchanan 2006; salamanders, Wise and Buchanan 2006, Perry et al. 2008). Disorientation due to artificial lighting obscuring the natural light of the horizon on developed beaches has proven to be a significant and well-documented hindrance to the ocean-finding ability of hatchling sea turtles (McFarlane 1963) and nesting female turtles (Witherington 1992); roads are often constructed along beaches, and are often well illuminated. Lighting from bridges affects small animal communities similarly to streetlights, and since the area under a bridge can serve as a passageway across the road, lighting that deters or disorients animals may have a disproportionately negative effect on habitat connectivity.

To reduce impacts, lighting can be modified in terms of intensity and schedule, and placed with careful consideration of minimizing disturbance to road segments critical to animal passage. This may be especially useful during times of the year when animal movement rates are high, thus providing a balance between road safety and conservation imperatives.

4.3.3. Microclimate

The microclimate of the managed road verge can be dramatically different from that of the landscape more distant from the road, especially in forested regions (for a recent review, see Pohlman et al. 2009). This is caused by two factors. First, vegetation management of the road verge usually favors low grass or herbaceous cover and little or no tree canopy cover. Second, the road pavement heats up considerably during daylight hours, especially on cloudless days. Some of this heat is transferred to the road verge soil. Thus, the conditions of the road and road verge relative to a forest include more intense solar radiation during the day, greater radiative cooling at night, less obstruction to air flow, and presence of a substrate (the road) that is significantly hotter than the surrounding soil. The degree of contrast in microclimate between the road verge and the surrounding landscape is a function of the width and material of the road, the width of the managed verge and the type of vegetation management, the land cover of the surrounding landscape, the intensity of solar radiation at the site, and regional climate. The verge is also windier than a forest, and relative humidity is lower. Soil moisture is lower near roads than nearby forest, because of the road microclimate and because compaction of roadside soils and fills promotes rapid surface runoff rather than infiltration (see Section 4.4.1).

Overall, the microclimate contrast between the roadside and surrounding landscape is much greater during the day than at night. The road surface is often dangerously hot during the day and warmer than the surrounding soil at night. An extreme microclimate contrast can deter some species from approaching a road, especially small forest vertebrates; they may not be able to tolerate the hot, desiccating conditions of the road verge, particularly during the daytime. So, for small diurnal animals that are sensitive to heat and low humidity, the roadside and road may be an impassable barrier to movement.

The heat of the roadside soil and surface may attract some animals for thermoregulation. Sun basking may occur on the low-vegetation substrate of the verge or on the road, especially in early hours after sunrise for active diurnal species. At night, nocturnal snakes and other reptiles and amphibians may move to the road to warm up via conduction from the heated road surface. Turtles move to the soil next to the road to nest, apparently attracted to the warm soil conditions found there (see Case Study 4.3.3.1). As a consequence of attraction to the road, animals are put at risk of becoming roadkill. Road mortality may become a significant problem on roads that are attractive as basking sites.

Vegetation management can be used to make roadside microclimate less of a barrier to small vertebrates, or more of a barrier, depending on whether management goals are to encourage road crossing or deter it. Vegetation management could be used creatively to make corridors of favorable microclimate that guide animals to passage structures or safe places to cross a road. Where nocturnal road basking is frequent, and there are alternative roads for motorists to use, nighttime road closures may be worth considering. Wildlife barrier fencing might be worthwhile if species of conservation concern are frequently attracted to a known, spatially restricted segment of road.

4.3.3.1. Case Study: Roadside Nesting
by Aquatic Turtles
Tom A. Langen

All turtles nest on land. Female turtles excavate a nest in the soil, deposit the eggs, cover them with soil, and abandon the nest. Eggs incubate in the soil for two months or more, and upon hatching the young of aquatic turtles leave the nest and move toward water.

Hatchlings of a few species, such as the painted turtle (*Chrysemys picta*), will actually overwinter in the nest in northern regions and then migrate to water in the spring (Packard and Packard 2001*a*).

Female turtles seek out places to nest where soil is warm, well drained, and easy to excavate. Preferred nest sites typically have low or absent vegetation, ample exposure to sunlight, and slight elevation above the surrounding landscape. Individual turtles may return to the same area to nest year after year (Kolbe and Janzen 2002).

The rate of embryo development is affected by temperature (warmer = faster). High soil humidity can reduce water loss in eggs, resulting in bigger, more viable hatchlings (Kolbe and Janzen 2001; Packard and Packard 2001*b*; Ackerman et al. 2008). In most turtle species, the sex of hatchlings is determined by egg temperatures during a critical period of embryo development—a characteristic known as environmental sex determination (reviewed in Valenzuela and Lance 2004; Steyermark et al. 2008). Species vary as to whether warmer temperatures produce males or females; there is commonly a gradient of temperature in turtle nests that results in clutches with hatchlings of both sexes. If temperatures are too high, however, there can be total mortality of embryos (Wilson 1998).

Roadside habitats provide most of the cues that, to turtles, indicate appropriate nesting sites; thus, female turtles frequently nest on roadsides (Figure 4.3). This attraction to roadsides for nesting habitat is most evident at causeways crossing wetland areas or in the vicinity of bridge abutments. Nests are placed near the pavement's edge. Because the soil is very compacted near a road, and because there is typically gravel or rock fill along the roadside in wetland areas, it can take a nesting female several attempts to excavate a nest chamber of adequate depth. Along unpaved roads, including compacted earth or gravel roads, turtles will attempt to nest in the middle of the road.

Females are killed when crossing roads at the time of nesting. Depending on the species and location, females may initiate nesting during the morning, the afternoon, or at night; the timing of nesting in relation to local traffic patterns affects how dangerous attempted road crossings will be. Adult females are more frequently struck by vehicles than males, at least for aquatic turtle species, and road mortality is more fre-

quent during the nesting period (Aresco 2005; Steen et al. 2006). However, Beaudry et al. (2010) estimated that mortality of both sexes associated with mate-searching movements also threatened the viability of Blanding's turtle (*Emydoidea blandingii*) populations in the northeastern United States. Wetlands near roads have more biased sex ratios favoring males than wetlands that are more distant from roads (Steen and Gibbs 2004). Based on demographic modeling and data on the numbers of females turtles killed on roads, several researchers have concluded that roadkill of nesting turtles may result in long term declines of turtle populations near roads (Gibbs and Shriver 2002; see also Practical Example 1). In addition, turtles nesting on the roadside are easily detected and convenient to collect, at least those that nest during the day (Tucker and Lamer 2004). Thus, roadside nesting may also result in losses of female turtles due to molestation, harvest for food, or the pet trade.

Roadsides appear to be dangerous places for females to nest, but what about the viability of the nests? About this we know surprisingly little. One study in northern New York State found that nest predation, a principle cause of nest failure, was lower at nests placed near roads than other nesting sites (Langen 2012). Roadside nests were more variable in temperature than other

Figure 4.3. Common snapping turtle (*Chelydra serpentina*) nesting alongside a highway. *Credit: Tom A. Langen.*

Figure 4.4. Wildlife barrier for turtles (a) installed along a causeway (b) by the New York State Department of Transportation (DOT). *Credit: Tom A. Langen.*

nesting sites, and peak temperatures were significantly higher, peaking at temperatures that resulted in egg inviability under constant temperature incubation conditions in the lab. Moreover, roadside nests were significantly colder in winter than other nest sites, dropping to temperatures that again affect hatchling viability in laboratory studies. Nothing is known about the effects of salt, metals, hydrocarbons, and other soil contaminants on embryo development or hatchling condition. Overall, we know little about the success of nests constructed near roads as compared to other sites, but there are numerous reasons to suspect that there are important differences between roadside nests and other habitat that turtles use for nesting. Given the frequency with which turtles use roadsides as nesting habitat, this should be a research priority for agencies concerned with turtle conservation.

Are there options to reduce mortality for nesting female turtles, or to prevent roadside nesting? Fencing can be effective at reducing trespass onto the road by turtles (Figure 4.4), and female turtles will nest behind fencing if the vegetation is kept low. One concern would be that predators or people could learn to patrol fences to locate females or their nests (Langen 2012). Fencing is a barrier to movement, and thus may have other negative consequences as well (Chapter 3).

Construction of alternative nesting sites away from a road is a potential long-term solution. However, to date there have been few attempts to intentionally create nesting habitat for aquatic turtles and to entice them to switch from one nesting area to another, but

this is a promising approach (Buhlmann and Osborn 2011). Since quite a bit is known about aquatic turtles' nest-site habitat preferences, there are some possibilities for "turtle-scale" habitat manipulation to attract turtles to safer and more productive nesting areas.

4.4. Habitat Modification

The roadside or road verge is the area bordering a road that is engineered and managed for the physical integrity of the road and motorist safety. On highways and other major roads, the verge is broad enough to be classified as a distinct habitat type that small vertebrates must traverse when approaching a road. The engineered conditions of the road verge affect animals' behavior when encountering a road and the kinds of environmental stresses experienced there. Topography, soil conditions, microclimate, and vegetation are modified in ways that can make a road verge a harsh and dangerous environment for many small vertebrates, although it may also provide attractive habitat for some.

4.4.1. Microtopography and Soil Conditions

The area immediately adjacent to a road typically consists of compacted fill. Roads tend to be slightly elevated from the surrounding topography, especially where precipitation is high, so that water runs off the road surface and flows away from the road; at the base of the fill slope (the bank sloping down from a road) there will typically be a drainage ditch or swale. Soil compaction reduces permeability so that less water can percolate through the soil; the ground near a road thus tends to be relatively dry. The surface transitions from compacted fill to more natural soil with increasing distance from the immediate border of the road.

Roadside soil and fill, besides being compacted and relatively impermeable, is also chemically different from soil distant from a road. Metals and hydrocarbons accumulate in roadside soils. Nutrient composition may be altered and various contaminants present in soils near roads (see Section 4.2). Soil organic matter is typically low, and soil microbial organisms and macroinvertebrates are less abundant than in soils further from roads (Langen et al. 2006). Overall, roadside soil conditions can inhibit the growth and lower the nutritive value of many plants. Road verges tend to have smoothed contours, and natural structure (e.g., rocks and coarse woody debris that provide cover) that is often removed to improve safety and simplify roadside mowing. As a consequence of the structural simplicity of a roadside and its typically sparse, low vegetation (see Section 4.4.2), the verge offers little protective cover for small vertebrates that venture onto it. However, some mammals, such as groundhogs and ground squirrels, may dig into the fill bank to build their burrows. Where there are steep slopes or problems with bank erosion, roadsides may be armored with rock fill. These sites provide protective cover for some small vertebrates, and may be colonized by ground squirrels, snakes, lizards, and other small animals that prefer rocky substrates.

The road verge can be managed to modify small vertebrate activities near roads. Maintaining the open, dry, exposed environment that is typical of highway verges may discourage many small animals from venturing toward the road, which may be desirable if the management goal is avoidance of roadkill. If, however, the goal is to maintain connectivity by encouraging road crossing or use of a crossing structure, corridors of high vegetation riddled with cover objects such as rocks and coarse woody debris may be effective. Detailed reviews of roadside microtopography and soil conditions include Spellerberg (1998), Trombulak and Frissell (2000), Forman et al. (2003), and Langen et al. (2006).

4.4.2. Vegetation Management and Conditions on the Roadside

Vegetation is managed along highways and other roads to increase visibility for drivers, provide a safe recovery zone in the event of a vehicle driving off the road, and to alter the type of vegetation present. This zone of vegetation management can be quite wide, often exceeding 10 m from the edge of the road, and of considerably different character from the surrounding landscape. Trees are removed near roads because they are obstacles to vehicles that run off the road and because they impede the field of view; grass height is also kept at a short length to reduce maintenance costs and improve visibility. Vegetation management, along with the other habitat alterations associated with the road verge, widens the barrier to small animal movement beyond the road surface.

The edge effect and the change in vegetation type favor some small animal species. High densities of rodents near roads can encourage risky hunting behavior for predators that venture into this area, causing increased mortality, especially for raptors such as barn owls (*Tytonidae* spp.; Borda-de-Água et al. 2014). Most small animals depend on cover to hide or escape from predators, and frequent mowing to manage road verges can be lethal not only due to the mowing itself, but also due to the reduction in cover.

If habitat features preferred by small animals of conservation concern are present in the clear zone (where vegetation is reduced), this area may benefit from management to promote habitat for the species, especially if vehicle-caused mortality is not a problem. For example, in ecosystems dependent on wildfire, prescribed burning can help maintain native plants in the clear zone as long as necessary precautions are in place to prevent spread of the fire beyond the intended area (Shochat et al. 2005). Conversely, vehicle-caused mortality may rise to levels of population concern if rare species are attracted to the ROW (right-of-way). An example is the federally listed Oregon silverspot butterfly (*Speyeria zerene hippolyta*) whose preferred nectar-producing plants are also uncommon and grow near a highway; mortality of these fragile insects can occur simply from the turbulence from passing vehicles (Zielin et al. 2012). Air turbulence from passing vehicles can also blast birds and small animals and cause fatal injuries, and barotrauma caused by air pressure waves from passing vehicles can cause fatal internal injuries to amphibians. If road mortality is a population concern, removing vegetation or mowing to reduce the attraction of vegetation may be a reasonable trade-off for a widened barrier effect. Managing habitat to make it less suitable near the highway can also be a potential solution, as it is for the threatened butterfly.

Similarly, in areas with declining populations of snakes due to mortality caused by basking on warm roadbeds, an inhospitable clear zone may be managed to hinder access to the warm road surface. Creating surfaces within the clear zone for snakes to bask may reduce mortality by providing alternative basking locations. Turtles are often found nesting along roadsides, where the construction materials favor soil textures amenable to digging nests. The attraction to the roadside comes with the risk of the adult females and hatchlings being killed while crossing. For turtle species with declining populations, creating suitable nesting substrates in locations where fewer road crossings are necessary would reduce the loss of these animals (see Case Study 4.3.3.1).

4.4.3. Arboreal Connectivity

Arboreal connectivity refers to tree crowns that overhang the road on both sides, closely enough that arboreal animals can cross the road above it. Arboreal connectivity is an important issue for road management in tropical regions, where many mammals, amphibians, reptiles, and other animals are highly arboreal. Although there are fewer arboreal animals in temperate forests than the tropics, arboreal connectivity may be beneficial for some species such as tree squirrels (*Sciurus* spp.), American pine marten (*Martes americana*), ratsnakes, and treefrogs. Many other semi-arboreal species such as raccoon (*Procyon lotor*) and Virginia opossum (*Didelphis virginiana*) may also use arboreal connections to cross roads when such opportunities arise. There is currently much work on the design and testing of arboreal crossing structures over tropical roads (see Case Study 4.4.3.1), but the utility of maintaining arboreal connectivity in other regions is relatively unknown, warranting further research.

An additional consideration in maintaining tree crowns that hang over a road is the effect on microclimate. The greater the extent of the overhang, the more natural the microclimate of the habitat below, along the road verge and road. Light and wind intensity will be lower, humidity and soil moisture will be higher, and temperatures will be less extreme than a typical road and verge. These microclimatic conditions will make it more likely that forest animals will enter a verge and, perhaps, attempt to cross a road.

4.4.3.1. Case Study: Small Vertebrates and Roads in the Tropics
Tom A. Langen

This book has emphasized small vertebrate issues associated with roads in developed countries within the temperate zone. Many of the critical issues and management options discussed in this book apply equally to developing countries in the tropics. However, there are several additional road-related issues that affect small vertebrates in developing tropical countries.

Although diverse in terms of economic conditions, many developing countries in the tropics are rapidly growing in population, and their wealth and resource consumption are also increasing; as a consequence, public infrastructure including roads is expanding explosively. Roads are being built into formerly inaccessible regions of these countries to facilitate long-distance transportation, resource extraction, development of isolated communities, and colonization (Laurance et al. 2009). The principal impacts on small vertebrates may include an immediate increase in exploitation (hunting, collection for the pet trade) along the road corridor, long-term loss of habitat, and increased habitat fragmentation as farms and communities are founded, developed, and expanded. Since control of homesteading or squatting is often weakly enforced (and sometimes officially encouraged), road corridors can be rapidly settled and cleared.

Some countries have been proactive about avoiding these impacts of road construction on environmentally sensitive areas. In the 1980s, Braulio Carillo National Park in Costa Rica was created along the corridor of a new highway to protect a vast highland cloud forest that had previously been inaccessible. This area surely would have been colonized after construction of the highway had it not been protected (Evans 1999).

In many developing countries, there are weak controls on poaching, and roads provide access to small vertebrate populations. In rural areas, traffic by foot, animal, or bicycle is common. Pedestrians are more likely to notice small vertebrates on or near a road than people in motor vehicles; this can elevate the risk that animals will be detected and killed or captured because they are perceived as dangerous (e.g., snakes), used as food (turtles, lizards, mammals), or kept as pets (amphibians and reptiles).

The higher diversity of tropical small vertebrate species, many with poorly documented natural histories, presents a challenge to the development of effective road management plans. There are at least four issues related to how small vertebrates interact with roads that are of greater concern in the tropics than the temperate zone.

The first is arboreality. Many small vertebrates in the tropics are arboreal or semi-arboreal; individuals of some species may never descend to the ground. For these species to maintain habitat connectivity across a road, there must be arboreal connectivity. Where roads are not too wide, this can be accomplished by allowing branches to grow over the road and intersect branches from the other side; this creates a "green tunnel" effect (Figure 4.5). For example, the Guanacaste Conservation Area in northwestern Costa Rica, in collaboration with the Costa Rica Ministry of Transportation, maintains arboreal connections over the Pan-American Highway where it passes through the conservation area.

Where roads are too wide, or branch overgrowth is not possible, inexpensive arboreal bridges can be installed (e.g., cables and lattice bridges can be suspended from between the canopy of trees on either side of the road), or else structures can be provided as perches or platforms for leaping and gliding animals. Such aerial wildlife crossing structures are effective for monkeys, sloths, marsupials, and other arboreal mammals, and are also likely to work for arboreal reptiles and amphibians. Recent successful examples from Australia include work completed by Weston et al. (2011) with rope bridges for arboreal mammals, and by Taylor and Goldingay (2012) with tall poles as stepping stones across roads for gliding mammals.

A second issue is the importance of riparian corridors for wildlife, including small vertebrates. Tropical rivers are often entrenched, and are usually bordered by evergreen riparian forest. In seasonally dry regions of tropical dry forest or savannah, these "gallery forests" are important refugia for wildlife in the dry season. In regions where evergreen forest has been extensively cleared for settlement and agriculture, gallery forest often remains. This provides a refugium for forest species, and often serves as a corridor between remnant forest patches. The roads and bridges that cross tropical

Figure 4.5. A "green tunnel" of tree canopy foliage and humid roadside verge on the Pan-American Highway, within the Guanacaste Conservation Area, Costa Rica. *Credit: Tom A. Langen.*

rivers are typically elevated above the river to avoid damage during floods; with care, much of the riparian vegetation can remain intact, and is thus appropriate for the occupancy and movement of animals.

A third issue is microclimate. Sunlight is more intense in the tropics than the temperate zone, and as a consequence, microclimate differences between a road and bordering forest can be much more extreme. In comparison to a surrounding forest, road verges are brighter and hotter in the day, usually cooler at night, and are drier and windier overall (see Section 4.3.3). This can make the road an extremely hostile environment and an effective barrier to movement for small vertebrates that are adapted to a forest's microclimate, especially for diurnal forest species. To reduce the microclimate and light-intensity contrast between a forest and road, tree branch overgrowth can be encouraged over a road. The Guanacaste Conservation Area's "green tunnel" policy along the Pan-American Highway, for example, along with providing arboreal connectivity, is also effective at maintaining cool, moist, still conditions along the highway's verge that are suitable for frogs and forest-interior rodents.

The final issue is that small forest vertebrates are often loathe to cross open gaps such as those created by roads; this is particularly true of evergreen forest species (Goosem et al. 2001). Animals may be deterred by changes in light and microclimate, the lack of cover, or the presence of humans associated with broad forest gaps such as those created by roads, as well as the roads themselves. In one Australian study, for example, some rainforest mice avoided crossing a narrow, unpaved forest tract (Goosem 2001, 2002). Appropriate vegetation management, cover objects, and underpass structures may work for enticing small vertebrates to cross a road (Goosem et al. 2001).

Research on the environmental impact of roads and road traffic on tropical wildlife lags far behind that of Europe and temperate North America. Given the explosive growth of road networks in developing tropical countries, research to understand region-specific problems and to design and implement appropriate solutions for small vertebrates and roads is imperative. Good general reviews of the environmental impacts of roads in the tropics include Goosem (2001) and Laurance et al. (2009).

4.4.4. Edge Effects and Roadside Habitat

In the field of landscape ecology (Forman 1995; Farina 2006), there are two general types of habitat edges—natural and artificial. Natural edges or ecotones occur where two or more vegetation communities meet. These edges may exhibit gradual or abrupt changes in types of vegetation, but primarily are composed of native species. Natural edges are generally irregular and random in form. Artificial edges occur at unnatural or human-influenced boundaries between vegetation communities. These edges often exhibit a sharp contrast in vegetation types and include native and non-native vegetation. Roadsides are typical examples of artificial edges.

The composition and structure of vegetation along edges are very different from those of core habitat. Within core habitat, there is often a high degree of uniformity in vegetation type and structure, such as in a forest with a relatively continuous closed canopy that provides protection from extreme environmental influences. Edges are typically abrupt, open, and exposed to an array of environmental effects:

- Microclimatic changes—more severe temperature extremes, lower humidity, higher wind velocities and solar incidence.
- Physical processes—increases in contaminant transport, surface sediment transport, erosion and deposition, and wind throw of trees.

- Ecological changes—altered forms and rates of predation, parasitism, competition, spread of invasive species.

Whereas natural edges are randomly shaped, complex (e.g., with curvilinear geometry and diverse composition of vegetation), and variable in extent (e.g., from a small treefall gap to an ecotone along a river floodplain), artificial edges are mostly nonrandom and linear in form. The complexity of natural edges dampens external environmental effects, whereas these effects are pronounced along artificial edges. These edge effects negatively influence the quality of habitat from tens to thousands of meters away from roads (Forman 1995; Forman et al. 2003).

When possible, the interface between the road and the adjacent habitat should include qualities of natural edges, such as increased complexity and irregularity of shape. For example, this might include planting a zone of shrub layers between the standard roadside groundcover and adjacent trees. Native species should be used rather than introduced species, and the diversity of plantings should emulate that of adjacent natural habitat.

4.4.5. Swales and Roadside Ditches

Drainage features near roads, such as swales, roadside ditches, or canals, provide artificial aquatic habitat. These features differ from natural water bodies both physically and biologically. Their size, shape, and depth play significant roles in their ability to provide long-term habitat for aquatic and semi-aquatic fauna. The complexity of vegetation and productivity of littoral zones are strongly influenced by the shape and depth of the underlying surface of the water feature. Roadside drainage features are typically engineered with steep slopes and relatively uniform depth, in contrast to the gradual slopes and complex depth profile of most natural water features, which support more productive littoral and benthic zones and therefore sustain more complex faunal assemblages.

Water quantity and hydroperiod are regulated primarily by frequency of rainfall, rate of percolation into underlying soil layers, and rate of runoff from the road. In contrast, natural water bodies may be fed by groundwater sources as well as from surface runoff, providing more stability in water levels. Hydroperiod is defined as the duration and frequency of inundation or saturation (Nuttle 1997). It plays a significant role in the character and composition of aquatic vegetation and suitability of the water feature to support populations of fish, amphibians, and reptiles (Wells 2007). Road drainage features, such as ditches, may act as "ecological traps" for some breeding amphibians, attracting them to deposit eggs despite hydroperiods that are too short for offspring to reach maturity (Babbitt 2005; Baber et al. 2003; Werner and McPeek 1994).

The water quality of roadside drainage features is strongly influenced by the road. Stormwater facilities are repositories for contaminants from vehicles, pavement surface treatments (e.g., road salt), and the pavement surface material itself (see Section 4.2); these contaminants reduce water quality and diminish suitability for aquatic and semi-aquatic fauna (Bishop, Struger, Barton, et al. 2000; Bishop, Struger, Shirose, et al. 2000).

When possible, stormwater retention areas should better resemble natural wetlands because their complex water features containing both shallow- and deepwater habitats reduce the impacts of contaminants (Lovell and Johnston 2009; Clark and Acomb 2008). Native aquatic vegetation in littoral zones and created wetlands should resemble that of local, natural water features. When ROW width allows, ditches and canals could be designed to be more similar to natural water courses in order to reduce flow velocities and improve habitat quality.

4.4.6. Erosion and Siltation

Runoff from paved surfaces can negatively affect the quality of adjacent habitat. The impervious road surface increases the volume of stormwater runoff for a given area. If unattenuated, the rate of runoff can be rapid, resulting in greater potential for soil erosion of adjacent areas and siltation in adjacent water features. Loss of surface soil or groundcover vegetation then eliminates potential habitat adjacent to roads. The rate of stormwater runoff can be reduced by the use of swales and more diverse plantings of native groundcover species (e.g., a mix of bunch grasses, short grasses, and other appropriate forbs) on road shoulders (Harper-Lore and Wilson 2000).

Siltation is known to reduce habitat suitability for fish and amphibians inhabiting water bodies adja-

cent to roads. In the Pacific Northwest, high-gradient streams experience the greatest sedimentation impacts from roads, resulting in decreased populations of various endemic amphibians (Corn and Bury 1989; Welsh and Ollivier 1998). In addition to direct reduction of amphibian densities due to mortality or reproductive failure, sedimentation from roads may reduce invertebrate prey densities in streams (Richter 1997; Welsh and Ollivier 1998; Semlitsch 2000). Careful placement and construction of roads can significantly reduce erosion problems. McCashion and Rice (1983) found that 24% of the erosion from logging roads in California (a major cause of sedimentation in forest streams) could be prevented with conventional engineering methods (e.g., proper grading, adequately sized culverts), and the remaining 76% could be attributed to site placement and conditions (see also Montgomery 1994).

4.5. Access

4.5.1. Human Movement and Access

A road is an element of public infrastructure that is designed to provide safe, efficient, and convenient access to some place, whether for the general public or a specific class of users. However, roads can also provide access to unintended places and for unintended users, and this can lead to habitat degradation or direct risks to small vertebrates. The presence of motor vehicles, bicyclists, and pedestrians deter some animals from approaching roads (see Section 4.6.1), and animals that do approach them may be targets of harassment—predators and game animals are legally or illegally shot from roads, and some animals are harvested for food or other purposes (e.g., Tucker and Lamer 2004).

Roads provide access for human entry into the surrounding landscape (Figure 4.6), which results in more frequent and widespread human-wildlife encounters and may also lead to habitat degradation. Perhaps the most severe result is increased levels of poaching and unregulated harvest near roads. This includes both direct harvest of animals and timber harvesting, which degrades habitat quality near roads. Given that in developed countries like the United States, much of the landscape is within a few hundred meters of a road, few regions are too remote for humans to easily access from a road (Riitters and Wickham 2003), and most animals encounter roads during their natural movements.

Figure 4.6. Viewed from within this wetland, the road lies just 3 m away. The litter in the foreground is proof of habitat degradation resulting from the road's proximity and consequent human access. This proximity can also mean that animals using the wetland are vulnerable to poaching or illegal harvest. *Credit: Steven P. Brady.*

Road networks intended for energy and timber extraction can inadvertently cause problems associated with increased access to wildlands when the roads are accessible to unintended users (see Case Study 4.5.1.1; Havlick 2002).

In fire-prone regions, road access can increase the frequency of wildfires. Many such fires are caused accidentally by carelessly discarded cigarettes along a roadside. Some fires may be accidental escapes of campfires made by people who have traveled by road. However, some fires are lit intentionally to clear vegetation from the land adjacent to a road, or by arsonists who can conveniently access a flammable target, start a fire, and leave the area before detection.

The most important solution to access issues is to minimize new road construction through environmentally sensitive wildlands. Where road networks are needed for resource extraction, however, locked gates or legal mechanisms should be used to prevent unintended users from exploiting the roads. There should be a contractual obligation to remove resource extraction roads or make them impassable once they are no longer needed. To reduce access to the adjacent landscape, safe parking places along the road should be restricted to those areas that are the least environmentally sensitive. Fencing can also be used to impede access to the landscape around the road.

4.5.1.1. Case Study: Road Proliferation due to Rapid Renewable Energy Development
Jeffrey E. Lovich

Renewable energy development, especially to harness wind and solar resources, is proceeding rapidly worldwide, particularly in the United States. The potential scale of road building associated with renewable energy development is significant by virtue of the large footprint of facilities (McDonald et al. 2009). For example, solar energy development is currently being considered on some portions of almost 40 million hectares of public land in 6 southwestern states (Lovich and Ennen 2011) including Arizona, California, Colorado, Nevada, New Mexico, and Utah. In addition, worldwide wind energy development exhibited a 15-fold increase in generating capacity between 1995 and 2006 (Golait et al. 2009), with 23% annual growth rates in the United States from 2000 to 2009 (Wilburn 2011). Associated with this surge in development is an increase in road building to provide access to wind turbines and solar arrays. Although the effects of roads on wildlife are well documented, little research has been conducted on their effects relative to renewable energy development (Lovich and Ennen 2013).

Since many renewable energy facilities are in remote areas, roads are built to provide access to both the general site and to specific infrastructure within a site. For example, roads are necessary to provide access to individual wind energy turbines for construction and maintenance. These roads can be paved, covered with gravel, or bladed into native soil and rocks. Water and wind erosion is a significant problem when wind energy development occurs in hilly or mountainous terrain (Wilshire and Prose 1987). Roads built on other than flat terrain require culverts to direct runoff away from the roadbed, and culverts by themselves can have adverse impacts on wildlife.

The various effects of roads on wildlife, including effects on mortality, behavior, and on the nearby environment, can all similarly result from roads at renewable energy sites. Agassiz's desert tortoise (*Gopherus agassizii*) is a federally threatened species that occurs in wind energy facilities near Palm Springs, California. Tortoises at the site have been the focus of ecological research since 1995, about 12 years after wind energy was established. In the mid-1990s, Lovich and Daniels (2000) concluded that burrows constructed by tortoises were located closer to dirt roads than random locations without burrows, and suggested that the herbivorous tortoises might use roads as foraging microenvironments because of roadside vegetation enhancement in the desert (Johnson et al. 1975). Although this association may appear to be beneficial, it also increases the probability of tortoise mortality from vehicle collisions (von Seckendorff Hoff and Marlow 2002; Lovich, Ennen, Madrak, Meyer, et al. 2011).

Additional research on tortoises at a Palm Springs wind energy facility with a dense network of roads found few differences in the growth rate, annual survivorship, or population structure of tortoises compared to more natural populations (Lovich, Ennen, Madrak, Meyer, et al. 2011). Nesting ecology is also comparable to that reported from populations in natural areas (Ennen et al. 2012), and nesting burrows were not located closer to turbines and roads. Continued research at the site suggests that tortoises no longer use areas near concentrations of turbines and roads as frequently as they did in the past (Lovich, unpublished data).

Although culverts are widely promoted as safe passages for wildlife under roads, they can also have negative consequences (Lovich, Ennen, Madrak, Grover, et al. 2011). In one case, a radio-tagged male tortoise used a culvert in a wind energy facility as a burrow surrogate. Heavy rains filled the culvert with a slurry of sand, entombing the tortoise for several weeks. After the tortoise was exhumed, it was released and lived less than 18 days before dying of cardiac and pulmonary complications associated with the burial. Ignition points for fire in the California deserts can also be directly associated with roads (Brooks and Matchett 2006) and are another source of wildlife injury and death.

Simple mitigation measures may ameliorate some of the negative effects of roads associated with renewable energy. Designing roads with larger concrete box culverts may reduce or eliminate the wildlife mortality occurring in small tubular 0.61 m steel culverts that are easily plugged by runoff. Slow speed limits and speed bumps at renewable energy facilities may further reduce vehicle collisions with slow-moving species like tortoises and rattlesnakes. Training facility personnel to be aware of wildlife on and around roads is another low-cost method with the potential to reduce vehicle impacts.

4.5.2. Invasive and Subsidized Species

Road verges are notorious as pathways for the spread of invasive species, including plants as well as animals (Mortensen et al. 2009). Verges are usually highly disturbed, especially during construction, allowing footholds for noxious weeds. Noxious weeds not only change native habitats that many small animals depend on, but these areas may become reservoirs of harmful chemicals used to treat noxious weeds, thus creating a secondary impact from toxicity. Roads can facilitate the spread of invasive plant species into habitats further from the road verge, potentially displacing native vegetation (Tyser and Worley 1992; Parendes and Jones 2000; Tikka et al. 2001; Mortensen et al. 2009).

In Australia, it was determined that roads and trail systems facilitated the range expansion of the invasive cane toad (*Rhinella marina*) across the country (Seabrook and Dettman 1996), which is now expanding its range at >50 km per year. In the southeastern United States, fire ants (*Solenopsis invicta*) have proliferated along roadsides (Stiles and Jones 1998) and will prey upon reptile nests (e.g., Allen et al. 2001; Buhlmann and Coffman 2001; Parris et al. 2002). In addition, subsidized predators, such as raccoons, are often found along roadsides or near other human-altered landscapes seeking food refuse and carrion (Hoffman and Gottschang 1977; Gehrt et al. 2002; Prange et al. 2004); reptile and bird nests (including both adults and young) in these areas are then potentially vulnerable to those predators' opportunistic depredation.

4.5.3. Development

Roads can facilitate development by providing improved access to agricultural, natural, or rural areas (Moon 1987; Ewing 2008). This leads to significant loss of habitat and increased fragmentation and barrier effects, resulting in loss of area-sensitive species and alteration of normal species movement and dispersal patterns (Trombulak and Frissell 2000). Development along new roads sited through pristine natural habitat is a severe problem in many regions in the tropics (see Case Study 4.4.3.1). While agricultural and rural areas can be designed and managed to maintain desirable habitat qualities for use by some species sensitive to disturbance (Scherr and McNeely 2007; 1000 Friends of Florida 2010), urban and suburban development transforms the landscape into an environment that favors species adapted to human-related disturbance and edge habitat (Noss and Cooperider 1994). To prevent undesirable development of the land adjacent to the road, appropriately designed and enforced land use regulations may be appropriate. Conservation easements can be used to restrict land use options on private property around a road.

Increased impervious surface area associated with development results in higher average temperatures, greater amounts of stormwater runoff and pollutant discharge, more traffic, more noise, and greater levels of many other disturbance and environmental degradation factors. Cumulative effects of these elements of the urban environment result in reduced cover and quality of habitat available to wildlife.

4.6. Animal Behavior and Roads

When considering how animal behavior contributes to ways in which animals are affected by roads and road traffic, one must consider two aspects of behavior. The first is whether animals are attracted to roads, avoid roads, or encounter roads randomly during movements. The second is how animals actually cross roads, and whether crossing behavior differs in the presence or absence of vehicle traffic. How animals behave near and on roads will both determine how roads affect wildlife populations and guide what mitigation measures to apply (Jaeger et al. 2005).

4.6.1. Animals That Avoid Roads

Many species avoid entering roads, some deterred solely by the road surface, and others also deterred from entering the strip of managed landscape along the road verge. For some species, the vegetation, topography, soil, microclimate, and other conditions

along roadsides reviewed in this chapter cause animals to avoid roads. For other species, the road itself may not be a deterrent, but the sights, sounds, smells, and vibrations of vehicles or other traffic along roads (e.g., pedestrians, bicycles) may be the deterrent. Many small vertebrates are deterred by roads (e.g., Andrews et al. 2006), including some rodents (Mader 1984; Goosem 2001, 2002; Ford and Fahrig 2008), birds (Laurance et al. 2009), reptiles (Shine et al. 2004; Andrews and Gibbons 2005; Hyslop et al. 2006; Plummer and Mills 2006; Shepard et al. 2008), and amphibians (Gibbs 1998; Madison and Farrand 1998; Marsh et al. 2005).

To evaluate the consequences of road avoidance and apply appropriate mitigation methods, it is essential that one identify why animals avoid roads. For species that are deterred by road traffic, roads with little traffic during the period that the animals are active will not be a serious barrier to movement, and road closures can be effective during periods of mass movements. Passageways under heavily trafficked roads may not be effective for these species if the stimuli that deter the animals (e.g., vibrations, vehicle lights) are detectable in the crossing structure.

For species that avoid roads because of the conditions on the road verge or the road itself, even roads with little or no traffic will serve as a barrier. Vegetation management, addition of cover structures, allowance of tree canopy overgrowth, and other measures could be used to make conditions more attractive at road locations where crossing is to be encouraged.

Many studies have found that populations of particular small vertebrate species are lower near roads than in similar habitat distant from roads (reviewed in Fahrig and Rytwinski 2009), even on lesser-traveled forest and logging roads (Semlitsch et al. 2007). Although this is sometimes interpreted as indicating that excessive road mortality near roads has resulted in a population decline in such species, it may also indicate road avoidance. Behavioral data are essential for resolving whether road mortality or behavioral avoidance causes reduced populations near roads, and are therefore also essential if effective mitigation measures are to be implemented.

4.6.2. Animals That Are Attracted to Roads

Although some animals are attracted to the road surface, many others are more often drawn to the managed road verge. Attraction to roads increases the risk of road mortality, exposure to toxic substances around roads, and harassment by humans. On the positive side, attraction to roads may make it more likely that habitat connectivity will be maintained if animals can safely cross the roads or have appropriate wildlife passages to use.

Reptiles and amphibians may be attracted to paved road surfaces for thermoregulation, especially those roads with infrequent traffic, and especially on cool nights following sunny days. Snakes are most frequently reported to use roads for thermoregulation (e.g., Klauber 1931; McClure 1951; Dodd et al. 1989), but there are many reports, mostly anecdotal, of various other species of herpetofauna encountered on roads while seemingly using the road surface as an aid in thermoregulation. Additionally, the warm microclimate of road verges is attractive for some reptiles for thermoregulation (e.g., Sartorius at al. 1999).

Small carrion-feeding mammals, and plausibly some reptiles, are attracted to road surfaces to feed on roadkill (Enge and Wood 2002; Smith and Dodd 2003; Antworth et al. 2005; Gerow at al. 2010). Mammals killed on the road are disproportionately composed of scavengers (Barthelmess and Brooks 2010).

Turtles and other reptiles are attracted to warm roadside soils for nesting (see Case Study 4.3.3.1; Practical Example 1; Hódar et al. 2000). Aquatic reptiles and amphibians are attracted to roadside ditches and swales, or large puddles and water-filled tire ruts on unpaved roads (e.g., Reh and Seitz 1990).

Roadsides provide attractive habitat for small mammals, reptiles, and amphibians that prefer to forage or move through grassy open areas. This includes grass-, forb-, or seed-eating rodents; small predators that feed on rodents; and insectivorous small vertebrates (Getz et al. 1978; Seabrook and Dettmann 1996; Adams and Geis 2002; Grilo et al. 2012). Hawks and owls are attracted to roadsides to hunt small vertebrates, and road mortality of these birds can be high enough in some cases to cause population declines (Borda-de-Água et al. 2014).

4.6.3. Behavior while Crossing a Road

For some small vertebrates, the road surface itself is a barrier, even in the absence of vehicle traffic. The width of a road, in terms of body lengths, number of paces, or time to cross, is much greater for small vertebrates than the large mammals that are typically the focus of concern on roads. The flat, featureless open expanse of the road, with a substrate surface temperature that differs from the surrounding landscape, may be a formidable obstacle to animals that prefer to remain near or under cover. Such animals may be deterred from attempting a road crossing, but should they attempt to cross, they move quickly and directly across the road to minimize time and distance on its surface (e.g., for snakes, Andrews and Gibbons 2005; turtles, Beaudry et al. 2008). Some small Australian rainforest rodents are hesitant to cross narrow unpaved logging roads, even in the absence of traffic (Goosem 2001, 2002). The propensity of snakes to cross a road can depend on the road surface (Shine et al. 2004) and canopy cover (see Chapter 9; Case Study 9.6).

Animals vary in important ways in their behavioral response to vehicle traffic on a road, which is important to consider when anticipating the potential effects of traffic volume on their populations. Amphibians and reptiles tend to exhibit antipredator behaviors in response to oncoming vehicles, meaning many of them immobilize in an attempt to blend in, which does not fare well with vehicular predators (e.g., Lima et al. 2014). Andrews and Gibbons (2005) have shown that even within one local assemblage of snakes, species differ dramatically in how they respond to roads and road traffic. Jacobson and Smith (2010) have provided a useful classification of behavioral responses of animals to road traffic. They classify animals into four categories: avoiders, pausers, nonresponders, and speeders.

Avoiders are hesitant about attempting to cross a road even when traffic volume is light—they regard the sights, sounds, and vibrations of moving vehicles as cues of danger. They may modify the timing of movements to avoid periods of elevated traffic volumes. Such species have relatively low risk of road mortality since they avoid crossing when traffic is present. Populations of such species may become isolated by roads if traffic volumes are constantly high, since these animals will avoid crossing under such conditions.

Pausers respond to oncoming vehicles by halting forward movement. The risk of road mortality increases disproportionately with road traffic volume, and the probability of a successful road crossing would be nearly nil when traffic volumes are high. Pausers include small vertebrates that respond to predators by using freezing and crypsis, thanatosis ("playing dead"), coiling (for snakes), or offensive responses to threats. Examples of pausers, which are frequently detected as roadkill, include skunks, porcupines, armadillos, opossums, turtles, and amphibians. Mazerolle et al. (2005) found that the strongest stimuli for immobilization behavior across six amphibian species were a combination of headlights and vibration. Andrews and Gibbons (2005) found a high rate of immobilization in response to a passing vehicle among three snake species, a rate high enough that these species would be unlikely to survive crossing a busy highway. At high traffic volumes, roads are complete barriers, whether because of the high risk of road mortality or deterrence from attempting to cross the road.

Nonresponders continue to cross regardless of traffic volume. Such animals behave as if they do not detect oncoming vehicles, or else do not have an appropriate behavioral response to the vehicles. Many salamander and frog species, for example, are nonresponders (e.g., Bouchard et al. 2009). Road mortality increases proportionately to increases in traffic volume, and traffic flow models of road mortality risk are appropriate (Hels and Buchwald 2001; Gibbs and Shriver 2005; van Langevelde and Jaarsma 2004). As traffic volume increases, roadkill rates increase and the probability of successfully crossing a road approaches zero, thus creating a complete barrier. Thus, populations of nonresponders are at risk of both excessive road mortality and barrier effects where traffic volume is high. However, in some instances, roadkill observations are highest at moderate traffic densities (Sutherland et al. 2010), but it is not clear whether this is a matter of detectability because carcasses are degraded and removed at a higher rate at high traffic densities, or whether more animals are crossing the road at reduced traffic volumes (Section 8.2.1.4; Section 12.4.3.1; K. M. Andrews, personal communication).

Speeders flee from perceived danger with increased speed and directed movements away from the threat. Speeders, such as some rapidly moving snakes or small carnivores, can respond to elevated traffic volumes by increasing speed to exploit traffic gaps. If traffic volumes are too high, gap distances become too small to safely pass, and the road becomes a complete barrier.

4.7. Key Points

- Winter salt management plans that aim to reduce the application and runoff of road deicing salts can better protect aquatic animal populations and their habitat, and reduction in road salt contamination is especially important where conservation of amphibians is a priority.

- Chloride is the element considered to be harmful to freshwater aquatic life, so replacing one chloride-based compound with another does not solve the problem for sensitive species. More recently developed deicers that contain an organic source (sugars or calcium magnesium acetate) show promise, because it is likely that these compounds will be consumed rapidly by microbes in wetlands and streams.

- Culverts installed to direct runoff downslope and away from wetlands and streams would help to reduce contamination of sensitive aquatic habitat. Legacy stormwater management designs that aim runoff at wetlands should be reconfigured.

- Redirecting runoff to a constructed wetland or through an infiltration system can result in a reduction in contaminant concentrations prior to runoff waters entering a natural habitat or groundwater. However, while constructed wetlands may appear to provide wildlife habitat, it is important to be aware that those with high contaminant burdens may function as ecological traps, endangering wildlife populations that are drawn to them.

- Most pollutant deposition occurs within 30 m of roads, so establishing appropriate 30 m wide vegetated strips along roads would help to contain the pollution and reduce impacts to the surrounding landscape.

- Community involvement seems to have had the greatest impact on reducing littering and dumping along roads. In rural areas where curbside waste pickup is unavailable, providing affordable and accessible collection stations for waste, including hazardous waste, may reduce clandestine dumping.

- Sound barriers can be effective tools to reduce road noise, although they may block animal passage, so options for passage should be considered. Natural vegetation can also be used to limit noise. Culverts and bridges can be equipped with materials to insulate vibration and dampen noise (Chapters 9 and 10).
- Lighting can be modified in terms of intensity and schedule, and placed with careful consideration of minimizing disturbance to road segments critical to animal passage, particularly where and when nocturnal animal movement rates are high.
- Vegetation management can be used to make roadside microclimate less of a barrier to small vertebrates or to provide a guide to passage structures, or else to create more of a barrier to deter road approach, depending on management goals. The interface between the road and the adjacent habitat should be as natural as possible, and include the qualities of natural edges, such as irregularity of shape, native plant species as opposed to exotic, and a high vegetation complexity (e.g., incorporation of a shrub layer). Retention or introduction of structural elements that can serve as cover objects, such as coarse woody debris and rocks, can also increase movement or occupancy of small animals along a road verge.
- Stormwater retention areas should also be as natural as possible and include more complex depth profiles (both shallow and deep), native aquatic vegetation, and water courses that mimic natural streams. These steps could reduce runoff and contamination and improve habitat quality for aquatic wildlife.
- To reduce human access to environmentally sensitive wildlands near roads, the most important step is to minimize new road construction through those areas. Where roads are needed, locked gates or legal mechanisms can keep unintended users off the roads, and there should be a contractual obligation to remove resource extraction roads or make them impassible once they are no longer needed. Parking restrictions and the use of fencing along the road in environmentally sensitive areas is also effective.
- To prevent undesirable development of land adjacent to a road, appropriately designed and enforced land use regulations may be suitable. Conservation easements can be used to restrict land use options on private property near a road.
- Renewable energy development is the impetus behind most new road construction in the United States currently (particularly in the western states). This development is resulting in wildlife mortality and landscape fragmentation of highly sensitive desert habitats and resident species.
- Species respond differently to roads and traffic, and the most effective solutions for reducing road impacts require an understanding of the behavior and natural history of the focal species of concern. For those species that will cross roads, knowledge of the volume and timing of traffic is also important.

LITERATURE CITED

1000 Friends of Florida. 2010. Wildlife habitat planning strategies, design features and best management practices for Florida communities and landowners. 1000 Friends of Florida, Tallahassee, Florida, USA. http://www.1000friendsofflorida.org/. Accessed 19 September 2014.

Ackerman. R. A., T. A. Rimkus, and D. B. Lott. 2008. Water relations of snapping turtle eggs. Pages 135–145 in Biology of the snapping turtle (*Chelydra serpentina*). A. C. Steyermark, M. S. Finkler, R. J. Brooks, and J. W. Gibbons, editors. Johns Hopkins Press, Baltimore, Maryland, USA.

Adams, L. W., and A. D. Geis. 2002. Effects of roads on small mammals. Journal of Applied Ecology 20:403–415.

Alexander, L., S. Lailvaux, and P. DeVries. 2012. Effects of salinity on early life stages of the Gulf Coast toad, *Incilius nebulifer* (Anura: Bufonidae). Copeia 2012:106–114.

Allen, C. R., E. A. Forys, K. G. Rice, and D. P. Wojcik. 2001. Effects of fire ants (Hymenoptera: Formicidae) on hatching turtles and prevalence of fire ants on sea turtle nesting beaches in Florida. Florida Entomologist 84:250–253.

American Association of State Highway and Transportation Officials. 2006. Roadside design guide. American Association of State Highway and Transportation Officials, Washington, DC, USA.

Amrhein, C., J. Strong, and P. Mosher. 1992. Effect of deicing salts on metal and organic matter mobilization in roadside soils. Environmental Science and Technology 26:703–709.

Andrews, K. M., and J. W. Gibbons. 2005. How do highways influence snake movement? Behavioral responses to roads and vehicles. Copeia 2005:772–782.

Andrews, K. M., J. W. Gibbons, and D. M. Jochimsen. 2006. Literature synthesis of the effects of roads and vehicles on amphibians and reptiles. Federal Highway Administration (FHWA), US Department of Transportation, Report No. FHWA-HEP-08-005. FHWA, Washington, DC, USA.

Antworth, R. L., D. A. Pike, and E. E. Stevens. 2005. Hit and run: using road surveys to detect avian and snake victims of road mortality. Southeastern Naturalist 4:647–656.

Aresco, M. J. 2005. The effect of sex-specific terrestrial movements and roads on the sex ratio of freshwater turtles. Biological Conservation 123:37–44.

Babbitt, K. J. 2005. The relative importance of wetland size and hydroperiod for amphibians in southern New Hampshire, USA. Wetlands Ecology and Management 13:269–279.

Baber M. J., K. J. Babbitt, and T. L. Tarr. 2003. Patterns of larval amphibian distribution along a wetland hydroperiod gradient. Canadian Journal of Zoology 81:1539–1552.

Bäckström, M. 2003. Grassed swales for stormwater pollution control during rain and snowmelt. Water Science and Technology 48:123–132.

Baker J. P., J. Van Sickle, C. J. Gagen, D. R. DeWalle, W. E. Sharpe, R. F. Carline, B. P. Baldigo, et al. 1996. Episodic acidification of small streams in the northeastern United States: effects on fish populations. Ecological Applications 6:422–437.

Barber, J. R., C. L. Burdett, S. E. Reed, K. A. Warner, C. Formichella, K. R. Crooks, D. M. Theobald, and K. M. Fristrup. 2011. Anthropogenic noise exposure in protected natural areas: estimating the scale of ecological consequences. Landscape Ecology 26:1281–1295.

Barber, J. R., K. R. Crooks, and K. M. Fristrup. 2010. The costs of chronic noise exposure for terrestrial organisms. Trends in Ecology and Evolution 25:180–189.

Barthelmess, E. L., and M. S. Brooks. 2010. The influence of body-size and diet on road-kill trends in mammals. Biodiversity and Conservation 19:1611–1629.

Beaudry F., P. G. deMaynadier, and M. L. Hunter Jr. 2008. Identifying road mortality threat at multiple spatial scales for semi aquatic turtles. Biological Conservation 141:2550–2563.

Beaudry, F., P. G. deMaynadier, and M. L. Hunter. 2010. Nesting movements and the use of anthropogenic nesting sites by spotted turtles (Clemmys guttata) and Blanding's turtles (Emydoidea blandingii). Herpetological Conservation and Biology 5:1–8.

Bellamy, P., R. Shore, D. Ardeshir, J. Treweek, and T. Sparks. 2000. Road verges as habitat for small mammals in Britain. Mammal Review 30:131–139.

Berthinussen, A., and J. Altringham. 2011. The effect of a major road on bat activity and diversity. Journal of Applied Ecology 42:82–89.

Birch, G., M. Fazeli, and C. Matthai. 2005. Efficiency of an infiltration basin in removing contaminants from urban stormwater. Environmental Monitoring and Assessment 101:23–38.

Bishop, C. A., J. Struger, D. R. Barton, L. J. Shirose, L. Dunn, A. L. Lang, and D. Sheperd. 2000. Contamination and wildlife communities in stormwater detention ponds in Guelph and the Greater Toronto Area, Ontario, 1997 and 1998. Part 1. Wildlife communities. Water Quality Research Journal of Canada 35:399–435.

Bishop, C. A., J. Struger, L. J. Shirose, L. Dunn, and G. D. Campbell. 2000. Contamination and wildlife communities in stormwater detention ponds in Guelph and the Greater Toronto Area, Ontario, 1997 and 1998. Part 2. Contamination and biological effects of contamination. Water Quality Research Journal of Canada 35:437–475.

Bishop, S. C. 1941. Salamanders of New York. New York State Museum Bulletin 324:1–365.

Blomqvist, G., and E.-L. Johansson. 1999. Airborne spreading and deposition of de-icing salt—a case study. Science of the Total Environment 235:161–168.

Boffetta, P., N. Jourenkova, and P. Gustavsson. 1997. Cancer risk from occupational and environmental exposure to polycyclic aromatic hydrocarbons. Cancer Causes and Control 8:444–472.

Bondello, M. C., and B. H. Brattstrom. 1979. The experimental effects of off-road vehicle sounds on three species of desert vertebrates. Unpublished Report. US Bureau of Land Management, Washington, DC, USA.

Boone, M. D., R. D. Semlitsch, E. E. Little, and M. C. Doyle. 2007. Multiple stressors in amphibian communities: effects of chemical contamination, bullfrogs, and fish. Ecological Applications 17:291–301.

Borda-de-Água, L., C. Grilo, and H. M. Pereira. 2014. Modeling the impact of road mortality on barn owl (Tyto alba) populations using age-structured models. Ecological Modelling 276:29–37.

Bouchard, J., A. T. Ford, F. E. Eigenbrod, and L. Fahrig. 2009. Behavioral responses of northern leopard frogs (Rana pipiens) to roads and traffic: implications for population persistence. Ecology and Society 14:23. http://www.ecologyandsociety.org/vol14/iss2/art23/. Accessed 28 April 2014.

Bowen, G. S., and M. J. Hinton. 1998. The temporal and spatial impacts of road salt on streams draining the Greater Toronto Area. Pages 303–310 in Proceedings of the groundwater in a watershed context symposium. A. R. Piggott, editor. Canadian Water Resources Association, Ontario, Canada.

Bradbury, J. W., and S. L. Vehrencamp. 2011. Principles of animal communication. Second edition. Sinauer, Sunderland, Massachusetts, USA.

Brady, S. P. 2012. Road to evolution? Local adaptation to road adjacency in an amphibian (Ambystoma maculatum). Scientific Reports 2:235–239.

Brady, S. P. 2013. Microgeographic maladaptive performance and deme depression in response to roads and runoff. PeerJ1:e163.

Breault, R. F., and G. E. Granato. 2000. A synopsis of technical issues of concern for monitoring trace elements in highway and urban runoff. US Geological Survey Open File Report 00-422. USGS, Washington, DC, USA.

Brooks, M. L., and J. R. Matchett. 2006. Spatial and temporal patterns of wildfires in the Mojave Desert, 1980–2004. Journal of Arid Environments 67:148–164.

Brown, M. G., E. K. Dobbs, J. W. Snodgrass, and D. R. Ownby.

2012. Ameliorative effects of sodium chloride on acute copper toxicity among Cope's gray tree frog (*Hyla chrysoscelis*) and green frog (*Rana clamitans*) embryos. Environmental Toxicology and Chemistry 31:836–842.

Bryson, G. M., and A. V. Barker. 2002. Sodium accumulation in soils and plants along Massachusetts roadsides. Communications in Soil Science and Plant Analysis 33:67–78.

Buchanan, B. W. 1993. Effects of enhanced lighting on the behavior of nocturnal frogs. Animal Behaviour 45:893–899.

Buchanan, B. W. 2006. Observed and potential effects of artificial night lighting on anuran amphibians. Pages 192–220 in Ecological consequences of artificial night lighting. C. Rich and T. Longcore, editors. Island Press, Washington, DC, USA.

Buhlmann, K. A., and G. Coffman. 2001. Fire ant predation of turtle nests and implications for the strategy of delayed emergence. Journal of the Elisha Mitchell Scientific Society 117:94–100.

Buhlmann, K. A., and C. P. Osborn. 2011. Use of an artificial nesting mound by wood turtles *Glyptemys insculpta*: a tool for turtle conservation. Northeastern Naturalist 18:315–334.

Calhoun, A., and P. deMaynadier. 2008. Science and conservation of vernal pools in northeastern North America. Third edition. CRC Press, New York, New York, USA.

Camponelli, K. M., R. E. Casey, J. W. Snodgrass, S. M. Lev, and E. R. Landa. 2009. Impacts of weathered tire debris on the development of *Rana sylvatica* larvae. Chemosphere 74:717–722.

Carls, M. G., S. D. Rice, and J. E. Hose. 1999. Sensitivity of fish embryos to weathered crude oil. Part I. Low-level exposure during incubation causes malformations, genetic damage, and mortality in larval Pacific herring (*Clupea pallasi*). Environmental Toxicology and Chemistry 18:481–493.

Chen, C. Y., R. S. Stemberger, B. Klaue, J. D. Blum, and P. C. Pickhardt. 2000. Accumulation of heavy metals in food web components across a gradient of lakes. Limnology and Oceanography 45:1525–1536.

Clark, M., and G. Acomb. 2008. Florida field guide to low impact development: enhanced stormwater basins. IFAS Extension, University of Florida, Gainesville, Florida, USA.

Colburn, E. A. 2004.Vernal pools: Natural history and conservation. McNaughton and Gunn, Saline, Michigan, USA.

Collins, S., and R. Russell. 2009. Toxicity of road salt to Nova Scotia amphibians. Environmental Pollution 157:320–324.

Corn, P. S., and R. B. Bury. 1989. Logging in western Oregon: responses of headwater habitats and stream amphibians. Forest Ecology and Management 29:39–57.

Corsi, S., D. Graczyk, S. Geis, N. Booth, and K. Richards. 2010. A fresh look at road salt: aquatic toxicity and water-quality impacts on local, regional, and national scales. Environmental Science and Technology 44:7376–7382.

Crain, D. A., and L. J. Guillette. 1998. Reptiles as models of contaminant-induced endocrine disruption. Animal Reproduction Science 53:77–86.

Davies, T. W., J. Bennie, and K. J. Gaston. 2012. Street lighting changes the composition of invertebrate communities. Biology Letters 8:764–767.

Davis, A. P., M. Shokouhian, and S. Ni. 2001. Loading estimates of lead, copper, cadmium, and zinc in urban runoff from specific sources. Chemosphere 44:997–1009.

Day, K. E., K. E. Holtze, J. L. Metcalfesmith, C. T. Bishop, and B. J. Dutka. 1993. Toxicity of leachate from automobile tires to aquatic biota. Chemosphere 27:665–675.

deMaynadier, P. G., and M. L. Hunter Jr. 1995. The relationship between forest management and amphibian ecology: a review of the North American literature. Environmental Review 3:230–261.

Denoël, M., M. Bichot, G. Ficetola, J. Delcourt, M. Ylieff, P. Kestemont, and P. Poncin. 2010. Cumulative effects of road de-icing salt on amphibian behavior. Aquatic Toxicology 99:275–280.

Dodd, C. K., Jr., K. M. Enge, and J. N. Stuart. 1989. Reptiles on highways in north-central Alabama, USA. Journal of Herpetology 23:197–200.

Duellman, W. E., and L. Trueb. 1986. Biology of amphibians. McGraw-Hill, New York, St. Louis, San Francisco, USA.

Endler, J. A. 1993. The color of light in forests and its implications. Ecological Monographs 63:2–27.

Enge, K. M., and K. N. Wood. 2002. A pedestrian road survey of an upland snake community in Florida. Southeastern Naturalist 1:365–380.

Ennen, J. R., J. E. Lovich, K. Meyer, C. Bjurlin, and T. R. Arundel. 2012. Nesting ecology of a desert tortoise (*Gopherus agassizii*) population at a utility-scale renewable energy facility in southern California. Copeia 2012: 222–228.

Environment Canada and Health Canada. 1999. Environmental Protection Act 1999. Priority Substances List Assessment Report—Road Salt. Ottawa, Canada.

Evans, S. 1999. The green republic. A conservation history of Costa Rica. University of Texas Press, Austin, Texas, USA.

Ewing, R. 2008. Highway-induced development: research results for metropolitan areas. Transportation Research Record 2067:101–109.

Fahrig, L., and T. Rytwinski. 2009. Effects of roads on animal abundance: an empirical review and synthesis. Ecology and Society 14:21. http://www.ecologyandsociety.org/vol14/iss1/art21/. Accessed 28 April 2014.

Farina, A. 2006. Principles and methods in landscape ecology: toward a science of landscape. Springer, Houten, The Netherlands.

Farmer, A. M. 1993. The effects of dust on vegetation—a review. Environmental Pollution 79:63–75.

Feist, B. E., E. R. Buhle, P. Arnold, J. W. Davis, and N. L. Scholz. 2011. Landscape ecotoxicology of coho salmon spawner mortality in urban streams. PLoS ONE 6:e23424.

Findlay, C. S., and J. Bourdages. 2000. Response time of wetland biodiversity to road construction on adjacent lands. Conservation Biology 14:86–94.

Finnie, W. C. 1973. Field experiments in litter control. Environment and Behavior 5:123–144.

Ford, A. T., and L. Fahrig. 2008. Movement patterns of eastern chipmunks (*Tamias striatus*) near roads. Journal of Mammalogy 89:895–903.

Forman, R.T.T. 1995. Land mosaics: The ecology of landscapes and regions. Cambridge University Press, Cambridge, UK.

Forman, R T.T., D. Sperling, J. A. Bissonette, A. P. Clevenger, C. D. Cutshall, V. H. Dale, L. Fahrig, et al. 2003. Road ecology. Science and solutions. Island Press, Washington, DC, USA.

Fort, D., P. Guiney, J. Weeks, J. Thomas, R. Rogers, A. Noll, and C. Spaulding. 2004. Effect of methoxychlor on various life stages of *Xenopus laevis*. Toxicological Sciences 81:454–466.

Gehrt, S. D., G. F. Hubert Jr., and J. A. Ellis. 2002. Long-term population trends of raccoons in Illinois. Wildlife Society Bulletin 30:457–463.

Gelbard, J. L., and J. Belnap. 2003. Roads as conduits for exotic plant invasions in a semiarid landscape. Conservation Biology 17:420–432.

Gerow, K., N. C. Kline, D. E. Swann, and M. Pokorny. 2010. Estimating annual vertebrate mortality on roads at Saguaro National Park, Arizona. Human-Wildlife Interactions 4:283–292.

Getz, L. L., F. R. Cole, and D. L. Gates. 1978. Interstate roadsides as dispersal routes for *Microtus pennsylvanicus*. Journal of Mammalogy 59:208–212.

Getz, L. L., L. Verner, and M. Prather. 1977. Lead concentrations in small mammals living near highways. Environmental Pollution 13:151–157.

Gibbs, J. P. 1998. Amphibian movements in response to forest edges, roads, and streambeds in southern New England. Journal of Wildlife Management 62:584–589.

Gibbs, J. P., and W. G. Shriver. 2002. Estimating the effects of road mortality on turtle populations. Conservation Biology 16:1647–1652.

Gibbs, J. P., and W. G. Shriver. 2005. Can road mortality limit populations of pool-breeding amphibians? Wetlands Ecology and Management 13:281–289.

Giesy, J. P., S. Dobson, and K. R. Solomon. 2000. Ecotoxicological risk assessment for Roundup Herbicide. Pages 35–120 *in* Reviews of environmental contamination and toxicology. Vol. 167. D. M. Whitacre, editor. Springer-Verlag, New York, New York, USA.

Golait N., R. M. Moharil, and P. S. Kulkarni. 2009. Wind electric power in the world and perspectives of its development in India. Renewable and Sustainable Energy Reviews 13:233–247.

Goosem, M. 2001. Effects of tropical rainforest roads on small mammals: inhibition of crossing movements. Wildlife Research 28:351–364.

Goosem, M. 2002. Effects of tropical rainforest roads on small mammals: fragmentation, edge effects and traffic disturbance. Wildlife Research 29:277–289.

Goosem, M., Y. Izumi, and S. Turton. 2001. Efforts to restore habitat connectivity for an upland tropical rainforest fauna: a trial of underpasses below roads. Ecological Management and Restoration 2:196–202.

Grilo, C., J. Sousa, F. Ascensao, H. Matos, I. Leitao, P. Pinheiro, M. Costa1, et al. 2012. Individual spatial responses towards roads: implications for mortality risk. PLoS ONE 7:e43811.

Grimmer, G., H. Brune, R. Deutsch-Wenzel, G. Dettbarn, and J. Misfeld. 1984. Contribution of polycyclic aromatic hydrocarbons to the carcinogenic impact of gasoline engine exhaust condensate evaluated by implantation into the lungs of rats. Journal of the National Cancer Institute 72:733–739.

Handa, T., T. Yamauchi, K. Sawai, T. Yamamura, and Y. Koseki. 1984. In situ emission levels of carcinogenic and mutagenic compounds from diesel and gasoline-engine vehicles on an expressway. Environmental Science and Technology 18:895–902.

Hankinson, O. 1995. The aryl-hydrocarbon receptor complex. Annual Review of Pharmacology and Toxicology 35:307–340.

Harless, M. L., C. J. Huckins, J. B. Grant, and T. G. Pypker. 2011. Effects of six chemical deicers on larval wood frogs (*Rana sylvatica*). Environmental Toxicology and Chemistry 30:1637–1641.

Harper-Lore, B., and M. Wilson. 2000. Roadside use of native plants. Island Press, Washington, DC, USA.

Harrison, R. M., R. Tilling, M.S.C. Romero, S. Harrad, and K. Jarvis. 2003. A study of trace metals and polycyclic aromatic hydrocarbons in the roadside environment. Atmospheric Environment 37:2391–2402.

Hautala, E. L., R. Rekilä, J. Tarhanen, and J. Ruuskanen. 1995. Deposition of motor vehicle emissions and winter maintenance along roadside assessed by snow analyses. Environmental Pollution 87:45–49.

Havlick, D. G. 2002. No place distant: Roads and motorized recreation on America's public lands. Island Press, Washington, DC, USA.

Hayes, T. B., A. Collins, M. Lee, M. Mendoza, N. Noriega, A. A. Stuart, and A. Vonk. 2002. Hermaphroditic, demasculinized frogs after exposure to the herbicide atrazine at low ecologically relevant doses. Proceedings of the National Academy of Sciences of the United States of America 99:5476–5480.

Hels, T., and E. Buchwald. 2001. The effect of road kills on amphibian populations. Biological Conservation 99:331–340.

Hódar, J. A., J. M. Pleguezuelos, and J. C. Proveda. 2000. Habitat selection of the common chameleon (*Chamaeleo chamaeleon*) (L.) in an area under development in southern Spain: implications for conservation. Biological Conservation 94:63–68.

Hoelker, F., T. Moss, B. Griefahn, W. Kloas, C. C. Voigt, D. Henckel, A. Haenel, et al. 2010. The dark side of light: a transdisciplinary research agenda for light pollution policy. Ecology and Society 15:13.

Hoffman, C. O., and J. L. Gottschang. 1977. Numbers, distribution, and movements of a raccoon population in a suburban residential community. Journal of Mammalogy 58:623–636.

Hoffman, R. W., C. R. Goldman, S. Paulson, and G. R. Winters. 1981. Aquatic impacts of deicing salts in the central Sierra Nevada Mountains, California. Water Resources Bulletin 17:280–285.

Hopwood, J. L. 2008. The contribution of roadside grassland restorations to native bee conservation. Biological Conservation 141:2632–2640.

Horvath, G., G. Kriska, P. Malik, and B. Robertson. 2009. Polarized light pollution: a new kind of ecological photopollution. Frontiers in Ecology and the Environment 7:317–325.

Hyslop, N. L., J. M. Meyers, and R. J. Cooper. 2006. Movement, survival, and habitat use of the threatened eastern indigo snake (*Drymarchon couperi*) in southeastern Georgia. Final Report (unpublished). Georgia Department of Natural Resources Nongame Endangered Wildlife Program, Forsyth, Georgia, USA.

Jackson, R. B., and E. G. Jobbágy. 2005. From icy roads to salty streams. Proceedings of the National Academy of Sciences of the United States of America 102:14487–14488.

Jacobson, S. L., and W. P. Smith. 2010. A conceptual framework for assessing barrier effects to wildlife populations using species group responses to traffic volume. Page 919 *in* Proceedings of the 2009 international conference on ecology and transportation. P. J. Wagner, D. Nelson, and E. Murray, editors. Center for Transportation and the Environment, North Carolina State University, Raleigh, North Carolina, USA.

Jaeger, J.A.G. 2012. Road ecology. Pages 344–350 *in* Ecosystem management and sustainability. R. K. Craig, J. C. Nagle, B. Pardy, O. J. Schmitz, and W. K. Smith, editors. Encyclopedia of sustainability, Vol. 5. Berkshire Publishing Group, Great Barrington, Massachusetts, USA.

Jaeger, J.A.G., J. Bowman, J. Brennan, L. Fahrig, D. Bert, J. Bouchard, N. Charbonneau, et al. 2005. Predicting when animal populations are at risk from roads: an interactive model of road avoidance behavior. Ecological Modelling 185:329–348.

Johnson, H. B., F. C. Vasek, and T. Yonkers. 1975. Productivity, diversity and stability relationships in Mojave Desert roadside vegetation. Bulletin of the Torrey Botanical Club 102:106–115.

Jones, J. A. 2000. Effects of roads on hydrology, geomorphology, and disturbance patches in stream networks. Conservation Biology 14:76–85.

Karraker, N. E. 2008. Impacts of road deicing salts on amphibians and their habitats. Pages 211–223 *in* Urban herpetology. R. E. Jung, and J. C. Mitchell, editors. Herpetological conservation Vol. 3. Society for the Study of Amphibians and Reptiles, Salt Lake City, Utah, USA.

Karraker, N. E., and J. P. Gibbs. 2011. Road deicing salt irreversibly disrupts the osmoregulation of salamander egg clutches. Environmental Pollution 159:833–835.

Karraker, N., J. Gibbs, and J. Vonesh. 2008. Impacts of road deicing salt on the demography of vernal pool-breeding amphibians. Ecological Applications 18:724–734.

Kaushal, S. S., P. M. Groffman, G. E. Likens, K. T. Belt, W. P. Stack, V. R. Kelly, L. E. Band, and G. T. Fisher. 2005. Increased salinization of fresh water in the northeastern United States. Proceedings of the National Academy of Sciences of the United States of America 102:13517–13520.

Kinnberg, K., H. Holbech, G. Petersen, and P. Bjerregaard. 2007. Effects of the fungicide prochloraz on the sexual development of zebrafish (*Danio rerio*). Comparative Biochemistry and Physiology. Part C, Toxicology and Pharmacology 145:165–170.

Klauber, L. M. 1931. A statistical survey of the snakes of the southern border of California. Bulletin of the Zoological Society of San Diego 8:1–93.

Klotzbach, J., and P. Durban. 2005. Metsulfuron methyl—human health and ecological risk assessment—Final Report. US Forest Service BPA: WO-01-3187-0150. US Forest Service, Washington, DC, USA.

Knight, R. L. 1997. Wildlife habitat and public use benefits of treatment wetlands. Water Science and Technology 35:35–44.

Kohl, R. A., C. G. Carlson, and S. G. Wangemann. 1994. Herbicide leaching potential through road ditches in thin soils over an outwash aquifer. Applied Engineering in Agriculture 10:497–503.

Kolbe, J. J., and F. J. Janzen. 2001. The influence of propagule size and maternal nest-site selection on survival and behaviour of neonate turtles. Functional Ecology 15:772–781.

Kolbe, J. J., and F. J. Janzen. 2002. Impact of nest-site selection on nest success and nest temperature in natural and disturbed habitats. Ecology 83:269–281.

Krein, A., and M. Schorer. 2000. Road runoff pollution by polycyclic aromatic hydrocarbons and its contribution to river sediments. Water Research 34:4110–4115.

Kucken, D. J., J. S. Davis, and J. W. Petranka. 1994. Anakeesta stream acidification and metal contamination—effects on a salamander community. Journal of Environmental Quality 23:1311–1317.

Lacki, M. J., J. W. Hummer, and H. J. Webster. 1992. Mine-drainage treatment wetland as habitat for herptofaunal wildlife. Environmental Management 16:513–520.

Langen, T. A. 2012. Design considerations and effectiveness of fencing for turtles: three case studies along northeastern New York state highways. Pages 521–532 *in* Proceedings of the 2011 international conference on ecology and transportation. P. J. Wagner, D. Nelson, and E. Murray, editors. Center for Transportation and the Environment, North Carolina State University, Raleigh, North Carolina, USA.

Langen, T. A., M. R. Twiss, T. C. Young, K. J. Janoyan, J. C. Stager, J. D. Osso Jr., H. Prutzman, and B. T. Green. 2006. Environmental impacts of winter road management at the

Cascade Lakes and Chapel Pond. Clarkson Center for the Environment Report No. 1. Clarkson University, Potsdam, New York, USA. https://www.dot.ny.gov/divisions /engineering/environmental-analysis/repository/cascade _lakes_final_report.pdf. Accessed 28 April 2014.

Laurance, W. F., M. Goosem, and S. G. Laurance. 2009. Impacts of roads and linear clearings on tropical forests. Trends in Ecology and Evolution 24:659–669.

Laursen, K. 1981. Birds on roadside verges and the effect of mowing on frequency and distribution. Biological Conservation 20:59–68.

Leblond, M., C. Dussault, J.-P. Ouellet, M. Poulin, R. Courtois, and J. Fortin. 2007. Management of roadside salt pools to reduce moose-vehicle collisions. The Journal of Wildlife Management 71:2304–2310.

Lee, M. A., L. Davies, and S. A. Power. 2012. Effects of roads on adjacent plant community composition and ecosystem function: an example from three calcareous ecosystems. Environmental Pollution 163:273–280.

Lefcort, H., K. A. Hancock, K. M. Maur, and D. C. Rostal. 1997. The effects of used motor oil, silt, and the water mold *Saprolegnia parasitica* on the growth and survival of mole salamanders. Archives of Environmental Contamination and Toxicology 32:383–388.

Lelong, B., C. Lavoie, Y. Jodoin, and F. Belzile. 2007. Expansion pathways of the exotic common reed (*Phragmites australis*): a historical and genetic analysis. Diversity and Distributions 13:430–437.

Lepori, F., and S. J. Ormerod. 2005. Effects of spring acid episodes on macroinvertebrates revealed by population data and *in situ* toxicity tests. Freshwater Biology 50:1568–1577.

Lima, S. L., B. F. Bradwell, T. L. DeVault, and E. Fernández-Juricic. 2014. Animal reactions to oncoming vehicles: a conceptual review. Biological Reviews. doi: 10.1111/brv.12093.

Long, E. R., D. D. Macdonald, S. L. Smith, and F. D. Calder. 1995. Incidence of adverse biological effects within ranges of chemical concentrations in marine and estuarine sediments. Environmental Management 19:81–97.

Lovell, S. T., and D. M. Johnston. 2009. Designing landscapes for performance based on emerging principles in landscape ecology. Ecology and Society 14:44. http://www.ecologyand society.org/vol14/iss1/art44/. Accessed 28 April 2014.

Lovich, J. E., and R. Daniels. 2000. Environmental characteristics of desert tortoise (*Gopherus agassizii*) burrow locations in an altered industrial landscape. Chelonian Conservation and Biology 3:714–721.

Lovich, J. E., and J. R. Ennen. 2011. Wildlife conservation and solar energy development in the desert southwest, United States. BioScience 61:982–992.

Lovich, J. E., and J. R. Ennen. 2013. Assessing the state of knowledge of utility-scale wind energy development and operation on non-volant terrestrial and marine wildlife. Applied Energy 103:52–60.

Lovich, J. E., J. R. Ennen, S. V. Madrak, and B. Grover. 2011. Turtles, culverts and alternative energy development: an

unreported but potentially significant mortality threat to the desert tortoise (*Gopherus agassizii*). Chelonian Conservation and Biology 10:124–129.

Lovich, J. E., J. R. Ennen, S. Madrak, K. Meyer, C. Loughran, C. Bjurlin, T. R. Arundel, et al. 2011. Effects of wind energy production on growth, demography and survivorship of a desert tortoise (*Gopherus agassizii*) population in southern California with comparisons to natural populations. Herpetological Conservation and Biology 6:161–174.

Lundmark, A., and B. Olofsson. 2007. Chloride deposition and distribution in soils along a deiced highway—assessment using different methods of measurement. Water, Air, and Soil Pollution 182:173–185.

Mader, H.-J. 1984. Animal habitat isolation by roads and agricultural fields. Biological Conservation 29:81–96.

Madison, D. M., and L. Farrand III. 1998. Habitat use during breeding and emigration in radio-implanted tiger salamanders (*Ambystoma tigrinum*). Copeia 1998:402–410.

Mahaney, P. A. 1994. Effects of freshwater petroleum contamination on amphibian hatching and metamorphosis. Environmental Toxicology and Chemistry 13:259–265.

Maltby, L., D. M. Forrow, A.B.A. Boxall, P. Calow, and C. I. Betton. 1995. The effects of motorway runoff on freshwater ecosystems: 1. Field-study. Environmental Toxicology and Chemistry 14:1079–1092.

Mangani, G., A. Berloni, F. Bellucci, F. Tatano, and M. Maione. 2005. Evaluation of the pollutant content in road runoff first flush waters. Water, Air, and Soil Pollution 160: 213–228.

Marsh, D. M., G. S. Milam, N. P. Gorham, and N. G. Beckman. 2005. Forest roads as partial barriers to terrestrial salamander movement. Conservation Biology 19:2004–2008.

Mazerolle, M. J., M. Huot, and M. Gravel. 2005. Behavior of amphibians on the road in response to car traffic. Herpetologica 61:380–388.

McBean, E., and S. Al-Nassri. 1987. Migration pattern of de-icing salts from roads. Journal of Environmental Management 25:231–238.

McCashion, J. D., and R. M. Rice. 1983. Erosion on logging roads in northwestern California: how much is avoidable? Journal of Forestry 81:23–26.

McClure, H. E. 1951. An analysis of animal victims on Nebraska's highways. Journal of Wildlife Management 15:410–420.

McDonald, R. I., J. Fargione, J. Kiesecker, W. M. Miller, and J. Powell. 2009. Energy sprawl or energy efficiency: climate policy impacts on natural habitat for the United States of America. PLoS ONE 4(8): e6802. doi:10.1371/journal .pone.0006802.

McFarlane, R. W. 1963. Disorientation of loggerhead hatchlings by artificial road lighting. Copeia 1963:153.

Meland, S., B. Salbu, and B. Rosseland. 2010. Ecotoxicological impact of highway runoff using brown trout (*Salmo trutta* L.) as an indicator model. Journal of Environmental Monitoring 12:654–664.

Miguel, A. H., T. W. Kirchstetter, R. A. Harley, and S. V. Hering. 1998. On-road emissions of particulate polycyclic aromatic hydrocarbons and black carbon from gasoline and diesel vehicles. Environmental Science and Technology 32:450–455.

Montgomery, D. R. 1994. Road surface drainage, channel initiation, and slope instability. Water Resources Research 30:1925–1932.

Moon, H. E., Jr. 1987. Interstate highway interchanges reshape rural communities. Rural Development Perspectives, October 1987:35–39.

Morgenstern, V., A. Zutavern, J. Cyrys, I. Brockow, U. Gehring, S. Koletzko, C. P. Bauer, et al. 2007. Respiratory health and individual estimated exposure to traffic-related air pollutants in a cohort of young children. Occupational and Environmental Medicine 64:8–16.

Morin, L. P. 2013. Nocturnal light and nocturnal rodents: similar regulation of disparate functions? Journal of Biological Rhythms 28:95–106.

Mortensen, D. A., E.S.J. Rauschert, A. N. Nord, and B. P. Jones. 2009. Forest roads facilitate the spread of invasive plants. Invasive Plant Science Management 2:191–199.

Motto, H. L., R. H. Daines, D. M. Chilko, and C. K. Motto. 1970. Lead in soils and plants: its relation to traffic volume and proximity to highways. Environmental Science and Technology 4:231–237.

Mudre, J. M., and J. J. Ney. 1986. Patterns of accumulation of heavy metals in the sediment of roadside streams. Archives of Environmental Contamination and Toxicology 15:489–493.

Mullaney, J. R., D. L. Lorenz, and A. D. Arntson. 2009. Chloride in groundwater and surface water in areas underlain by the glacial aquifer system, northern United States. US Geological Survey Scientific Investigations Report 2009-5086. USGS, Washington, DC, USA.

Munguira, M., and J. Thomas. 1992. Use of road verges by butterfly and burnet populations, and the effect of roads on adult dispersal and mortality. Journal of Applied Ecology 29:316–329.

Mungur, A., R. Shutes, D. Revitt, and M. House. 1995. An assessment of metal removal from highway runoff by a natural wetland. Water Science and Technology 32:169–175.

Nash R. F., G. G. Gallup Jr., and M. K. McClure. 1970. The immobility reaction in leopard frogs (Rana pipiens) as a function of noise-induced fear. Psychonometric Science 21:155–156.

Nega, T., C. Smith, J. Bethune, and W.-H. Fu. 2012. An analysis of landscape penetration by road infrastructure and traffic noise. Computers, Environment and Urban Systems 36:245–256.

Nielsen, T., H. E. Jørgensen, J. C. Larsen, and M. Poulsen. 1996. City air pollution of polycyclic aromatic hydrocarbons and other mutagens: occurrence, sources and health effects. Science of the Total Environment 189–190:41–49.

Noss, R. F., and A. Y. Cooperider. 1994. Saving nature's legacy: Protecting and restoring biodiversity. Defenders of Wildlife and Island Press, Washington, DC, USA.

Nriagu, J. O., and J. M. Pacyna. 1988. Quantitative assessment of worldwide contamination of air, water and soils by trace metals. Nature 333:134–139.

Nuttle, W. K. 1997. Measurement of wetland hydroperiod using harmonic analysis. Wetlands 17:82–89.

Orton, F., J. Carr, and R. Handy. 2006. Effects of nitrate and atrazine on larval development and sexual differentiation in the northern leopard frog Rana pipiens. Environmental Toxicology and Chemistry 25:65–71.

Packard, G. C., and M. J. Packard. 2001a. The overwintering strategy of hatchling painted turtles, or how to survive in the cold without freezing. BioScience 51:199–207.

Packard, G. C., and M. J. Packard. 2001b. Environmentally induced variation in size, energy reserves and hydration of hatchling painted turtles, Chrysemys picta. Functional Ecology 15:481–489.

Parendes, L. A., and J. A. Jones. 2000. Role of light availability and dispersal in exotic plant invasion along roads and streams in the H. J. Andrews Experimental Forest, Oregon. Conservation Biology 14:64–75.

Parris, K. M., and A. Schneider. 2008. Impacts of traffic noise and traffic volume on birds of roadside habitats. Ecology and Society 14:29. http://www.ecologyandsociety.org/vol14/iss1/art29/. Accessed 28 April 2014.

Parris, K. M., M. Velik-Lord, and J.M.A. North. 2009. Frogs call at a higher pitch in traffic noise. Ecology and Society 14:25. http://www.ecologyandsociety.org/vol14/iss1/art25/. Accessed 28 April 2014.

Parris, L. B., M. M. Lamont, and R. R. Carthy. 2002. Increased incidence of red imported fire ants (Hymenoptera: Formicidae) presence in loggerhead sea turtle (Testudines: Cheloniidae) nests and observations of hatchling mortality. Florida Entomologist 85:514–517.

Patz, J. A., N. Geller, T. K. Graczyk, and A. Y. Vittor. 2000. Effects of environmental change on emerging parasitic diseases. International Journal for Parasitology 30:1395–1405.

Paul, M. J., and J. L. Meyer. 2001. Streams in the urban landscape. Urban Ecology 32:207–231.

Perera, F. P., K. Hemminki, D. Brenner, G. Kelly, T. L. Young, and R. M. Santella. 1988. Detection of polycyclic aromatic hydrocarbon-DNA adducts in white blood cells of foundry workers. Cancer Research 48:2288–2291.

Perera, F. P., K. Hemminki, E. Gryzbowska, G. Motykiewicz, J. Michalska, R. M. Santella, T. L. Young, et al. 1992. Molecular and genetic damage in humans from environmental pollution in Poland. Nature 360:256–258.

Perry, G., B. W. Buchanan, R. N. Fisher, M. Salmon, and S. E. Wise. 2008. Effects of artificial night lighting on urban reptiles and amphibians. Pages 239–246 in Urban herpetology. R. E. Jung and J. C. Mitchell, editors. Herpetological Conservation Vol. 3. Society for the Study of Amphibians and Reptiles, Salt Lake City, Utah, USA.

Petranka, J. W., and E. J. Doyle. 2010. Effects of road salts on the composition of seasonal pond communities: can the use of road salts enhance mosquito recruitment? Aquatic Ecology 44:155–166.

Plummer, M. V., and N. E. Mills. 2006. *Heterodon platirhinos* (Eastern hognose snake). Road crossing behavior. Herpetological Review 37:352.

Pohlman, C. L., S. M. Turton, and M. Goosem. 2009. Temporal variation in microclimatic edge effects near powerlines, highways and streams in Australian tropical rainforest. Agricultural and Forest Meteorology 149:84–95.

Prange, S., S. D. Gehrt, and E. P. Wiggers. 2004. Influences of anthropogenic resources on raccoon (*Procyon lotor*) movements and spatial distribution. Journal of Mammalogy 85:483–490.

Ramwell, C. T., A.I.J. Heather, and A. J. Shepherd. 2002. Herbicide loss following application to a roadside. Pest Management Science 58:695–701.

Reeves, M. K., C. L. Dolph, H. Zimmer, R. S. Tjeerdema, and K. A. Trust. 2008. Road proximity increases risk of skeletal abnormalities in wood frogs from National Wildlife Refuges in Alaska. Environmental Health Perspectives 116:1009–1014.

Reh, W., and A. Seitz. 1990. The influence of land use on the genetic structure of populations of the common frog *Rana temporaria*. Biological Conservation 54:239–249.

Relyea, R. A. 2004. Synergistic impacts of malathion and predatory stress on six species of North American tadpoles. Environmental Toxicology and Chemistry 23:1080–1084.

Relyea, R. A. 2005. The lethal impact of roundup on aquatic and terrestrial amphibians. Ecological Applications 15:1118–1124.

Relyea, R. A., and J. Hoverman. 2006. Assessing the ecology in ecotoxicology: a review and synthesis in freshwater systems. Ecology Letters 9:1157–1171.

Relyea, R. A., and N. Mills. 2001. Predator-induced stress makes the pesticide carbaryl more deadly to gray treefrog tadpoles (*Hyla versicolor*). Proceedings of the National Academy of Sciences of the United States of America 98:2491–2496.

Rich, C., and T. Longcore, editors. 2006. Ecological consequences of artificial night lighting. Island Press, Washington, DC, USA.

Richter, K. O. 1997. Criteria for the restoration and creation of wetland habitats of lentic-breeding amphibians of the Pacific Northwest. Pages 72–94 in Wetland and riparian restoration: Taking a broader view. K. B. Macdonald, and F. Winmann, editors. US EPA, Region 10, Seattle, Washington, USA.

Riitters, K. H., and J. D. Wickham. 2003. How far to the nearest road? Frontiers in Ecology and the Environment 1:125–129.

Rodriguez-Flores, M., and E. Rodriguez-Castellon. 1982. Lead and cadmium levels in soil and plants near highways and their correlation with traffic density. Environmental Pollution Series B, Chemical and Physical 4:281–290.

Rogge, W. F., L. M. Hildemann, M. A. Mazurek, G. R. Cass, and B.R.T. Simoneit. 1993. Sources of fine organic aerosol, 3. Road dust, tire debris, and organometallic brake lining dust—roads as sources and sinks. Environmental Science and Technology 27:1892–1904.

Rohr J. R., T. Sager, T. M. Sesterhenn, and B. D. Palmer. 2006. Exposure, postexposure, and density-mediated effects of atrazine on amphibians: breaking down net effects into their parts. Environmental Health Perspectives 114:46–50.

Rowe, C. L., O. M. Kinney, and J. D. Congdon. 1998. Oral deformities in tadpoles of the bullfrog (*Rana catesbeiana*) caused by conditions in a polluted habitat. Copeia 1998:244–246.

Ruter, J. M. 2003. Selections for the salty coast. American Nurseryman, September 1, 2003:38–40.

Salt, D. E., M. Blaylock, N.P.B.A. Kumar, V. Dushenkov, B. D. Ensley, I. Chet, and I. Raskin. 1995. Phytoremediation: a novel strategy for the removal of toxic metals from the environment using plants. Biotechnology 13:468–474.

Sandahl, J., D. Baldwin, J. Jenkins, and N. Scholz. 2007. A sensory system at the interface between urban stormwater runoff and salmon survival. Environmental Science and Technology 41:2998–3004.

Sanzo, D., and S. J. Hecnar. 2006. Effects of road de-icing salt (NaCl) on larval wood frogs (*Rana sylvatica*). Environmental Pollution 140:247–256.

Sartorius, S. S., L. J. Vitt, and G. R. Colli. 1999. Use of naturally and anthropogenically disturbed habitats in Amazonian rainforest by the teiid lizard *Ameiva ameiva*. Biological Conservation 90:91–101.

Scanlon, P. F. 1979. Ecological implications of heavy metal contamination of roadside habitats. Pages 136–145 in Proceedings of the 33rd annual conference of the southeastern association of fish and wildlife agencies. R. W. Dimmick, J. D. Hair, and J. Grover, editors. Southeastern Association of Fish and Wildlife Agencies 33:136–145.

Scherr, S. J., and J. A. McNeely. 2007. Biodiversity conservation and agricultural sustainability: towards a new paradigm of "ecoagriculture" landscapes. Philosophical Transactions of the Royal Society B 363:477–494.

Schwartz, J. J., B. W. Buchanan, and H. C. Gerhardt. 2001. Female mate choice in the gray treefrog (*Hyla versicolor*) in three experimental environments. Behavioral Ecology and Sociobiology 49:443–455.

Schwartz, J. J., and K. D. Wells. 1983. An experimental study of acoustic interference between two species of neotropical treefrogs. Animal Behaviour 31:181–190.

Seabrook, W. A., and E. B. Dettman. 1996. Roads as activity corridors for cane toads in Australia. Journal of Wildlife Management 60:363–368.

Semlitsch, R. D. 2000. Principles for management of aquatic-breeding amphibians. Journal of Wildlife Management 64:615–631.

Semlitsch, R. D., T. J. Ryan, K. Hamed, M. Chatfield, B. Drehman, N. Pekarek, M. Spath, and A. Watland. 2007. Salamander abundance along road edges and within abandoned logging roads in Appalachian forests. Conservation Biology 21:159–167.

Sezgin, N., H. K. Ozcan, G. Demir, S. Nemlioglu, and C. Bayat. 2004. Determination of heavy metal concentrations in street dusts in Istanbul E-5 highway. Environment International 29:979–985.

Shepard, D. B., A. R. Kuhns, M. J. Dreslik, and C. A. Phillips. 2008. Roads as barriers to animal movement in fragmented landscapes. Animal Conservation 11:288–296.

Sheridan, G. J., and P. J. Noske. 2007. A quantitative study of sediment delivery and stream pollution from different forest road types. Hydrological Processes 21:387–398.

Shine, R., M. Lemaster, M. Wall, T. Langkilde, and R. Mason. 2004. Why did the snake cross the road? Effects of roads on movement and location of mates by garter snakes (*Thamnophis sirtalis parietalis*). Ecology and Society 9:9. http://www.ecologyandsociety.org/vol9/iss1/art9/. Accessed 28 April 2014.

Shochat, E., D. H. Wolfe, M. A. Patten, D. L. Reinking, and S. K. Sherrod. 2005. Tallgrass prairie management and bird nest success along roadsides. Biological Conservation 121:399–407.

Siver, P. A., R. W. Canavan, C. K. Field, L. J. Marsicano, and A. Lott. 1996. Historical changes in Connecticut lakes over a 55-year period. Journal of Environmental Quality 25:334.

Smith, L. L., and C. K. Dodd Jr. 2003. Wildlife mortality on U.S. highway 441 across Payne Prairie, Alachua County, Florida. Florida Scientist 66: 128–140.

Solomon, K. R., D. B. Baker, R. P. Richards, D. R. Dixon, and S. J. Klaine. 1996. Ecological risk assessment of atrazine in North American surface waters. Environmental Toxicology and Chemistry 15:31–74.

Sparling, D., C. Matson, J. Bickham, and P. Doelling Brown. 2006. Toxicity of glyphosate as Glypro and LI700 to red-eared slider (*Trachemys scripta elegans*) embryos and early hatchlings. Environmental Toxicology and Chemistry 25:2768–2774.

Spellerberg, I. F. 1998. Ecological impacts of roads—a literature review. Global Ecology and Biogeography Letters 17:317–333.

Steen, D. A., M. J. Aresco, S. G. Beilke, B. W. Compton, C. K. Dodd Jr., H. Forrester, J. W. Gibbons, et al. 2006. Relative vulnerability of female turtles to road mortality. Animal Conservation 9:279–273.

Steen, D. A., and J. P. Gibbs. 2004. Effects of roads on the structure of freshwater turtle populations. Conservation Biology 18:1143–1148.

Steyermark, A. C., M. S. Finkler, and R. J. Brooks. 2008. Biology of the snapping turtle (*Chelydra serpentina*). Johns Hopkins University Press, Baltimore, Maryland, USA.

Stiles, J. H., and R. H. Jones. 1998. Distribution of the red imported fire ant, *Solenopsis invicta*, in road and powerline habitats. Landscape Ecology 335:335–346.

Sutherland, R. W., P. R. Dunning, and W. M. Baker. 2010.

Amphibian encounter rates on roads with different amounts of traffic and urbanization. Conservation Biology 24:1626–1635.

Svartz, G., J. Herkovits, and C. Perez Coll. 2012. Sublethal effects of atrazine on embryo-larval development of *Rhinella arenarum* (Anura: Bufonidae). Ecotoxicology 21:1251–1259.

Tanner, C. C. 1996. Plants for constructed wetland treatment systems—a comparison of the growth and nutrient uptake of eight emergent species. Ecological Engineering 7:59–83.

Taylor, B. D., and R. L. Goldingay. 2012. Restoring connectivity in landscapes fragmented by major roads: a case study using wooden poles as "stepping stones" for gliding mammals. Restoration Ecology 20:671–678.

Tikka, P. M., H. Högmander, and P. S. Koski. 2001. Road and railway verges serve as dispersal corridors for grassland plants. Landscape Ecology 16: 659–666.

Trinchella, F., M. Cannetiello, P. Simoniello, S. Filosa, and R. Scudiero. 2010. Differential gene expression profiles in embryos of the lizard *Podarcis sicula* under in ovo exposure to cadmium. Comparative Biochemistry and Physiology. Part C, Toxicology and Pharmacology 151:33–39.

Trombulak, S. C., and C. A. Frissell. 2000. Review of ecological effects of roads on terrestrial and aquatic communities. Conservation Biology 14:18–30.

Tucker, J. K., and J. T. Lamer. 2004. Another challenge in snapping turtle (*Chelydra serpentina*) conservation. Turtle and Tortoise Newsletter 8:10–11.

Tunnel, K. D. 2008. Illegal dumping: large and small scale littering in rural Kentucky. Southern Rural Sociology 23:29.

Turer, D. G., and B. J. Maynard. 2003. Heavy metal contamination in highway soils. Comparison of Corpus Christi, Texas and Cincinnati, Ohio shows organic matter is key to mobility. Clean Technologies and Environmental Policy 4:235–245.

Turer, D., J. B. Maynard, and J. J. Sansalone. 2001. Heavy metal contamination in soils of urban highways comparison between runoff and soil concentrations at Cincinnati, Ohio. Water, Air, and Soil Pollution 132:293–314.

Turtle, S. L. 2000. Embryonic survivorship of the spotted salamander (*Ambystoma maculatum*) in roadside and woodland vernal pools in southeastern New Hampshire. Journal of Herpetology 34:60–67.

Tyser, R. W., and C. A. Worley. 1992. Alien flora in grasslands adjacent to road and trail corridors in Glacier National Park, Montana (U.S.A.). Conservation Biology 6:253–262.

Urbonas, B. R. 1999. Design of a sand filter for stormwater quality enhancement. Water Environment Research 71:102–113.

US EPA (United States Environmental Protection Agency). 1998. Illegal dumping prevention handbook. EPA 905-B-97-001. US EPA, Washington, DC, USA.

Valenzuela, N., and V. A. Lance. 2004. Temperature-dependent sex determination in vertebrates. Smithsonian Press, Washington, DC, USA.

van Bohemen, H. D., and W.H.J. van de Laak. 2003. The influence of road infrastructure and traffic on soil, water, and air quality. Environmental Management 31:50–68.

Vandenabeele, W., and O. Wood. 1972. The distribution of lead along a line source (highway). Chemosphere 1:221–226.

van Langevelde, F., and C. F. Jaarsma. 2004. Using traffic flow theory to model traffic mortality in mammals. Landscape Ecology 19:895–907.

Vaze, J., and F.H.S. Chiew. 2002. Experimental study of pollutant accumulation on an urban road surface. Urban Water 4:379–389.

Viskari, E.-L., and L. Karenlampi. 2000. Roadside Scots pine as an indicator of deicing salt use—a comparative study from two consecutive winters. Water, Air, and Soil Pollution 122:405–419.

Viskari, E. L., R. Rekilä, S. Roy, O. Lehto, J. Ruuskanen, and L. Kärenlampi. 1997. Airborne pollutants along a roadside: assessment using snow analyses and moss bags. Environmental Pollution 97:153–160.

von Seckendorff Hoff, K., and R. W. Marlow. 2002. Impacts of vehicle road traffic on desert tortoise populations with consideration of conservation of tortoise habitat in southern Nevada. Chelonian Conservation and Biology 4:449–456.

Wang, W. 1987. Factors affecting metal toxicity to (and accumulation by) aquatic organisms—overview. Environment International 13:437–457.

Weis, J. S., and P. Weis. 2004. Metal uptake, transport and release by wetland plants: implications for phytoremediation and restoration. Environment International 30:685–700.

Wells, K. D. 2007. The ecology and behavior of amphibians. University of Chicago Press, Chicago, Illinois, USA.

Welsh, H. H., Jr., and L. M. Ollivier. 1998. Stream amphibians as indicators of ecosystem stress: a case study from California's Redwoods. Ecological Applications 8:1118–1132.

Werner, E. E., and M. A. McPeek. 1994. The roles of direct and indirect effects on the distributions of two frog species along an environmental gradient. Ecology 75:1368–1382.

Weston, N., M. Goosem, H. Marsh, M. Cohen, and R. Wilson. 2011. Using canopy bridges to link habitat for arboreal mammals: successful trials in the wet tropics of Queensland. Australian Mammalogy 33:93–105.

Wik, A., and G. Dave. 2009. Occurrence and effects of tire wear particles in the environment—a critical review and an initial risk assessment. Environmental Pollution 157:1–11.

Wilber, W. G., and J. V. Hunter. 1977. Aquatic transport of heavy metals in the urban environment. Journal of the American Water Resources Association 13:721–734.

Wilburn D. R. 2011. Wind energy in the United States and materials required for the land-based wind turbine industry from 2010 through 2030. US Geological Survey Scientific Investigations Report 2011-5036. USGS, Washington, DC, USA.

Wilshire H., and D. Prose. 1987. Wind energy development in California, USA. Environmental Management 11:13–20.

Wilson, D. S. 1998. Nest-site selection: microhabitat variation and its effects on the survival of turtle embryos. Ecology 79:884–1892.

Wise, S. E., and B. W. Buchanan. 2006. Influence of artificial illumination on the nocturnal behavior and physiology of salamanders. Pages 221–251 in Ecological consequences of artificial night lighting. C. Rich, and T. Longcore, editors. Island Press, Washington, DC, USA.

Witherington, B. E. 1992. Behavioral responses of nesting sea turtles to artificial lighting. Herpetologica 48:31–39.

Wood, T. M. 2001. Herbicide use in the management of roadside vegetation, western Oregon, 1999–2000: effects on the water quality of nearby streams. US Geological Survey and Oregon Department of Transportation, Water-Resources Investigation Report 01-4065. USGS, Portland, Oregon, USA.

Woodsen, W. M. 2003. Avoiding salt damage to trees along roadsides. American Nurseryman, March 15, 2003:10.

Yang, S.Y.N., D. Connell, D. Hawker, and S. Kayal. 1991. Polycyclic aromatic hydrocarbons in air, soil and vegetation in the vicinity of an urban roadway. Science of the Total Environment 102:229–240.

Zakaria, M., H. Takada, S. Tsutsumi, K. Ohno, J. Yamada, E. Kouno, and H. Kumata. 2002. Distribution of polycyclic aromatic hydrocarbons (PAHs) in rivers and estuaries in Malaysia: a widespread input of petrogenic PAHs. Environmental Science and Technology 36:1907–1918.

Zehetner, F., U. Rosenfellner, A. Mentler, and M. Gerzabek. 2009. Distribution of road salt residues, heavy metals and polycyclic aromatic hydrocarbons across a highway-forest interface. Water, Air, and Soil Pollution 198:125–132.

Zielin, S. B., C. E. de Rivera, W. P. Smith, and S. L. Jacobson. 2012. A synopsis of the case study: targeting ecological investigations to prioritize management for reducing vehicle-caused butterfly mortality. Pages 105–109 in Proceedings of the 2011 international conference on ecology and transportation. P. J. Wagner, D. Nelson, and E. Murray, editors. Center for Transportation and the Environment, North Carolina State University, Raleigh, North Carolina, USA.

Zielinska, B., J. Sagebiel, J. McDonald, K. Whitney, and D. Lawson. 2004. Emission rates and comparative chemical composition from selected in-use diesel and gasoline-fueled vehicles. Journal of the Air and Waste Management Association 54:1138–1150.

5

Engaging the Public in the Transportation Planning Process

Julia Kintsch,
Kari E. Gunson,
and Tom A. Langen

5.1. Introduction

Road ecology researchers and project managers often cite a need for greater public education and understanding of conservation issues, particularly where amphibians, reptiles, or small mammals are involved. While, in general, there has been a gradual shift over recent decades in public attitudes about wildlife apart from management and use for strictly human benefit (Manfredo et al. 2003), public understanding of wildlife conservation, and especially the needs of small animals, remains low. Likewise, despite recognition that wildlife may present hazards for drivers along some stretches of road, few among the motoring public understand the deleterious impacts of public infrastructure and traffic on wildlife populations.

To improve public understanding of these issues, dedicated outreach strategies are on the rise, including tools such as roadside signs, kiosks, posters, and informational brochures. Public outreach may be integrated into any research, mitigation, or monitoring project, offering a much-needed mechanism for engaging the public in small animal conservation. Building public support for transportation projects early in the planning process is an essential ingredient in the funding and implementation of mitigation measures, particularly for these species that otherwise may be easily ignored.

Public awareness efforts are also commonly built into citizen science programs. Such efforts are growing as a means of increasing data collection capacity while engaging citizens in research and conservation.

Long-term datasets on wildlife activity and movement patterns are essential in siting, designing, and evaluating the effectiveness of wildlife crossing structures and other mitigation measures (Clevenger and Waltho 2000; Ng et al. 2004; Clevenger and Huijser 2011). However, the high costs and extended time periods generally needed to produce these data are often prohibitive (e.g., Au et al. 2000). In particular, long-term data sets are generally lacking for herpetofauna and small mammals whose movement needs must be identified to support informed decision making on where and how to implement mitigation measures. Drawing on enthusiastic volunteer field assistants to aid in time- and labor-intensive field data collection tasks offers a low-cost solution to fulfilling this need.

Citizen science, generally defined as public participation in scientific research (Trumbull et al. 2000), was once regarded dubiously among academic researchers (Stokes et al. 1990; Darwall and Dulvy 1996; Heiman 1997; Engel and Voshell 2002), but is now considered more reliable, particularly when strong protocols and quality controls are employed and when there is frequent communication between scientists or managers and their volunteers (Lee 2012). Citizen engagement in data collection can offer many benefits. In addition to new or expanded datasets, it may facilitate an improved comprehension of the scientific process (Trumbull et al. 2000; Krasny and Bonney 2005; Fink et al. 2009) and a heightened awareness of conservation or management issues (Bromenshenk and Preston 1986; Gouveia et al. 2004). In turn, enhanced awareness may result in behavioral changes among the volunteers (e.g., drivers slowing down for wildlife; Lee et al. 2010). Further, this awareness can reach beyond the citizen scientists themselves to the broader community through organizational and media coverage as well as friend-to-friend networking, where volunteers share their citizen scientist experiences with their family, friends, and colleagues.

Developing a citizen science program requires careful consideration of its specific goals, as a public awareness-oriented project must be considered differently than a data-intensive research study. For both, clear research and education goals stated from the outset allow for a tailored program designed to meet those goals. Now, and in the future, the role of citizen scientists is expected to become increasingly important as a strategy for overcoming resource restrictions and providing increased capacity for research. This strategy greatly expands the scientific and management community's ability to monitor wildlife over longer time frames than traditional research using professional biologists or trained graduate students (Cooper et al. 2007; Cohn 2008), and promotes public understanding of wildlife and roads issues.

5.2. Building Public Support and Awareness

Public understanding of road effects on wildlife is instrumental in building support and generating funding for mitigation, such as wildlife crossing structures and other strategies. Early engagement with the public in the transportation planning process when mitigation measures for wildlife are being considered, planned, and designed is worthwhile, spawning broad understanding and public support for projects that otherwise may become easy targets during political and budgetary disputes. Building an understanding of the need for and benefits of mitigation among local communities and public officials, alongside a clear demonstration of the wise use of funding, can go a long way toward smooth project implementation. For example, while in some cases new crossing structures are required to accommodate target species, in others, existing structures may be modified to better accommodate wildlife passage (Kintsch and Cramer 2011; Chapters 9 and 10). Identifying where these lower-cost retrofit opportunities are possible may help in building public con-

fidence and support for mitigation. Yet even the best public education campaigns struggle to engage people who either aren't inherently interested in wildlife or are actively antagonistic to wildlife conservation, leaving campaign organizers grappling with how to reach the entire motoring public. Identifying the target audience is a first, crucial step in designing an effective public outreach campaign. Public outreach strategies encompass a broad array of tools and techniques for targeting different audiences, such as specific communities, motorists (including local drivers and commuters, truckers, tourists, or other pass-through drivers), public officials, or other agencies.

5.2.1. Education and Awareness Tools
Signage
Roadside signage can be a useful tool for generating greater awareness about the potential for wildlife-vehicle collisions. Well-placed and well-designed informational roadside signage is a behavior-based mitigation strategy that is meant to encourage motorists to slow down and drive with caution in order to avoid collisions with wildlife. Additionally, the installation of new signage may provide a media opportunity to garner public support from a wider audience than those driving particular roadways.

Historically, wildlife warning signage along roads has been used to alert drivers to the risk of dangerous collisions with large animals, especially ungulates. Although collisions with small animals have little risk of causing human injury or property damage, and little is known about the effectiveness of signs at reducing road mortality (Huijser et al. 2007), warning signs for these animals increasingly are being used to raise public awareness.

Placement of wildlife signs for amphibians began in Switzerland in the late 1960s and their use quickly spread across much of Europe (Chapter 1). Today, amphibian signs can also be found in Germany, Austria, the Netherlands, United Kingdom, Belgium, and Italy. In the 1990s, wildlife crossing signs began being used in North America, where they have been more commonly used for turtles (Figure 5.1). Some signs in Europe serve a dual purpose of bringing attention to both wildlife and people on roads. For example, signs are often placed where mass migrations for certain amphibians occur, and on warm spring nights concerned

Figure 5.1. Examples of crossing signs in Canada for turtles, for general (a) and seasonal awareness (b). *Credits: Kari E. Gunson.*

citizens actively carry animals across the road (Langton 1989; Section 9.2.3). In Ontario, Canada, wildlife signs for turtles are meant to draw attention to both their declining populations as well as the threats they face from road traffic (Gunson et al. 2012).

To maximize the effectiveness of using road signage for small animals in order to generate environmental awareness, standardized protocols are needed (Gunson et al. 2012). Ideally, signs should be placed within defined seasonal periods when the target species are most active. This may reduce driver desensitization to signage and heighten their awareness to periods of peak wildlife crossing activity. The use of standardized images on signs may help them to be more widely recognized and understood, and may also result in a decrease in the theft associated with more novel signs. As

an example, in Ontario, Canada, 27% of turtle warning signs were stolen or vandalized within one year of placement (Gunson et al. 2012). Another concern is that signage may draw attention to the locations of populations of certain sensitive species that may be prone to poaching, harassment, or persecution (Ashley et al. 2007).

While standardized signage may be less prone to theft, specialized wildlife signage may be more effective in catching a driver's attention. Such specialized signage may contain more information than just a message of caution. For example, in the Netherlands signs have been used to tally roadkill numbers of toads, frogs, and salamanders (Figure 5.2). Other eye-catching informational signs may direct motorists to slow down for specified distances, or may draw their attention to certain species or crossing hotspots (Figures 5.3 and 5.4). Careful consideration of sign placement in locations with moderate to high traffic volumes (Gunson et al. 2012), and the use of theft-reduction methods,

Figure 5.2. Informational roadside sign tallying roadkill of toads, frogs, and salamanders in the Netherlands. *Credit: Kari E. Gunson.*

Figure 5.3. Informational roadside sign for snakes in Manitoba, Canada. *Credit: Manitoba Conservation and Water Stewardship.*

Figure 5.4. Crossing sign for endangered red squirrels (*Sciurus vulgaris*) in Scotland. *Credit: Kari E. Gunson.*

such as greased nuts, can help reduce vandalism of these unique signs.

Public Awareness and Education

Wildlife warning signage, speed reduction signs, and other measures to encourage motorists to slow down, such as speed bumps, have been shown to be more effective if used in combination with a public awareness and education campaign (Joyce and Mahoney 2001; Reed 2008). The idea is that if a motorist understands why roadkill is an issue for a particular species they will be more aware of the roadside signs and speed bumps and respond accordingly. In addition, educational materials can improve the effectiveness of mitigation

measures and facilitate public support and advocacy for efforts to prevent road mortality. An informed public may also be more likely to become involved in citizen science efforts (Section 5.3) that enhance the planning process for mitigation measures, including their design and placement or to monitor their effectiveness following construction.

Conveniently located and engaging kiosks may be installed at rest areas or scenic pull-offs to generate awareness among drivers. Such displays offer a low-cost means of informing motorists about road effects on wildlife and existing or potential mitigation strategies. Billboards may also be used to generate public awareness. Two examples from the United States can be found in the states of Washington and Colorado. In Washington, the winning entry from a children's drawing contest to educate youth about wildlife crossings was displayed on a billboard along Interstate 90. In Colorado, a private company sponsored a billboard along Interstate 70 advertising a citizen science wildlife reporting website (Figure 5.5).

Beyond the road, educational materials and displays may also be useful as part of a comprehensive public outreach strategy to inform the public about the effects of roads on wildlife. These include informational displays at museums (e.g., Whyte Museum of the Canadian Rockies in Banff National Park), zoos (e.g., Toronto Zoo), and visitor centers where interested individuals are likely to take notice. In addition, some

Figure 5.5. Billboard on Interstate 70 in Colorado, United States, advertising a website for reporting wildlife observations. *Credit: Rocky Mountain Wild.*

nonprofit organizations are designing and distributing flyers, brochures, and articles through targeted venues such as car rental companies, visitor centers, and insurance company newsletters. Other organizations, such as PARC (Partners in Amphibian and Reptile Conservation), have created websites and publications providing information and management guidelines that include the effects of roads (and other management actions) on small animals.

Competitions

Design competitions are an emerging forum for stimulating new ideas and innovation in road ecology; though few, if any, have focused on wildlife crossing designs for smaller animals. The 2010 Animal Road Crossing (ARC) International Wildlife Crossing Infrastructure Design Competition engaged professional interdisciplinary design teams of landscape architects, engineers, transportation, and ecological disciplines to design a new generation of wildlife crossing structures. The competition focused on designing a cost-effective, multispecies overpass on an interstate highway in Colorado that would blend with the ecosystem and use technology adaptable to diverse locations. In addition to inspiring new concepts for designing wildlife bridges, the competition created a platform for educating professionals outside of ecology and conservation about wildlife crossing needs, bringing new awareness to potential solutions for reducing the impacts of roads on wildlife (ARC Solutions 2012).

In contrast to the interdisciplinary engagement among professionals generated by the ARC Competition, the 2011 conference of the Infra Eco Network Europe (IENE) in Velence, Hungary, hosted a children's art competition focusing on new concepts for reducing wildlife mortality on roads. Not only did the competition stimulate new ideas for crossing designs, but professionals were inspired by the creativity and artistic nature of the children's submissions.

5.2.2. Collaboration and Information Sharing
Agency Support and Collaboration

Increasingly, nongovernmental organizations (NGOs), government agencies, academia, and conservation practitioners are supporting and promoting road ecology initiatives. Several NGOs across North America have received grant funding for road ecology projects.

Often, these projects include collaborations with state and provincial wildlife or transportation agencies, lending added credibility and expertise. Such projects, in turn, help generate greater support for these initiatives within the agencies themselves. For example, the Colorado Wildlife on the Move Coalition is a consortium that includes the Colorado State Patrol, wildlife and transportation agencies, and NGOs; also included is an association of insurance companies that releases biannual press releases reminding Colorado drivers to pay attention to wildlife on the road, particularly during migration seasons.

Interorganizational collaborations offer multiple benefits to road ecology and transportation projects. Government transportation agencies seek information about planning, designing, and monitoring road mitigation solutions, while universities are performing much-needed research that feeds into the decision-making process. Practitioners then collate all the information and implement solutions on the ground. One such example of a formalized collaboration is the Ontario Road Ecology Group, an interdisciplinary collaboration of government and nongovernment scientists, educators, and transportation planners. Since its inception in 2008 at the Toronto Zoo, the group has promoted public education about road ecology in Ontario, supported a website for reporting roadside observations of wildlife, disseminated information about road ecology projects in the province, and promoted new research in road ecology (Ontario Road Ecology Group 2010).

Knowledge Dissemination

Sharing knowledge beyond small academic and professional circles is an important step in building widespread awareness across disciplines and across organizations. Formal meetings and conferences offer people from diverse entities and geographies to convene, share information, and create partnerships. Biennial conferences focused on road ecology in Europe (IENE) and North America (International Conference on Ecology and Transportation [ICOET]) provide important gathering places for such dissemination. In recent years, technical presentations formerly dominated by talks on large animals now increasingly incorporate research on smaller animals, including special sessions on topics related to mitigating the impacts of roads on small vertebrates.

Continued research on the effects of roads on small animals and their habitats is paramount for developing enhanced, scientifically defensible, and adaptable mitigation strategies. A new generation of professional engineers, transportation planners, and biologists will need to be trained to fill knowledge gaps across professions in order to develop and implement solutions to mitigate impacts on small animals. Several United States universities have recognized this need and have established programs in road ecology (e.g., Montana State University and the University of California, Davis) or offer dedicated courses (e.g., Portland State University). Short courses for professionals have also been offered through the University of Colorado, Denver, and the Western Transportation Institute. Undergraduate internships, co-op programs, and postgraduate research are other opportunities to foster scientific research in this highly applied, fast-evolving, and yet underrepresented area of science.

5.3. Citizen Science

For much of its history, the use of citizen scientists in ecological research has been dominated by ornithological surveys, including the Christmas Bird Count (Droege 2007), the Breeding Bird Survey (Sauer et al. 1997), and eBird (Sullivan et al. 2009). Initially, increased data collection capacity was the sole objective of involving nonscientists in research studies. As the environmental movement grew in the 1960s, it became clear that direct engagement by individuals could also lead to increased public support for the burgeoning movement, and citizen science blossomed beyond ornithological surveys into diverse fields of study.

Today, citizen science encompasses a wide variety of research projects, ranging from those that ask people to report on what they see in their own backyards (e.g., Cooper et al. 2007) to those that enlist trained volunteers to visit specifically designated sites, such as monitoring transects (e.g., Przybyl 2003). Some projects direct students from multiple institutions to collect standardized data about the environment around their school campuses, while others engage specific user groups, such as anglers, birders, or hikers, to collect data in the areas where they are traveling. Broad participation in citizen science is increasingly facilitated by technological advances, including web-based and mobile applications that tally wildlife observations made by drivers along specified or unspecified routes. Some United States examples include the California Roadkill Observation System, the I-70 Wildlife Watch (Colorado), and Roadkill Garneau (nationwide).

Ecological research and monitoring are long-term endeavors that require substantial investments of time, labor, and money (Trumbull et al. 2000; Clevenger and Huijser 2011; Chapter 12). Scientists have long recognized the power of collaborative research across university labs, states, or continents for task and cost sharing, and for generating research at broader scales or greater depth. In the field of conservation biology, datasets encompassing large temporal and spatial scales are essential in understanding complex systems and detecting trends in species' responses to environmental change (Fink et al. 2009). In studies of wildlife crossing use by large mammals, it may take several years for local wildlife populations to adapt to using new structures; this adaptation process may only be captured via long-term monitoring (Clevenger and Waltho 2003; Dodd et al. 2009). With citizen scientists, the reach of a given study may extend further and at lower cost than traditional research studies. Conceivably, road ecology research compiled by citizen scientists from different locations could be compared to detect trends and behavioral variations across a vast geography. However, no standardized protocols for monitoring wildlife movement have been established to date (Paul 2007), hampering systematic large-scale analyses.

Despite the potential of citizen science, few road ecology research projects incorporate volunteers, and among those that do, even fewer involve small mammals, amphibians, and reptiles. Road ecology projects to date have enlisted volunteers to collect baseline information to inform the placement of wildlife mitigation measures, or to monitor wildlife use of new crossing structures. A handful of citizen science projects are dedicated to collecting information exclusively on small mammals, amphibians, or reptiles. Among those that do are the North American Amphibian Monitoring Program (http://www.pwrc.usgs.gov/naamp), where volunteers identify amphibians by their calls (i.e., conduct call surveys) along assigned roadside routes (though not related to road crossings); the Carolina Herp Atlas (www.carolinaherpatlas.org), a website where volunteers contribute observations of salamanders, frogs, turtles, lizards, snakes, and alligators from across the Carolinas; and the Front Range Pika Project in Colorado in which "pika patrollers" are trained to collect data on American pika (*Ochotonoa princeps*) habitat occupancy (www.pikapartners.org).

The relative scarcity of citizen science projects involving road ecology and small animals may be attributed to the general lack of attention that these species receive relative to larger and more visible mammals (Glista et al. 2008), which are also more associated with safety concerns due to vehicular damage and human injuries in the event of a collision. The relative lack of citizen science projects targeting amphibians and reptiles, in particular, also may be due to the cryptic nature of these species, making them more difficult to observe, as well as the aversion that some people feel toward them (Price and Dorcas 2011). Small mammals tend not to generate as strong an emotional response with the public and remain less noticeable than their larger counterparts and, accordingly, are less studied. While road ecology research using citizen scientists has captured incidental small mammal activity with camera traps, for example, few are deliberately designed to monitor small mammal movement. There is much room for growth in citizen science projects in road ecology research, particularly for small animals.

5.3.1. Building a Citizen Science Program in Road Ecology

The classic citizen science program engages participants to follow established protocols and to submit data to a central location. Models for citizen science projects have become as diverse as the research topics themselves. Community-based projects typically involve area residents in a project with local appeal or relevance, often with the purpose of advancing a specific conservation concern. These types of projects require community involvement and direct interaction with a dedicated program manager to keep volunteers engaged and energized over the long term. In these programs, volunteers typically submit data forms to the managing organization, which is then responsible for data entry (if data are not submitted online), quality control, analysis, and reporting results. Other research studies benefit from a dispersed network of volunteers who, collectively, compile information across a large landscape, continent, or even the globe. In this model, participants follow prescribed protocols and submit their data to a website through which researchers from various institutions may access and analyze the data for research and publication. Another model is a network of schools (K-12 or universities) that use students to collect data using standardized protocols under the direction of a trained instructor. Data from participant schools are typically posted on the web at a common project archive.

An effective citizen science program must be carefully tailored to meet both its research and its public awareness goals. Data collection goals and education or awareness-building goals will define how and where volunteers are recruited, what sort of training events and materials are needed, how to communicate with participants to ensure their long-term commitment, and how to catalyze broader public and media interest in the program and its conservation objectives.

As the use of volunteer field assistants grows, citizen science is increasingly recognized as a credible means of expanding data collection capacity as well as a valuable tool for giving people direct scientific experience and engendering in them a heightened level of concern and understanding of conservation issues. Those considering the establishment of a citizen science program must carefully evaluate whether such a framework offers the best mechanism for collecting useful data while providing an effective means of reaching out to target audiences or the general public. Any citizen science program requires appropriate infrastructure for its administration and sustenance. As citizen science encompasses a broad array of community-based research activities, programmatic structures may vary widely across the spectrum of citizen science research. However, all citizen science programs must include components for recruiting and managing volunteers, establishing scientific credibility through defined research objectives and protocols, maintaining rigorous data quality controls and expertise for conducting analyses and reporting results, and managing systems for communicating results to volunteers and to a broader public audience (Cooper at al. 2007; Lee 2012). Despite the current rarity of citizen science programs targeting small animals, we can learn a great deal from other citizen science efforts, particularly those addressing roads and wildlife issues.

Program Objectives

Facilitating scientific inquiry is a primary goal of citizen science. Yet in most cases, program objectives extend beyond data collection and may include goals of educating volunteers, increasing public awareness, building support and inspiring action toward specific conservation actions, or inspiring young people to study science and math. For example, the Citizen Science Wildlife Monitoring Program at Vail Pass was spearheaded in 2006 by a Colorado-based nonprofit organization for the dual purposes of jump-starting efforts to establish baseline data on wildlife populations and activity, and building awareness and support among local communities regarding the need for wildlife crossing structures to improve passage across a heavily trafficked interstate.

For amphibians, reptiles, and small mammals, data collection and public education and awareness goals are arguably of equal importance—successful conservation of these animals will ultimately depend on both an enhanced understanding of the species as well as greater public concern for the health of their populations. Though still small in number, the existing citizen science programs focusing on these species demonstrate that there are volunteers interested in projects that offer opportunities for their direct engagement (Price and Dorcas 2011). Emerging citizen science programs can be encouraged by these trends. Citizen science programs may also act as mechanisms for generating broader interest and support for a conservation issue as both funders and the media may be attracted by community involvement in a project. The positive feedback loop may continue by bringing new volunteers to the program and leveraging additional funding for citizen science research and, potentially, for mitigation actions.

The most successful citizen science programs are those that address a tangible conservation question that resonates with volunteers and motivates participation. This is most effective when there is a clear statement of both the research objectives and education and awareness goals. Developing these objectives is integral to the scientific process, ensures an appropriate application of citizen science to the research question at hand, and provides the foundation for communicating program milestones to funders, volunteers, and the public at large.

Research questions that a citizen science framework may address despite limited training of volunteers should be focused on presence and location of species or species groups with respect to certain habitat or human-made features such as roads or existing and potential crossing structures. Such programs may direct volunteers to check and maintain camera traps on a periodic basis, for example, or require them to become proficient in identifying species and signs of wildlife activity. Some programs may necessitate volunteer training and performance evaluations to ensure volunteers are equipped with the skills to identify animals and their tracks and signs, or conduct more complex research activities.

Volunteer Recruitment and Retention

Local partners and nonprofit organizations can help recruit volunteers from their membership. Recruitment efforts may also target residents of local communities, including neighborhood organizations or specific user groups. Some groups, such as those for hikers, anglers, or amateur naturalists, have members who are already inclined to participate in outdoor activities and have cultivated an understanding of the natural world. In addition, many companies encourage employees to participate in volunteer activities and will support a certain amount of work hours being applied for this purpose. Either after-school student programs or those that are classroom based may also be recruited into a citizen science program, although their participation may be restricted by the length of the school year or summer program. Not all citizen science programs will be appropriate for volunteers who are only able to provide a short-term commitment, and recruitment strategies should be designed accordingly. A preliminary public education component may be an important early step in developing a program oriented to small animals. This effort may help to generate interest among potential volunteers, as well as attract funders to support efforts focusing on these less visible species whose ecology and conservation needs are not widely recognized.

A fundamental component of a successful volunteer program is setting forth clear expectations regarding volunteer responsibilities, including the time commitment, prompt data submission, and how to access research sites (and how strenuous it may be to access certain sites). Stating these expectations clearly and repeatedly helps prevent the recruitment of well-intentioned but uncommitted volunteers who drain staff time and energy. Citizen science programs may also require applicants to sign a volunteer contract, simply as a means of reinforcing the expected commitment. Volunteers who are faithful to their research tasks are essential to the data collection process and overall research quality.

Retaining good volunteers demands regular communication, prompt replies and feedback, as well as recognition for their contributions. Retaining volunteers from year to year lends a program credibility among new volunteers (Genet and Sargent 2003). This retention also strengthens the program by creating leaders within the volunteer base whose feedback can lead to an adaptive approach in the overall program and whose experience can position them to train new recruits (Cooper et al. 2007).

Most importantly, the volunteer experience should be fun and interesting, with volunteers empowered as meaningful contributors to a conservation program. By setting up a dedicated webpage, listserv, online discussion board, or social networking site, program managers can report back with timely results, respond to questions, and provide a forum where volunteers may share experiences and photographs. The more engaging the volunteer experience is, the more likely the volunteers are to share their experiences with family and peers, which ultimately can lead to enhanced public support for the fundamental goals of the program.

Volunteer Training and Support

Volunteer training is essential for establishing conformity to research protocols. A secondary objective of a training session is to reinforce the volunteer's commitment and the gravity of their role in helping to achieve program objectives. A "face-to-face" training session serves as an introduction to the program, confirms the volunteer's relationship with the organizations involved in the research, further educates them regarding the conservation issue at hand, and establishes or reinforces connections within the volunteer community. Ongoing training can serve as a skills refresher at the start of each field season (Genet and Sargent 2003) and reinforce the volunteer's relationships and commit-

ment to the program. Volunteers should leave a training session—whether in person or online—confident in their ability to carry out the research activities, and secure in the knowledge that they will receive support as needed to accomplish these tasks, whether it be in confirming a species identification or clarifying how to take a measurement.

Many citizen science programs require volunteers to go out in the field in pairs or teams. This is both for safety reasons and as a quality control measure whereby team members verify one another's work and ensure data forms are completed properly. Static teams over the course of a field season also help volunteers to build personal relationships, further deepening the citizen scientist experience. Programs such as the Cascade Wildlife Monitoring Project assign leaders who have received additional training in track identification to serve as the team's resident expert and decision maker (Moskowitz et al. 2007). Each volunteer or volunteer group must be equipped with the necessary field equipment, including data forms, field guides and any project-specific equipment such as measurement tools, Global Position System (GPS) devices, batteries, or spare parts.

Volunteer Management

Volunteer management requires a significant amount of time and attention from a program manager (Krasny and Bonney 2005). The program manager must be responsive to volunteer questions and requests, and provide scheduled feedback regarding the program. Regular communication and volunteer feedback may also help to refine and improve the data collection process. For example, midseason feedback from citizen science teams in the Cascade Wildlife Monitoring Project in Washington State led to the addition of "ambiguous" as a species category so that volunteers would not be obligated to record a species or obscured tracks that they could not identify with certainty (Moskowitz et al. 2007). Depending on the nature of the research activities, volunteers may schedule their own field outings or they may be assigned to specific dates when the research tasks are time sensitive. In either case, a volunteer calendar is a useful tool in committing volunteers to the program and allowing them sufficient leeway in arranging their schedule to honor that commitment.

Volunteer safety is of utmost importance for any program, particularly when volunteers are asked to conduct research tasks adjacent to high-traffic roadways, or park on road shoulders to access research sites. Developing safety protocols in coordination with the local authorities or district transportation offices is mandatory, as is setting clear guidelines for the use of safety gear (i.e., hardhats, reflective or blaze orange vests, cell phones, flashing vehicle lights) and access points along road shoulders where other off-road access is not available. For projects along rights-of-way, permits from the responsible authority may be required, and it can be prudent to alert local law enforcement and conservation officers, especially if fieldwork is conducted at night.

Data Collection Protocols

Strong and clear data collection protocols are an essential component in ensuring that citizen science data are accurate, complete, and standardized so that they can be used to answer the research questions as intended. Thoroughly training volunteers in an explicit and easy-to-follow study regimen helps prevent errors from entering the dataset. With the increase in online access and the growing power of the internet, more and more citizen scientists are entering data online, rather than submitting hard copy data forms. Online data submission removes the burden of data entry from program staff and the need for staff to decipher hand writing and cryptic markings on hard copy forms that can result in erroneous data. Citizen science programs are also increasingly integrated into mobile applications that utilize GPS technology and data dictionaries for data entry on cell phones and other handheld devices (Ament et al. 2012). Although such applications may be limited by a lack of cell service in remote field locations, they have the potential to standardize and provide more accurate wildlife data.

Quality Assurance and Quality Controls

Variability in data collected by volunteers may be remedied by implementing rigorous quality controls. Several evaluations of citizen science programs that compared the accuracy and completeness of data collected by volunteers versus those collected by experts have generally concluded that citizen science data are reliable, and other studies have demonstrated that even elementary-aged children can collect useable, high-quality data (Krasny and Bonney 2005). Genet and Sargent (2003) determined that volunteers were able to assess species presence reliably using amphibian call surveys, but subjective determinations of abundance indices were not dependable. Problems that have arisen in citizen science programs include the improper use of equipment, greater inconsistencies in the data due to the larger number of observers (Bromenshenk and Preston 1986), and nonintentional data biases (Stokes et al. 1990; Price and Dorcas 2011); the last includes volunteers who may be more likely to report unique or rare species rather than multiple observations of more common ones. However, these concerns may be largely alleviated by proper study design and adequate and ongoing volunteer training.

Careful photo documentation of the species or wildlife sign provides an important voucher for confirming identifications made by citizen scientists (Moskowitz et al. 2007; Price and Dorcas 2011). For wildlife such as amphibians and reptiles, which may be easily misidentified, this step can be the difference between a reliable data point and one that must be discarded due to uncertainty. Price and Dorcas (2011) further recommend prompt staff review of all submitted observations so that any questionable entries may be quickly addressed before a volunteer's memory of the outing begins to fade; a confidence code for each sighting can then be assigned to reflect the level of certainty in the observation's accuracy.

Spurious observations may also be removed from a dataset, typically erring on the side of caution (i.e., removing inaccurate data points, such as those that fall outside of the study area or records of species that are not known to be present in the region). In the case of a wildlife observation website or roadkill tally, protocols must be put in place to avoid duplicate records by multiple surveyors, either by devising a field marking system or by imposing decision rules that define duplicate records according to set spatial and temporal limits (e.g., Huijser et al. 2008; Singer et al. 2011).

Disseminating Results

Citizen science, like any research, benefits from publication in peer-reviewed journals as a means of authenticating and sharing research methods and results with a wider audience. Programs with an academic partnership will almost certainly pursue publication, yet all citizen science programs benefit from the peer-review process and exposure garnered through publication in a scientific journal.

Equally important for many programs are forums for communicating research results with the media and public at large to facilitate broader awareness of the conservation issue and research implications (Cooper et al. 2007). Periodic newsletters and updates issued throughout the research season are useful tools for sharing interim results and photographs with the volunteer researchers themselves, keeping them motivated and engaged. These updates may also be released to an organization's membership or the media to foster greater public understanding and awareness.

Measuring Effectiveness

Measuring program effectiveness is an important step in both refining citizen science programs so that they better achieve their goals and building widespread support for citizen-based research. The effectiveness of a citizen science research program should be evaluated in terms of its ability to meet the research objectives as well as the education and outreach objectives. This includes evaluating the accuracy and completeness of the citizen-compiled data. In addition, an assessment of the structure of the volunteer program and research protocols may lead to refinements that enhance the ability of a volunteer to collect useful data. Finetuning of methods is common to any research study and is equally important for adaptive management in a citizen science scenario (Lee et al. 2010), allowing iterative refinements to maximize program effectiveness. Although most citizen science programs have not developed mechanisms for measuring progress toward the program's education and awareness goals, volunteer surveys are an effective way of evaluating participants' learning about conservation issues and the scientific processes (Krasny and Bonney 2005).

Interorganizational Collaborations

Citizen science abounds with opportunities for creating collaborative relationships between NGOs and agencies or universities. Such relationships greatly enhance the collective ability to gather and analyze wildlife data, which can then be used to inform decision making and advise placement of crossing structures. Departments of transportation (DOTs) do not typically initiate citizen science research programs, though several state DOTs have become involved in road ecology research (Section 5.2.2).

State or provincial, county, and municipal transportation departments, in general, are ill equipped to manage citizen science research, and partnerships with state wildlife agencies and local NGOs offer an effective alternative. Many state fish and wildlife agencies have established volunteer networks that could be engaged for road ecology research, and in the United States both the National Forest Service and the National Park Service have been long-time promoters of citizen science (Cohn 2008). In most cases, citizen science research focused on road ecology has been spearheaded by NGOs concerned with promoting habitat connectivity and wildlife crossings. These organizations tap into citizen science as a means of compiling baseline data for which there may otherwise be no funding, mobilizing their membership and other concerned citizens around a particular roadway of concern.

Nongovernmental organizations that team up with agencies and academia are well positioned to create a robust and credible citizen science program. Universities can offer program training support, quality controls and academic legitimacy, lab analysis services, and a stable institutional home for long-term research, data analysis, publication, and storage (Savan et al. 2003). A collaborating nonprofit partner can undertake the overall program and volunteer management, and enhance public education goals (Krasny and Bonney 2005). A multiparty effort will typically lend greater credibility to the data and the volunteer effort itself, as demonstrated by the eBird partnership, a collaboration between the National Audubon Society and the Cornell Lab of Ornithology. By engaging in these partnerships, agencies greatly expand their data collection capacity, which is an oft-cited limitation by DOTs considering mitigation measures for wildlife.

5.3.2. Case Study: Maine Audubon Endangered Species Road Watch

Barbara Charry and Fraser Shilling

In Maine, three state endangered and threatened species are known to be at serious risk of extinction due to mortality caused by roads. These include state endangered Blanding's turtles (*Emydoidea blandingii*) and northern black racers (*Coluber constrictor constrictor*), and state threatened spotted turtles (*Clemmys guttata*). Very little is known about the road interactions of New England cottontails (*Sylvilagus transitionalis*), though populations divided by Interstate 95 have been shown to be genetically distinct (Fenderson et al. 2014); this species is also endangered in Maine. To learn more about each of these species, their interactions with roads, and the habitat and roadway characteristics where they may attempt to cross a road, Endangered Species Road Watch was launched in spring 2012 by Maine Audubon in collaboration with the Maine Department of Inland Fisheries and Wildlife, Maine DOT and the University of California, Davis, Road Ecology Center. Volunteers across two counties in southern Maine collect field data, which is then submitted on an interactive wildlife observation reporting website (Figure 5.6; www.wildlifecrossing.net/maine/).

Collecting the amount of data needed to answer the research questions above would not be possible without volunteer effort. In addition, the project is generating new citizen advocates for these species, which helps to raise public awareness about the threats they face and potential mitigation solutions. Local champions can raise the awareness and concern needed to develop support for implementing solutions such as signs, traffic-calming measures, wildlife crossings, and improved transportation and land-use planning. As part of the project, volunteers are asked to share the project findings with their town officials who are responsible for the roads that make up nearly 66% of the state's road network. Broader public awareness and media attention that ripples from citizen involvement in the project will also be instrumental in increasing support for ultimately implementing strategies to reduce road mortality of these species.

Volunteer recruitment, training, and support are the most labor-intensive aspects of the Road Watch program. Recruitment activities include information sessions, press releases, radio interviews, posted flyers, social media, and extensive email communications to civic groups and outdoor organizations, as well as personal phone calls. To date, these efforts have resulted in 100 people who have either attended a presentation or contacted Road Watch to express interest in the

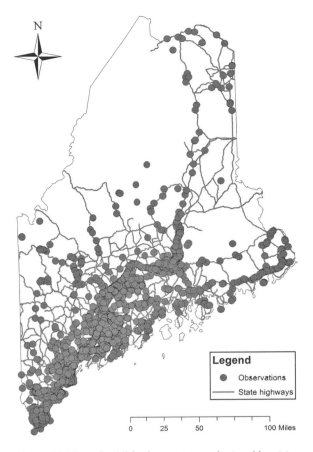

Figure 5.6. Map of wildlife observations submitted by citizen scientists to the Maine Audubon Wildlife Road Watch website during the inaugural season, from spring through summer 2012, in the United States. *Credit: Maine Audubon and University of California, Davis.*

project; 40 of these people have committed to becoming volunteers. Indirectly, the recruitment effort has also brought added visibility to the project.

Given Road Watch's previous experience with a volunteer project to monitor road segments using only online training materials, it was already known that training volunteers in person for this project would be critical to its success. At the training sessions volunteers learned why this project is necessary, how to identify species and collect standardized data, and how enter their data on the website.

Project protocols were developed collaboratively by the project partners. In addition to the targeted endangered species, more common and likely to be observed proxy species (those that use similar habitats) were included in the study. Volunteers were asked to report and photograph all dead and live wildlife observations made while walking a 1 mi (1.61 km) long route. Assigning routes that could be safely walked was critical because it is impossible to see small species such as frogs, turtles, and snakes while driving. Surveys were conducted once a week from mid-April through mid-May, and then once a month from June through September, for a total of eight surveys. Survey routes were identified through a Geographic Information Systems (GIS) analysis of previously identified GIS-modeled habitat connectors for the four target species that either intersected species occurrence data or Maine DOT priorities for future construction projects. Volunteers were provided datasheets with an aerial photo of their route, a ruler for scale in photographs, and a safety vest. Once volunteers collected the data, they entered it on the Maine Audubon Wildlife Road Watch website. The power of this website is the collection of a location using a simple interactive map for GIS analysis, making it unnecessary for all volunteers to have GPS units or learn how to use them. Once they have submitted their data, they may view the compiled data points on an interactive map; this allows them to visualize the results of their surveying and encourages their ongoing participation.

Several analyses may be conducted using these wildlife observations, particularly as more data are compiled about these routes and new routes are added to the study. The analyses include the following: (1) "hotspot" analysis, which is useful in identifying particular stretches of road (if any) where wildlife are attempting most frequently to cross; (2) investigating proximate causes of wildlife-vehicle collisions, such as adjacent land cover, traffic patterns and speed, and road characteristics; (3) testing whether or not previously mapped connectors are good predictors of wildlife presence and movement activity; and (4) comparing surveyed routes for similarities and differences to reveal if certain roads are more dangerous to wildlife than others.

As the Endangered Species Road Watch progresses, it will produce an extensive collection of wildlife observations that will be useful for understanding the impacts of roads and the changing distributions of wildlife due to land-use and climate change impacts. The Maine DOT plans to use this information in their project planning process to avoid and reduce impacts on these en-

dangered species. At the same time, Maine Audubon will continue working with volunteers to share the results with their local communities and implement conservation measures to reduce road mortality.

5.4. Conclusions

Stimulating public education and awareness about road effects on small animals through a variety of informative and interactive tools and techniques is an important step in generating support for transportation planning efforts that incorporate road ecology research, mitigation, and monitoring. Campaigns to engage and mobilize citizens have significant roles in influencing decision makers and the distribution of funds for research and application. The strategy for any public awareness campaign must be customized to meet specific education and awareness objectives such that these messages are heard and absorbed by the intended audiences.

Citizen science, in turn, provides a cost-effective means for both promoting public awareness about road effects on small animal populations and collecting much-needed data. The boundaries between scientists and the public are diminishing, not only making scientific inquiry and understanding more accessible, but creating mutually beneficial relationships where citizens become important contributors to scientific understanding of conservation issues. An effective citizen science program requires substantial management, but if this management recognizes the value of volunteer field assistants and is specifically tailored for them, it can reap widespread benefits that are well worth the effort. Regardless of the program structure, the involvement of citizen scientists allows for increased effort in data collection due to broader participation among volunteers or students, rather than relying solely on paid scientists. Indeed, many research studies would be unsustainable without volunteer participation (Cohn 2008). Digital databases and the internet have greatly facilitated the growth of citizen science projects and expanded participation among a broader array of volunteers (e.g., eBird network, Sullivan et al. 2009).

Both public outreach campaigns and citizen science have important roles in the future of road ecology. Practitioners involved in all aspects of such campaigns will benefit from proactive efforts to bring greater awareness, understanding, and funding to reduce the threat of roads on small mammals and herpetofauna. Ultimately, this increased public support and engagement can lead to the regular practice of incorporating mitigation measures for small animals in transportation projects.

5.5. Key Points

- Identify the target audience as a first, crucial step in designing an effective public outreach campaign.
- Use roadside signs within a defined season when target species are most active; this reduces driver desensitization to signage and heightens awareness during peak activity.
- Incorporate public education displays, such as information kiosks or billboards, in mitigation projects to educate the public and increase support for mitigation measures.
- Consider incorporating a citizen science program as a means of collecting valuable road ecology data while generating public awareness, support, and funding for mitigation measures.
- Select a tangible conservation question that will resonate with volunteers and motivate their participation; as part of this, ensure that research goals as well as education and outreach goals are clearly stated from the outset.
- State volunteer expectations clearly and repeatedly, including responsibilities, time commitment, and data collection and submission requirements, to help ensure that volunteers know exactly what they are agreeing to do.
- Develop training sessions that equip volunteers with the skills and tools they will need to be confident in their ability to carry out the research tasks.
- Develop safety protocols in coordination with local authorities, including clear guidelines and expectations.
- Rigorous quality-assurance and quality-control protocols can minimize variability in data collected by volunteers.
- Build a collaborative team including departments of transportation, university, and nonprofit partners to maximize the program's effectiveness in collecting reliable data.
- Both public engagement and citizen science have important roles in the future of transportation

planning and road ecology; proactive efforts to raise public awareness, understanding, and funding may ultimately lead to the public support for, and standard practice of, incorporating mitigation measures for small animals in transportation projects.

LITERATURE CITED

Ament, R., D. Galarus, D. Richter, K. Bateman, M. Huijser, and J. Begley. 2012. Roadkill observation collection system (ROCS): phase III development. Publication No. FHWA-CFL/TD-12-001. Federal Highway Administration, Central Federal Lands Program, Lakewood, Colorado, USA.

ARC Solutions. 2012. ARC Solutions. www.arc-solutions.org. Accessed 9 August 2014.

Ashley, E. P., A. Kosloski, and S. A. Petrie. 2007. Incidence of intentional vehicle-reptile collisions. Human Dimensions of Wildlife 12:137–143.

Au, J., P. Bagchi, B. Chen, R. Martinez, S. A. Dudley, and G. J. Sorger. 2000. Methodology for public monitoring of total coliforms, Escherichia coli and toxicity in waterways by Canadian high school students. Journal of Environmental Management 58:213–230.

Bromenshenk, J., and E. Preston. 1986. Public participation in environmental monitoring: a means of attaining network capability. Environmental Monitoring and Assessment 6:35–47.

Clevenger, A. P., and M. P. Huijser. 2011. Wildlife crossing structure handbook: design and evaluation in North America. Report No. FHWA-CFL/TD-11-003. Federal Highway Administration, Washington, DC, USA.

Clevenger, A. P., and N. Waltho. 2000. Factors influencing the effectiveness of wildlife underpasses in Banff National Park, Alberta, Canada. Conservation Biology 14:47–56.

Clevenger, A. P., and N. Waltho. 2003. Long-term, year-round monitoring of wildlife crossing structure and the importance of temporal and spatial variability in performance studies. Pages 293–302 in Proceedings of the 2003 international conference on ecology and transportation. C. L. Irwin, P. Garrett, and K. P. McDermott, editors. Center for Transportation and the Environment, North Carolina State University, Raleigh, North Carolina, USA.

Cohn, J. P. 2008. Citizen science: can volunteers do real research? BioScience 58:192–197.

Cooper, C. B., J. Dickinson, T. Phillips, and R. Bonney. 2007. Citizen science as a tool for conservation in residential ecosystems. Ecology and Society 12:11.

Darwall, W.R.T., and N. K. Dulvy. 1996. An evaluation of the suitability of non-specialist volunteer researchers for coral reef fish surveys, Mafia Island, Tanzania—a case study. Biological Conservation 78:223–231.

Dodd, N. L., J. W. Gagnon, S. Boe, K. Ogren, and R. E. Schweinsburg. 2009. Effectiveness of wildlife underpasses in minimizing wildlife-vehicle collisions and promoting wildlife permeability across highways: Arizona Route 260. Final project Report 603. Arizona Transportation Research Center, Arizona Department of Transportation, Phoenix, Arizona, USA.

Droege S. 2007. Just because you paid them doesn't mean their data are better. Pages 13–26 in Citizen science toolkit conference. C. McEver, R. Bonney, J. Dickinson, S. Kelling, K. Rosenberg, and J. Shirk, editors. Cornell Laboratory of Ornithology, Ithaca, New York, USA.

Engel, S. R., and J. R. Voshell Jr. 2002. Volunteer biological monitoring: can it accurately assess the ecological condition of streams? American Entomologist 48:164–177.

Fenderson, L. E., A. I. Kovach, J. A. Litvaitis, K. M. O'Brien, K. M. Boland, and W. J. Jakubas. 2014. A multiscale analysis of gene flow for the New England cottontail, an imperiled habitat specialist in a fragmented landscape. Ecology and Evolution 4:1853–1875.

Fink, D., W. M. Hochachka, M. J. Iliff, C. L. Wood, B. L. Sullivan, K. V. Rosenberg, M. A. Munson, and M. Riedewald. 2009. Obtaining new insights for biodiversity conservation from broad-scale citizen science data. Nature Proceedings. doi:10.1038/npre.2009.3967.1.

Genet, K. S., and L. G. Sargent. 2003. Evaluation of methods and data quality from a volunteer-based amphibian call survey. Wildlife Society Bulletin 31:703–714.

Glista, D. J., T. L. DeVault, and J. A. DeWoody. 2008. Vertebrate road mortality predominantly impacts amphibians. Herpetological Conservation and Biology 3:77–87.

Gouveia, C., A. Fonseca, A. Camara, and F. Ferreira. 2004. Promoting the use of environmental data collected by concerned citizens through information and communication technologies. Journal of Environmental Management 71:135–154.

Gunson, K. E., F. W. Schueler, and J. Middleton. 2012. Placement of species at risk crossing signs. Final Report. Ontario Ministry of Transportation, Toronto, Canada.

Heiman, M. K. 1997. Science by the people: grassroots environmental monitoring and the debate over scientific expertise. Journal of Planning Education and Research 16:291–299.

Huijser, M. P., A. V. Kociolek, P. McGowen, A. Hardy, A. P. Clevenger, and R. Ament. 2007. Wildlife-vehicle collision and crossing mitigation measures: a toolbox for the Montana Department of Transportation. Final Report submitted. Western Transportation Institute, Bozeman, Montana, USA.

Huijser, M. P., A. V. Kociolek, L. Oechsli, and D. E. Galarus. 2008. Wildlife data collection and potential highway mitigation along State Highway 75, Blaine County, Idaho. Board of Blaine County Commissioners, Hailey, Idaho, USA.

Joyce, T. L., and S. P. Mahoney. 2001. Spatial and temporal distributions of moose-vehicle collisions in Newfoundland. Wildlife Society Bulletin 29:281–291.

Kintsch, J., and P. C. Cramer. 2011. Permeability of existing structures for terrestrial wildlife: a passage assessment

system. Research Report No. WA-RD 777.1. Washington Department of Transportation, Olympia, Washington, USA.

Krasny, M., and R. Bonney. 2005. Environmental education through citizen science and participatory action research. Pages 291–318 in Environmental education or advocacy: Perspectives of ecology and education in environmental education. E. A. Johnson and M. J. Mapin, editors. Cambridge University Press, Cambridge, UK.

Langton, T.E.S., editor. 1989. Amphibians and roads: proceedings of the toad tunnel conference. ACO Polymer Products, Shefford, United Kingdom.

Lee, T., M. Quinn, and D. Duke. 2010. Road watch in the pass: web-based citizen involvement in wildlife data collection. Pages 95–101 in Proceedings of the 2009 international conference on ecology and transportation. P. J. Wagner, D. Nelson, and E. Murray, editors. Center for Transportation and the Environment, North Carolina State University, Raleigh, North Carolina, USA.

Lee, Y. 2012. Draft conceptual framework of recommendations for monitoring amphibians and reptiles using non-calling surveys and volunteers. Final Report for US Fish and Wildlife Service Competitive State Wildlife Grant U-3-R-1 (MI-U-14-R-1), Amphibian and Reptile Conservation Needs. Michigan Natural Features Inventory, Lansing, Michigan, USA.

Manfredo, M., T. Teel, and A. Bright. 2003. Why are public values toward wildlife changing? Human Dimensions of Wildlife 8:287–306.

Moskowitz, D., M. Clarke, R. Ashton, and S. Kachel. 2007. Cascade wildlife monitoring project: winter 2006–2007 field season report. Unpublished Report. Wilderness Aware School, Duvall, Washington, USA.

Ng, S. J., J. W. Dole, R. M. Sauvajot, S.P.D. Riley, and T. J. Valone. 2004. Use of highway undercrossings by wildlife in southern California. Biological Conservation 115:499–507.

Ontario Road Ecology Group. 2010. A guide to road ecology in Ontario. Toronto Zoo, Ontario, Canada.

Paul, K. S. 2007. Auditing a monitoring program: can citizen science document wildlife activity along highways? Master's thesis, University of Montana, Missoula, Montana, USA.

Price, S. J., and M. E. Dorcas. 2011. The Carolina Herp Atlas: an online citizen science approach to document amphibian and reptile occurrences. Herpetological Conservation and Biology 6:287–296.

Przybyl, J. 2003. Management model and training program for Sky Island Alliance's volunteer-based wildlife monitoring program. Master's thesis, Prescott College, Prescott, Arizona, USA.

Reed, H. 2008. Evaluating turtle road mortality mitigation: identifying knowledge gaps and public attitudes. Undergraduate honor's thesis, Dalhousie University, Halifax, Nova Scotia.

Sauer, J. R., J. E. Hines, G. Gough, I. Thomas, and B. G. Peterjohn. 1997. The North American breeding bird survey results and analysis. Version 96.4. Patuxent Wildlife Research Center, Laurel, Maryland, USA.

Savan, B., A. J. Morgan, and C. Gore. 2003. Volunteer environmental monitoring and the role of universities: the case of Citizens' Environment Watch. Environmental Management 31:561–568.

Singer, P., A. Huyett, and J. Kintsch. 2011. Eco-Logical monitoring and I-70 Wildlife Watch report. Unpublished Report. Rocky Mountain Wild, ECO-resolutions, Colorado Department of Transportation, Denver, Colorado, USA.

Stokes, P., M. Havas, and T. Brydges. 1990. Public participation and volunteer help in monitoring programs: an assessment. Environmental Monitoring and Assessment 15:225–229.

Sullivan, B. L., C. L. Wood, M. J. Iliff, R. E. Bonney, D. Fink, and S. Kelling. 2009. eBird: a citizen-based bird observation network in the biological sciences. Biological Conservation 142:2282–2292.

Trumbull, D. J., R. Bonney, D. Bascom, and A. Cabral. 2000. Thinking scientifically during participation in a citizen-science project. Science Education 84:265–275.

6

STEPHEN TONJES, SANDRA L.
JACOBSON, RAYMOND M.
SAUVAJOT, KARI E. GUNSON,
KEVIN MOODY, AND
DANIEL J. SMITH

The Current Planning and Design Process

There are two states to any large project—
too early to tell and too late to stop.
—Ernest Fitzgerald 2010

6.1. Introduction

Road projects are like trains—slow to get rolling, averse to getting off track, and hard to slow down. As a result, the best opportunities to incorporate wildlife accommodations into a project are in the early stages of project development, where deadlines are more flexible and the larger issues of road corridor impacts on regional landscapes are usually considered. However, people concerned about the impacts of road projects are often frustrated trying to learn exactly what is being proposed, and how they can have their concerns addressed. Project managers are under pressure to deliver their projects on time and within budget, and the purpose and need of transportation projects are almost always defined solely in terms of the transportation goals. Hence, integrating ecological infrastructure into road planning, design, and operations requires a basic appreciation of the decision-making processes and administrative procedures that transportation agencies use.

Even if project managers are sympathetic to the impacts of roads on wildlife, they must be able to justify spending time and transportation money on elements that do not contribute directly to the transportation goals; therefore, it is helpful to ensure they are aware of environmental laws and public interests related to accommodating wildlife. Chapter 5 discusses various ways to generate public understanding of, and support

for, wildlife issues. The road project development process is structured to obtain public input, and provides formal and informal opportunities for the public and resource professionals to offer education and demonstrate support for wildlife concerns; however, public input will have much more influence on project outcomes if it is provided to the right people at the right time.

Road projects usually proceed through distinct phases, distinguished by the kinds of information gathered and the professional disciplines involved. In the United States, development of state and federal transportation systems consists of three major phases: (1) preliminary planning; (2) project delivery (preliminary engineering / conceptual design and detailed design); and (3) construction, maintenance, and monitoring (discussed in Chapters 11 and 12). This chapter describes the first two phases and highlights the points at which people concerned about wildlife can most effectively engage with these efforts. The phases may have different names in different agencies and different countries, and may be combined or shortened depending on the complexity of the project, but most projects follow a similar sequence. Public input is most effective when it is addressed to the aspects of the project under consideration at the time.

6.2. Initial Considerations in Planning and Design

In the preliminary planning phase for federal and state highways in the United States, current and anticipated deficiencies are identified in the regional road network, and individual projects are proposed to correct them. This phase is usually conducted by professional planners working in the state departments of transportation (DOTs) and in regional transportation planning organizations, as described in 6.2.1. To most effectively integrate ecological infrastructure into road planning and design, any proposed mitigation measures should be well supported by documents, such as:

- scientific data,
- site-specific photos, and
- shared management goals and objectives among the agency and other relevant stakeholders.

When the above information is provided to decision makers, the chances are greater that measures will be adopted into the project plan. As part of this documentation and justification, information necessary to weigh trade-offs includes (1) what would result from no action (i.e., no mitigation measures implemented) versus what would result from implementation of preferred alternatives, including the ideal one; (2) what uncertainties exist; and (3) potential ways to assess effectiveness (i.e., performance measures). Throughout this process, the level and type of detail in the documentation and justification materials provided must be in line with the level of detail that the target audience can meaningfully integrate.

Decisions to integrate ecological infrastructure with transportation infrastructure should start with a discussion of the natural resources present in the transportation corridors. Ideally, the discussion will also include information about community areas (e.g., either a state or locally designated bird sanctuary or a wildlife viewing area would be considered a community value) and notable features (e.g., a natural wetland or forested area). Being equipped with this information, particularly if the proposed measure to maintain ecological infrastructure also connects community areas and notable features, can support the justification for the proposal offered to decision makers. The justification is further strengthened when community areas or notable features are associated with performance indicators (e.g., increases in wildlife diversity or maintenance of fish or wildlife populations as a result of the proposed mitigation measure).

6.2.1. Allocation of Funding

Federal transportation funding originates from federal gas tax and general fund tax dollars. These funds support the Federal-aid Highway Program (FAHP), and the Federal Lands Highway Program (FLHP). Summarized simply, the FAHP distributes tax dollars for the construction, operation, and maintenance of state highway systems, including interstates, primary highways, and secondary local roads (http://www.fhwa.dot.gov/federal-aidessentials/federalaid.cfm), and the FLHP provides funding for public roads on federally managed lands (http://flh.fhwa.dot.gov/).

FAHP funds are distributed to states according to a formula established by Congress. For large metropolitan areas, states are required to establish regional transportation planning organizations, which are intended to provide state DOTs with a balanced assessment of regional transportation needs and avoid arbitrary or inequitable distribution of funds. These regional organizations are known variously as metropolitan planning organizations, transportation planning organizations, transportation planning authorities, transportation councils, or councils of governments, among others. They are comprised of elected officials from the municipalities in the region and are supported by staff and by citizen and technical advisory committees. These planning organizations develop long range (20 year) transportation plans (LRTPs), and short range (5 year) transportation improvement programs (TIPs), both of which establish the region's transportation priorities. The state DOTs generally follow the priorities in these plans to develop their own long-range and short-range plans (work programs).

These plans are updated annually by adding new projects and adjusting priorities of projects from the previous plan. New projects usually are added into the later years of the plan, which gives the agencies time to adjust their workloads. Sometimes, however, under political pressure, projects slated for later years may be initiated sooner, and new projects also may be crowded into earlier years, making it more difficult for agencies to give all of these projects the full time and resources they need.

6.2.2. Integrating Wildlife and Natural Resources Information during Planning

Wildlife and ecology professionals should participate in the development of these large-scale, proactive planning efforts. Planners need to know how the road network will interact with the surrounding ecological landscape. Interested citizens can work directly with the elected officials who are members of the regional transportation planning organization, or participate on a citizen advisory committee, to ensure that the regional transportation plan takes ecological infrastructure into account when roads are initially planned. Unfortunately, few people interested in environmental issues are even aware of the existence of the regional transportation planning agencies and the opportunities they offer.

In addition to public input, the early planning phase should include consultation with environmental resource agencies, such as the US Fish and Wildlife Service (USFWS) and state wildlife agencies, and should take into account the many sources of environmental resource information available in most states (e.g., state wildlife action plans, multispecies habitat conservation plans, natural heritage databases, land use plans, green infrastructure planning, watershed plans, historic resource inventories, special area management plans, GIS maps of resource corridors). But there is always political pressure to shorten the transportation development process, and early review by resource agencies is often a step that is compromised.

In 2005, the federal transportation funding bill mandated that states develop a formal process to be approved by the Federal Highway Administration (FHWA) in order to incorporate consultation with environmental resource agencies during early transportation planning. In response, the state of Florida developed a plan that was essentially a waiver based on their previously established Efficient Transportation Decision Making (ETDM) process. ETDM gathers and displays the best information available about the area that would be affected by a proposed road project. This information is distributed to resource agencies for formal comment and is also available to the public. Florida initiated the ETDM process because in the absence of early resource reviews, time and money were being invested in detailed development of projects that later became bogged down when controversial environmental impacts were identified. However, even this process is sometimes rushed by local pressure to complete a particular project; therefore, even when tools are available, it is important for citizens to insist on their effective use.

6.3. Project Delivery

After a project has been scheduled in the planning phase and adopted into a state DOT work program, it is assigned to internal or external consulting engineers for design. In most state DOTs, design occurs in two separate phases. The first stage of the project delivery phase is called preliminary engineering, or conceptual design; the second is called detailed design. In preliminary engineering, the scope of needed improvements and the way they will fit into the landscape are determined based largely on readily available information, such as topography, soil maps, and original plans (if the improvements are to an existing road). Conceptual plans are produced that show the general dimensions and locations of the proposed improvements. This preliminary engineering stage is where the research and documents required under the National Environmental Policy Act (NEPA) are generated (Section 6.3.1). In detailed design, extensive elevation and location surveys, soil testing, and other site data are obtained and made available to design project team members in several different engineering specialties (e.g., road, drainage, structures, and pavement). These engineers then design their respective elements and fit them together into a set of detailed plans that can be used for construction. Detailed design is covered in Section 6.7.

6.3.1. Preliminary Engineering

All actions, approvals, and funding issuing from federal agencies are governed by NEPA. This legislation established a formal process to evaluate all the environmental impacts of federal or federally funded actions, and to disclose the results of its evaluation to the public. Whenever a federal agency proposes to fund, approve, or build a road project, it must undertake a study and prepare a document in which it develops alternatives that balance the traffic benefits, costs, and impacts to the environment (both the human and natural resource environment) of the proposed project. It is actually the state DOT that designs the road and prepares the document, but the project must be developed according to explicit federal rules and the document must be approved and signed by FHWA as satisfying the requirements of NEPA.

Depending on the significance of the expected impacts, the primary NEPA document may be (1) an environmental impact statement (EIS); (2) an environmental assessment (EA), leading either to an EIS or to a finding of no significant impact (FONSI); or (3) a type II, type I, or programmatic categorical exclusion (CE), in descending order of complexity (Box 6.1). The process of assigning the class of action to a project is not always consistent among or even within agencies. Building a major highway in a new location generally requires an EIS, and upgrading traffic signals on an existing road generally is categorically excluded, but for projects of an intermediate nature, establishing the significance of impacts has been debated since NEPA was enacted. Therefore, a project that would be developed with an EA in one jurisdiction may go straight to an EIS in another. Nevertheless, the process of evaluating project alternatives for environmental impacts is the same, and the difference in the documents is only in the degree of detail and extent of review by other agencies and the public.

NEPA has produced many benefits, including the requirement for the agencies to consider a much wider range of factors than cost and engineering. It also acts as an "umbrella" to ensure compliance with a number of laws protecting many different kinds of resources. A similar process has been adopted in many states for projects that have no federal involvement (e.g., California Environmental Quality Act [CEQA] in California, and State Environmental Impact Report [SEIR] in Florida).

The opportunity for stakeholders to provide input is a unique aspect of NEPA. The EIS is the only document for which a formal public hearing is required by NEPA, but many state DOTs include public involvement, with or without formal hearings, for all projects at or above the level of a type II CE to identify potential controversy that could delay the project later. However, large road projects are very complex and before a project can be formally presented to the public, considerable effort and money are usually expended in developing and evaluating proposed alternatives, and in preliminary coordination with resource and permitting agencies. The public hearing is the official opportunity to comment on the adequacy of the alternatives analysis and bring to light impacts and public sentiment that might not have been identified or given sufficient attention. Projects have indeed been redirected as a result of input from the public hearing provided by NEPA. However, any substantial changes at this stage have the

> **Box 6.1. Key Terms from the National Environmental Protection Act (NEPA)**
>
> *Environmental Impact Statement (EIS)*
> A full disclosure document that details the process through which a federal project was developed (e.g., actions planned and decisions made), includes consideration of a range of reasonable alternatives, analyzes the potential impacts resulting from the alternatives, and demonstrates compliance with other applicable environmental laws and executive orders (adapted from 40 CFR 1502.1; see also 23 CFR 771.123–125).
>
> *Categorical Exclusion (CE)*
> A category of [federal] actions that do not individually or cumulatively have a significant effect on the human environment ... and ... for which, therefore, neither an environmental assessment nor an environmental impact statement are required (adapted from 40 CFR 1508.4; see also 23 CFR 771.117).

disadvantage of requiring the revision of work in which time and effort have already been invested. Effective avoidance and mitigation of environmental impacts is better assured if the potential impacts have been identified during the planning phase, and if wildlife experts are involved in development of the alternatives before they are presented at the public hearing, as discussed in the next section.

6.3.1.1. Early Involvement with the Project Team during the NEPA Study

A more effective way to influence a project during the NEPA study is to become involved early in the development of the environmental analyses. It has become more common for project teams to hold informal public workshops or information meetings (prior to the official public hearing) early in the process of alternatives analysis. If these opportunities are not offered, then individuals or organizations should contact the road project manager as early as possible and ask how they can be involved.

6.3.1.2. Early Involvement with Wildlife Resource Agencies during the NEPA Study

For wildlife issues, it may be most effective for wildlife experts to get involved in the transportation project through the state and local fish and wildlife divisions and other environmental or natural resource agencies. By law, the project sponsor and lead transportation agencies must invite the participation of relevant agencies (including wildlife and natural resource agencies) during the development of various portions of the reports (http://www.fhwa.dot.gov/hep/section6002/1.htm). External stakeholder input, particularly when supported by data, can be a valuable supplement to that of the state fish and wildlife agencies.

Reviewers at the agencies are likely to welcome input from specialists in road ecology and experts on the species that will be affected by the project. The biologists who present the transportation agency's assessment of wildlife impacts and the biologists who are consulted for comment at the resource agencies come from a variety of backgrounds and specialties and may or may not be current with the rapidly evolving field of road ecology. Therefore, not all biologists may understand how to translate the biology of organisms to the need for accommodations or the specific requirements of the target species in relation to engineering and design specifications. Furthermore, as discussed below, not all biologists are likely to be familiar enough with the engineering process in the detailed design stage (Section 6.7), which follows the NEPA study, to be able to suggest commitments in the NEPA documents that are specific enough to ensure that wildlife accommodations will be designed to function as intended.

Specialized biological impact analyses and other supporting technical documents (e.g., noise study, cultural resources assessment, contamination assessment) are produced in support of the primary NEPA document (EIS, EA/FONSI, or CE). These technical documents are reviewed by specialists on staff at FHWA along with the primary document, but only the primary NEPA document is routinely consulted by the state DOT engineers in the subsequent design phase. Therefore, any recommendations made in the biological impact analyses must be translated into specific commitments, or agreed-upon actions, in the primary NEPA document. In making this translation, ecologists and engineers must be aware of important differences in the approaches of their respective sciences, which are discussed in more detail in Section 6.7.

For example, when wildlife scientists recommend crossing structures, they may assume that fencing to guide the animals to the structures will be included as part of a good design, but most engineers in the design phase will be surprised at this "extra" request if fencing has not explicitly been specified in the NEPA documents. There are also misconceptions about where wildlife fencing may be installed. Sometimes a wildlife corridor occurs on both public and private land. If the private land use will continue to allow wildlife movement, some states have routinely installed fencing in the road right-of-way adjacent to both the public and private land, coordinating with the land owner to maintain access with gates or other devices. In other states, engineers are reluctant to install fencing adjacent to any private land, even next to physical features already precluding access (S. Tonjes, Florida DOT, personal observation).

Design errors that compromise the functioning of wildlife accommodations are common (Sections 6.8 and Section 1.3.4). The final NEPA decision document should identify the target species, and the conceptual design should include the locations, types, and dimensions of the wildlife crossing structures, any special characteristics of the substrate and landscaping leading up to the structures, and locations and types of fencing. These become "commitments" that are similar to contractual agreements. The commitments should require continued involvement in, and review of, both concepts and detailed plans by wildlife experts during the design phase, and during pre- and post-construction monitoring and adaptive management as described in Chapter 12 (Section 12.2.2).

6.3.1.3. Case Study: Proactive Planning and Interagency Collaboration

Stephen Tonjes and Daniel J. Smith

State Road (SR) 40 runs east and west through the middle of Ocala National Forest (ONF) in central Florida and connects major cities in the eastern and central parts of the state, as well as two interstates. Established in 1908 as one of the first national forests east of the Mississippi River, ONF protects many significant archaeological, historical, geological, and botanical resources, including the world's largest contiguous sand pine scrub forest. It supports a diverse wildlife community, including the threatened Florida scrub jay (*Aphelocoma coerulescens*), Florida sand skink (*Plestiodon reynoldsi*), and wide-ranging species such as the Florida black bear (*Ursus americanus floridanus*) and the threatened eastern indigo snake (*Drymarchon couperi*).

To accommodate projected traffic volumes, the closest alternative east-west connectors outside ONF would require major upgrades with their own environmental impacts and would substantially lengthen the travel distances between the cities now served by SR 40. The Florida Department of Transportation (FDOT) began the preliminary engineering process to add lanes to SR 40 three times between 1988 and 1998. However, several state and federal resource agencies had identified the Greater Ocala National Forest Area as being especially susceptible to the negative consequences of fragmentation and habitat loss as a result of the widening of SR 40. Each attempt led to a stalemate over the impact of widening a road that was already perceived as splitting ONF in half.

In 2003, FDOT finally convened a task force consisting of all stakeholders who had an interest in the road, and who wanted ONF to take a broader perspective and discuss what alternatives might be feasible to improve transportation and to avoid or mitigate the wildlife impacts of the road. The task force meetings were professionally facilitated and aerial photography and Geographic Information Systems (GIS) data were used to develop consensus among the stakeholders concerning the general locations and priority of all wildlife crossings, and concerning other measures that would be needed to satisfactorily mitigate effects of widening the road. These conclusions were documented and used as the starting point of another preliminary engineering NEPA study in 2006.

The preliminary engineering study continued to involve the stakeholders in several advisory committees, including a Wildlife Crossings Task Force that met frequently to decide on the sizes and locations of the crossing structures. A commitment was made in the NEPA document to reconvene this task force during design to review the precise placement of the structures and other parameters (e.g., substrates, lighting, landscaping, and interior cover).

Because of these commitments, the engineering firms that were selected for the design phase had added in-house or subcontracted wildlife ecologists to their usual design teams; these scientists worked with the engineers from the beginning. They collaborated early and met in person to discuss concepts before time and money was spent developing finished plans. Further, the teams looked beyond the usual boundaries of their own project phase by developing a maintenance protocol for each crossing to ensure that important features such as substrates, interior cover, fencing, and landscaping, which could vary among crossings and change over time, would be maintained properly. The advantages of including wildlife experts early in the detailed design process are discussed in more detail in Section 6.7. The benefits of establishing specific maintenance protocols are discussed in Chapter 11.

This case study is an example of effective collaboration between a transportation agency and a range of stakeholders to develop a major road-widening project through a sensitive ecosystem. It illustrates the value of the following:

- Stakeholder and wildlife expert participation in the planning and preliminary engineering phases. Large, complex projects have a much greater chance of successful outcomes if they are planned and designed under the guidance of a formal process to engage all stakeholders, collect and analyze all relevant information, facilitate and document decisions, and monitor implementation and results. (See Section 6.4.2 for examples of formal models to guide early stakeholder collaboration.)
- Strong and specific commitments in the NEPA documents.
- Participation and review by wildlife experts in the detailed design phase. The need for collaboration with wildlife experts in the latter stages of project development is not as well recognized as the need to do so in the early stages, but it is just as important, as illustrated in Section 6.7.

6.4. Public Lands

When taken as a whole, federally owned lands encompass nearly 25% of the United States and provide habitat for almost two-thirds of all endangered and threatened species (White and Ernst 2004). While public lands are thought to be scenic landscapes and pristine sanctuary for wildlife, they are not exempt from the many impacts of roads (Andrews et al. 2006). The US Bureau of Transportation Statistics (www.transtats.bts .gov, 2004) defines an urban area as "a municipality . . . with a population of 5,000 or more." By this definition, many national parks and wildlife refuges have daily visitation levels equivalent to that of small urban areas and during months of peak visitation have traffic volumes comparable to some cities (www.nps.gov, 2004). Furthermore, an estimated 10% of all roads in the United States occur in national forests (~611,420 km, Forman 2000), a length that could encircle the earth approximately 15 times. In an analysis of road fragmentation in national parks by Schonewald-Cox and Buechner (1992), even the largest parks (up to 9,000 km^2) encompass few areas that lie more than 10 km from roads. Additionally, 30% of the land area within highly fragmented parks is within 1 km of a road, and all land parcels comprising 100 km^2 were within 500 m of roads.

6.4.1. Environmentally Sensitive Areas on Public Lands—Roads in National Parks

Most federal agencies that manage public land also have legal mandates and policies that delineate procedures for allowing significant developments on their land, including roads. Thus, in addition to the standard federal requirements of NEPA and other broadly applicable federal laws (e.g., Endangered Species Act [ESA], Clean Water Act [CWA]), specific agency mandates and policies can have significant influence on if, when, where, and how roads are developed. Understanding these policies and procedures, and participating early in the planning and development process, can result in important design improvements for roads in environmentally sensitive areas on public lands.

From the early years of the national park system and particularly during the 1950s and 1960s, facilitating access to national parks was an important priority and a driving factor in the creation of improved visitor facilities and infrastructure. Park visitation expanded rapidly as more campgrounds, lodges, hotels, visitor centers, and other services became available. Through this development, roads became the key arteries for bringing people into the parks, and the design of roads was even seen as a way to enhance visitor experience.

National park roads therefore not only facilitated access, but also contributed to the park experience due to stunning and dramatic roads developed at sites of impressive vistas and other notable natural features. In general, however, little knowledge or consideration was given to the potential wildlife impacts of roads and their role in fragmenting habitat.

Bumper-to-bumper traffic is now a common situation in many national parks. Most federal parks have a low speed limit of about 45 mph (70 kph), and because people are typically looking around for wildlife, traffic jams caused by animals near the roads are common. This fairly slow speed and attention to wildlife does not guarantee a lack of vehicle-caused mortality, and for small animals that are essentially invisible to drivers, the mortality impact can be comparable with other roads in urban areas. The high traffic volume of many parks can lead to a complete barrier for many small species (anurans, Hels and Buchwald 2001; snakes, Andrews and Gibbons 2005; turtles, Aresco 2005).

The National Park Service (NPS) is part of the US Department of the Interior and provides an example of how agency specific policies can influence road development and provide opportunities to reduce impacts on wildlife. Nearly 400 units are managed by the NPS in the United States, including well-known sites (e.g., Yellowstone and Yosemite National Parks) as well as those lesser known (e.g., Lewis and Clark National Historical Park). Overall, the NPS owns and operates close to 8,900 km (5,500 mi) of paved public roads and over 1,400 associated structures such as bridges, culverts, and tunnels (NPS and FHWA 2008). Funds to develop, maintain, and improve park roads come from many sources, including federal appropriations from transportation programs, NPS facility maintenance funds, and even from private and nonprofit donations. In the last 20 to 30 years, NPS road developments have received NEPA reviews consistent with other federal agencies, and similarly were required to abide by other federal environmental laws and regulations. However, as an agency focused on resource preservation with a clear mandate to protect those values in perpetuity, additional standards and expectations have become incorporated into the planning and design expectations for NPS roads. In general, infrastructure priorities for individual park units are articulated in their General

Management Plans and in site-specific designs, such as Development Concept Plans for visitor use areas or park access routes. The mission of the agency becomes a key driving factor in influencing these planning documents and development priorities, and activities that have the potential to permanently impair resource values are prohibited by law and policy.

As knowledge of road ecology and the threats posed by habitat fragmentation have advanced, NPS road planning and design have become more responsive to reducing impacts on wildlife. For example, transportation planners at the NPS Denver Service Center, which provides design and engineering support for national parks, have been compiling road and wildlife mitigation materials when road projects are contemplated or envisioned. Consequently, new opportunities to influence road developments in parks are advancing, motivated less by government-wide environmental regulations (e.g., NEPA) and more by the mission, policies, and mandates of the agency. Because of this, commitments to wildlife-oriented road designs can be incorporated at earlier levels of infrastructure design and planning, from the broadest management policies of the agency at the national level, to park-specific General Management Plans and location-specific development plans and designs.

To help ensure that heightened levels of wildlife sensitivity are included in park roadway plans, designs, and construction, the NPS developed a reference manual in collaboration with the FHWA. The manual articulates environmental review procedures, mission goals, and policies of the NPS so that road development projects under the FLHP (Section 6.2.1) proceed in a manner consistent with NPS-specific requirements (NPS and FHWA 2008). In addition to providing substantial program administration details, the reference manual lays out communication expectations to promote mutual understanding between NPS and FHWA when working together on road projects in national parks. This approach has far-reaching implications for park roads because most major road projects in the parks are implemented as part of FLHP using federal transportation funds. Indeed, FLHP projects have occurred or are underway in many national parks, from the old growth forests of Mount Rainier to the deserts of Death Valley, with formal agreements in place that combine NPS-specific resource preservation priorities with park visitor transportation needs. Individual projects implemented in this way are also subject to review by stakeholders and the public through the NEPA process, so significant opportunities are available to incorporate wildlife accommodations with FHWA and NPS.

6.4.2. Public Highways through Public Lands

A congressional report on wildlife-vehicle collisions found that rural, two-lane highways with speed limits of 55 mph (90 kph) or greater have the highest rate of wildlife-vehicle collisions (Huijser, McGowen, Fuller, et al. 2008). Although this finding was based on large rather than small animals, these roads probably constitute a formidable threat to small animals as well. Rural highways of this description are the most common roads on public lands. These roads traverse the best remaining wildlife habitat in the nation, often in large core tracts. As described above for the NPS, agency-specific missions, mandates, and goals can supplement NEPA requirements to help promote road designs that reduce impacts to wildlife.

Planning for wildlife features on highways through public lands, however, is complicated by the need to know whose jurisdiction covers the highway. Often it can be federal, state, and local, with different authorities for different aspects of planning, construction, and maintenance. In areas of the rural west, the US Forest Service (USFS) and US Bureau of Land Management (BLM) manage vast acreages of contiguous lands, so working on a highway development project may have fewer stakeholders than highways across federal lands in the eastern states where private inholdings are common. Highway projects through public lands are usually led by the state DOT, and the federal land management agencies take a supportive role. Although it is becoming more common for land management agencies to take a more assertive approach toward highway development on their land, it is highly inconsistent across the country and some state and federal agencies have yet to understand the issues and available mitigation measures for wildlife.

All of the federal land and resource management agencies must follow federal laws, including NEPA, during highway development projects. However, different agencies have different regulations for implementing

Box 6.2. The Rapid Assessment Process
Daniel J. Smith and William Ruediger

If cost and time are significant limiting factors to conducting an in-depth ecological planning assessment, a Rapid Assessment Process (RAP) can be a practical and effective approach (see www.rapidassessment.net for more information). In many instances, particularly when early engagement did not occur, the timeline available for conducting more extensive (multi-year or replicated) ecological assessments is insufficient. Here, the choice is to neglect the ecological planning aspect of the project or to employ a more rapid, yet still integrated, assessment. To understand the expected outcomes from a RAP, it is important to recognize the different perspectives of the stakeholders (e.g., project managers, engineers, scientists, regulatory agencies, nongovernmental organizations [NGOs]) and their knowledge of the elements involved in the problem and evaluation of effective and feasible solutions. For example, a project manager might be focused on types, quantities, and the quality of information needed to properly guide an assessment of linkages, structures, or other mitigation measures. Transportation engineers will seek what information is available or critical for decision making and will be acutely aware of cost elements and

the practicality of which design structures will work best with the immediate landscape. Biologists can assist with gathering existing information, informing stakeholders, and providing recommendations for additional ecological assessments. All stakeholders need to focus on how to minimize the amount of funds spent on (a) data collection and assessment, and (b) structures or other mitigation measures without compromising the integrity and objective of the project.

Given the varying backgrounds of stakeholders, it is important that all of these perspectives convene for the RAP. Specifically, a comprehensive list of what relevant information exists should be compiled with the project manager. This information should then be used as guidance in assessing what and how much information is needed given the available time and financial resources. For example, a valid need for new information could include confirming the presence of target species or groups, or collecting data on hydrology that would affect design specifications of road infrastructure features. In other words, new information is considered necessary if it directly influences the timing or design of the construction project.

NEPA. This difference in approach and pace can lead to misunderstandings whereby land management and transportation agencies, with quite different missions, and personnel with different academic training, find their agencies at odds. For example, the USFS has an agency-specific list of "sensitive" species that must be analyzed in its environmental documents; consultants who have worked for state transportation agencies but not on projects on national forest lands may be surprised to find they have an additional list of species to assess in a schedule that does not allocate the time to do an adequate biological evaluation.

To be effective, stakeholders and the transportation agency must be informed about the resource conservation policies of the agency, understand the

infrastructure planning and decision-making process, and participate in the process with this knowledge in mind in order to ensure that road plans and designs reflect commitments to wildlife conservation. The project development process should include regular meetings with consistent points of contact and careful documentation of decisions. A formal model to guide collaboration among stakeholders involved with transportation projects is provided below as an example, and there are also others available (Brown 2006; CEE 2009; CRAFT 2012). By engaging at the earliest conceptual stages of infrastructure planning (e.g., the General Management Plan stage for NPS) when project-specific design features such as wildlife crossing structures are proposed (e.g., as part of an EIS process), the value of such measures will already be established as important and consistent with agency policies and priorities.

6.5. County, Municipal, Industrial, Commercial, or Residential Roads

Local roads can have substantial impacts on wildlife. Many local roads have traffic volumes comparable to major state and federal highways, and even roads with low volumes have been documented as a formidable threat for smaller and slower animals (Hels and Buchwald 2001; Chapters 3 and 4; Andrews et al. 2008).

6.5.1. Development Process

Local public and private roads are subject to most of the same laws protecting natural resources, such as the Endangered Species Act, that govern state and federal highways. However, since NEPA applies only to federal actions, there is usually no comparable umbrella environmental process that comprehensively monitors for compliance and provides opportunities for public input. An important exception is a local road that receives federal funding or has another federal nexus, such as US Army Corps of Engineers permits or other impacts on a federal interstate highway. Some larger municipalities in the United States also may have a more comprehensive local process to engage the public and document compliance with environmental laws.

In Canada, the regional and local governments have a process similar to NEPA, known as Class Environmental Assessments (CEA). CEAs are divided into three schedules: A, B, and C. They vary in scale, potential impacts on the environment, and level of consultation required. Schedule A projects are typically small-scale routine projects, such as maintenance activities on roads. Schedule B projects are required to undertake a screening process that involves mandatory contact with the public, individuals, and entities who are directly affected; relevant government agencies are also contacted to ensure that all are aware of the project and that their concerns are addressed. An example of a schedule B project is improvements to an existing road, such as addition of culverts. Schedule C projects have the potential for significant environmental effects and must proceed under the full planning and documentation procedures specified in the Class EA document, which includes the development of an Environmental Study Report. Road extensions and upgrades are generally Schedule C projects.

6.5.2. Incorporating Wildlife Concerns into Local Projects

For roads not subject to NEPA or a similar process, concerned citizens or environmental groups who want to integrate wildlife concerns with road plans must first learn about the local procedure for planning and designing roads, along with agencies that have review and permitting authority. For example, some counties and large municipalities have their own long range transportation plans that are periodically advertised for public comment. An interested citizen or agency could identify potential points of conflicts where planned county roads might affect wildlife resources; they could do this either when the transportation plan is advertised, or, even better, in early conversations with staff in the county planning department. This opportunity is similar to that available for federally and state funded projects through the regional planning organizations described in Section 6.2.1. However, when county projects begin the engineering phase, there is no formal opportunity similar to NEPA for public review of individual project designs. As discussed below, this process requires finding and meeting with appropriate staff members.

As with the NEPA process, earlier is better. Most municipalities have a planner or a planning department with staff who should be able to explain how public road improvements are planned, how private roads are permitted, and where the opportunities are to address environmental concerns. Some municipalities may be more progressive, and may have an existing environmental review process or may be more open to addressing ways to maintain ecological infrastructure. Opportunities do not have to be formal. In fact, it may be more effective to work with a sympathetic staff person who will help navigate the planning phase and then perhaps identify a helpful contact in the road engineering department when the project is ready for design.

6.5.3. The Role of Resource Agencies

Like state and federal highways, any local road project that affects wetlands will require permits, and wildlife habitat is an important consideration in wetland permit reviews. The US Army Corps of Engineers (COE) reviews projects with impacts to wetlands that fall under their jurisdiction, and in fact must follow the NEPA process in order to issue its federal wetland permit. Many states, some counties, and other municipalities may also have their own wetland permit requirements or ordinances in addition to COE compliance requirements. The COE and the state wetland permitting agencies issue public notices for individual permit applications. These permits are another opportunity for input from wildlife experts, especially if the municipality planning the road has been unresponsive. Sometimes municipal wetland ordinances can also present opportunities for proactive measures.

6.5.4. Case Study: Effective Community Design to Maintain Ecological Infrastructure
David A. Mifsud

The growth and development of the City of Novi located in Oakland County, Michigan, has been driven by environmentally focused planning that protects and reduces impacts to wildlife habitat. The City of Novi was a pioneer in developing local-level regulations and innovative community planning to protect their natural resources. They were one of the first municipalities with a local wetland ordinance (adopted in 1986) and are still one of the few communities in Michigan with a woodland protection ordinance. The City of Novi was also one of the first communities to develop conservation easements on sensitive land, and wetland setback ordinances. Through proactive conservation strategies and incentives for responsible planning that promote road minimization, this community has preserved large tracts of land, reduced wildlife habitat fragmentation, and promoted the desired growth of their community.

Armed with these guidelines and a firm strategy for environmentally sensitive site design, the City of Novi Planning Department and their team of environmental consultants guided a residential development project in 2001 to avoid fragmentation of critical wildlife habitat. The proposed Maybury Park Estates was a 54.63 ha parcel located between high-end subdivisions, single-family homes, active agriculture, and a state park. Prior to development, the site contained portions of active agricultural land and natural areas. During the City's Environmental Review and Site Plan Approval process, the area was identified as an impor-

tant community natural resource because it was part of a large designated wildlife linkage in the municipality and part of a core habitat reserve. Further investigation revealed several high-quality beech, maple, and oak-hickory woodlands and an assortment of wetlands (emergent marsh, open water ponds, and numerous vernal pools) that supported a diverse assemblage of amphibians and reptiles (i.e., herpetofauna). Herpetofauna richness was high for the area, with 17 observed species and 5 additional species likely present based on habitat availability. Herpetofauna observed at this site and in the immediate vicinity included five species of salamanders, nine species of frogs, seven species of snakes, and one species of turtle.

Due to the high quality of portions of the site and the historic analysis conducted by the city identifying this region as part of a critical habitat corridor, measures were taken to minimize negative ecological effects from the proposed development. The project was submitted under the Residential Unit Development category (RUD), one of several ecologically responsible development options offered by the city, which allows the developer to create smaller lot sizes and increase the number of lots available if habitat preservation and impact minimization criteria are fulfilled. This strategy also reduces negative impacts to habitat by encouraging smaller overall development footprints. These and similar planning options encourage developers to minimize impacts and improve wildlife habitat by increasing potential profits. The developer and the City of Novi worked together to capitalize on this mutually beneficial opportunity, resulting in a total of 106 lots on the 54.63 ha site, with approximately 38.8% of the site (20.04 ha) permanently preserved (Figure 6.1).

Habitat fragmentation and mortality resulting from roads bisecting wildlife habitat can significantly affect herpetofauna and other wildlife populations (Chapter 4 and references therein). Roads create permanent barriers and introduce environmental disturbances that can reduce a biological community's diversity, increase genetic isolation, and increase species competition, thereby resulting in potential extirpations (e.g., Fahrig et al. 1995; Findlay and Bourdages 2000; Gibbs and Shriver 2002; Beaudry et al. 2010; Gunson et al. 2012; Sarver 2012).

One of the goals of development planning was to maintain the connectivity and ecosystem integrity,

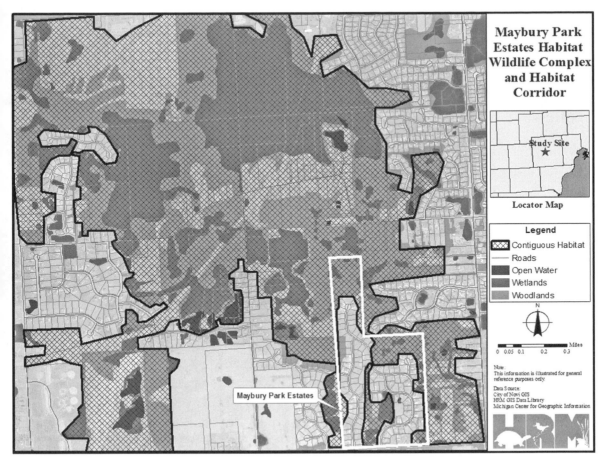

Figure 6.1. Location of Maybury Park Estates in proximity to contiguous protected habitat complex and adjacent residential developments. *Credit: Herpetological Resource and Management, LLC.*

thus maintaining ecological infrastructure. The road network within Maybury Park Estates was configured to avoid the highest-quality habitat in order to meet habitat connectivity goals, while also addressing traffic flow and emergency vehicle access. This alignment allowed for public infrastructure needs to be met without disregarding ecological infrastructure. To further reduce potential fragmentation resulting from road placement, low rolled curbs with gradual slopes, which small frogs and turtles could easily traverse, were used throughout the subdivision (Figure 6.2; Section 9.2.7). Additionally, to help minimize road-related mortality at critical herpetofauna movement corridors, wildlife signs were designed and located to alert residents of locations where wildlife would likely cross. Use of wildlife road-crossing culverts was also evaluated, but due to the road grade, these structures were not suitable for this site.

As a condition of plan approval, the developer enhanced existing wetlands and restored several wetlands onsite that had been invaded by exotic invasive plants or littered with decades of residential and agricultural debris. Restoration plans were created to remove invasive plants and establish a long-term management plan for control of these species. At the time of development (2001), the usual practice was to vegetate detention ponds intended for stormwater collection with turf grass and possibly some landscape plants along the border to provide screening; however, these ponds then had very little ecological value. Maybury Park Estates was different; the detention basins were vegetated with native plants to maximize the functionality of the landscape as both a means of flood mediation and wildlife habitat. In addition, the aquatic plants were selected for their ability to remove nutrients and improve water quality. Maybury Park Estates also supported a

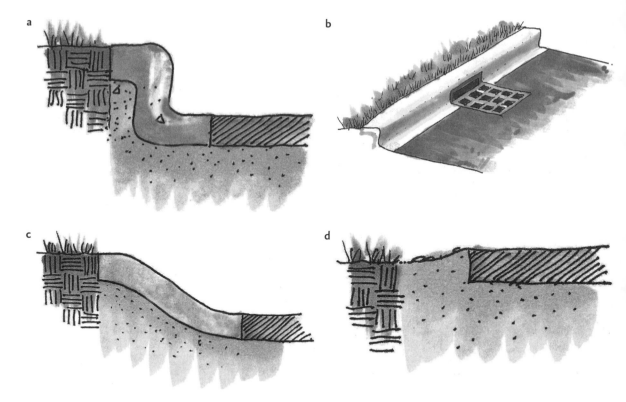

Figure 6.2. Traditional curb systems (a) typically limit or inhibit movement of many species of amphibian and reptile and can serve as a wildlife barrier and sink. This is especially true when curb and gutter systems are in place (b). When possible, modifying design for a more gradual slope (c) or elimination of a curb (d) can help facilitate small animal passage and reduce road-related mortality. *Credit: Herpetological Resource and Management, LLC.*

regional detention pond designed to address larger storm events and provide wildlife and water quality benefits. Best management practices (BMPs), including establishment of the shoreline in native aquatic plants and placement of logs for reptile basking, were incorporated into the pond design (Mifsud 2014). The regional detention pond was also situated adjacent to existing wetlands to enhance connectivity and expand the mosaic of habitat types available for wildlife. Incorporating these landscape features between existing high-quality habitat (i.e., ecological infrastructure) and road infrastructure reduced the impacts of road pollution on wildlife.

The willingness of the developer to work with the City of Novi to implement novel, ecologically friendly design solutions helped move the project through planning phases quickly in comparison to similar projects with complex environmental issues. These expedited design and review phases allowed for earlier project completion and higher sales. Lots adjacent to natural features sold for 20% more than interior lots, demonstrating that natural features are desirable and homebuyers are willing to pay more for these aesthetics. Additionally, the project was also recognized by the city for its environmentally sound design.

Maybury Park Estates is an excellent example of how thoughtful long-term community planning and environmentally responsible design can facilitate cost-effective and profitable development, or public infrastructure, while preserving large tracts of wildlife habitat, or ecological infrastructure, and avoiding increased wildlife habitat fragmentation from roads. This project continues to serve as a benchmark for ecologically sound, community-driven development and demonstrates the importance of strategic road planning to minimize the negative effects of development on wildlife while facilitating community growth. As with any project the mark of success is demonstrated through careful evaluation. Follow-up assessments would be warranted to determine how the species are faring and how the overall ecosystem is responding 10 years later.

6.6. Correcting Deficiencies When No Project Is Planned

One of the permissible uses of federal road funds is to initiate projects to mitigate environmental damage from existing roads. Wildlife advocates could request such projects in the transportation planning phase (Section 6.2.2), either through the citizen advisory committee, their elected representatives serving on the regional planning body, or through the state DOT. However, a mitigation project would need strong public support in order to compete against road capacity projects.

Support and credibility could be gained by coordinating with the resource agencies and by using any of the conservation tools mentioned in Section 6.2.2 to demonstrate where roads are in conflict with natural systems. Also, projects might be identified from deficiencies discovered in monitoring and adaptive management of previous projects, as described in Chapter 12. As discussed in Chapter 7, other funding sources are available and partial support from another source might be useful as leverage to encourage decision makers to allocate funding for these deficiencies.

6.7. Detailed Design

6.7.1. Integrating Wildlife Accommodations into the Detailed Design Process

The integration of road ecology into the transportation development process has naturally brought together two scientific disciplines—ecology and engineering. However, there are fundamental differences between how practitioners in these two disciplines think and apply practical solutions to a road project. Engineers focus on cost, motorist safety, and hydrology, while ecologists focus on the abiotic and biotic environments in relation to a road. Engineers make recommendations based on well-developed algorithms that have been utilized and tested throughout a long history of road building, whereas road ecologists make recommendations that are still undergoing testing and validation in an un-

derfunded research arena. An excellent example here is using the openness ratio for design of wildlife crossing structures (see Clevenger and Huijser 2011). Ideally, engineers desire a well-defined height, width, and length of a structure for each target wildlife species; however, this relationship has not been experimentally tested, especially for small animals, and enhancements such as light through openings in the upper structure may or may not compensate for longer tunnels.

Complicating matters further, an engineer has well-established protocols and design drawings to build a road in challenging landscapes, and often a road ecologist will add layers of complexity to a well-established procedure. In addition, a road ecologist desires more flexibility in the road building process to allow for an adaptive and experimental approach for mitigation. Luckily, because roads are often built in phases due to funding constraints, lessons learned from environmental measures on one road segment could be applied to others if the agency can be persuaded to entertain the changes.

Budgets also greatly affect the way ecologists and engineers think and operate. Engineers typically have large budgets for proven technologies, while ecologists are grappling for ways to measure the effectiveness of new and untested mitigation designs. Road ecology research is often underfunded because of a disparity in how research is directed. Most agencies that manage or administer funding for land or facilities (e.g., local and state wildlife and conservation agencies, FHWA, state DOTs, USFS, NPS, BLM) respond more favorably when funding is directed toward solving a concrete issue. Conversely, regulatory and review agencies tend to focus on the collection, analysis, and review of scientifically defensible data to justify mitigation requirements (e.g., US Environmental Protection Agency and USFWS). Universities and others involved in research and data collection tend to emphasize these latter aspects also. Too often, scientific findings are not translated back to management agencies in ways that can be readily applied or implemented. This pattern has resulted in ecological research being deemed a lower priority, and these agencies consequently have fewer avenues for funding scientific investigation.

The contrasts between ecological and engineering approaches are most evident in the detailed design

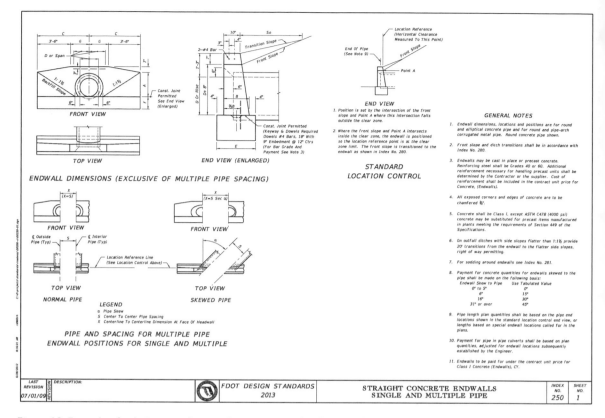

Figure 6.3. Example of a design specification for projects in Florida, United States. *Credit: Florida Department of Transportation.*

phase. In the state transportation agencies, preliminary engineering and design are even separate departments. Consultant engineering firms sometimes have engineers who do both, but rarely on the same project. When transportation agencies hire consulting engineers for a project, preliminary engineering and design are each established through separate contracts. In preliminary engineering, the engineers evaluate the surrounding environment and consider alternatives to find a balance between engineering and environmental constraints, whereas detailed designs are governed by formally established standards, as noted previously, which prescribe every detail of each individual structural element that goes into the road (for an example, see Figure 6.3). Designs from each specialty are more or less independently incorporated into a set of draft plans, which are then circulated among the different specialized professionals and among peers for review and comment. The comments are then incorporated into the next, more detailed, iteration of plans until a final set is approved for construction. Each discipline

can work independently on these standard designs because only structural elements that have been proven to work in combination with others over years of application become standards in the first place; therefore, different combinations of the elements, such as pavements, bridges, and drainage structures, may be mixed and matched with confidence.

Unfortunately, wildlife accommodations don't lend themselves to this typical approach. Incorporating structural wildlife accommodations into road projects is relatively new in the United States, so there are very few standards available. It's also more difficult because the behavior and requirements of wildlife are much more complex and unpredictable than the behavior and requirements of vehicles and flowing water, and the functioning of wildlife structures can be seriously compromised by neglecting any of a multitude of details that do not have to be considered in road structures (e.g., substrate, light, noise, moisture, cover, approaches). There have been several helpful manuals and best practices guides published in the last few years

(Carsignol 2005; SETRA 2005; Huijser, McGowen, Clevenger, et al. 2008; Road and Delivery Performance Division 2010; Clevenger and Huijser 2011), but none that could be applied without the help of trained wildlife biologists. In fact, these guidelines are still changing as more research is done. An important example is the wire mesh size for wildlife fencing specified in Clevenger and Huijser 2011; in "Hot Sheet" 13, they recommend a "standard" mesh size for small vertebrates of 0.5 in (1.27 cm) or larger, depending on the target species. But in preliminary trials of small animal containment by different mesh sizes conducted in Florida, two different species of snakes entangled themselves in 0.5 in (1.27 cm) mesh and could only be extricated by cutting the mesh away (Smith and Noss 2011). Therefore, even if a larger mesh size would be adequate for a chosen target species, a smaller size should be used to avoid a hazard to smaller fauna. Clevenger and Huijser's handbook (2011) also recommends a height of 1.25 feet (0.38 m) for drift fences leading to amphibian and reptile tunnels, but in the Florida trials above, 3 feet (0.91 m) of height was needed to effectively contain most frogs and snakes. The fencing was largely ineffective for lizards because of their climbing ability. These examples demonstrate that there is no universal set of standards that can be uniformly applied to all locations and species. Geographic and biological variability require the adaptation of standards for particular places and species. Even if there were established standards effective for all small animals, engineers could not be expected to apply them appropriately to a project, just as drainage engineers would not be expected to apply structural or pavement engineering standards.

Planning and preliminary engineering have been evolving toward greater stakeholder and biologist participation, but engineers in the detailed design phase are not accustomed to including biologists when developing detailed plans. Hence, design teams seldom involve biologists with adequate experience in road ecology or the target wildlife species to guide detailed design of wildlife accommodations. Biologists are often employed in design, but only for the purpose of acquiring permits for wetland impacts or for mitigation of impacts to individual protected species.

At the permit agencies, the details of project drainage structures are scrutinized by agency drainage engineers to ensure that they have been properly designed to perform their stormwater functions. However, the details of wildlife mitigation measures seldom are examined to ensure that they have been properly designed to perform their wildlife functions. As in the project design teams, there are few biologists in the permit agencies who have enough specialized knowledge to comment on design details, or who even understand that critical details may be neglected during the usual design process. Similarly, post-construction monitoring reports are almost always required by permit conditions when projects include wetland creation or enhancement, and these reports are scrutinized by the permitting agency for progress in meeting specified success criteria, but post-construction monitoring requirements for wildlife accommodations are rare.

In Section 6.8, we describe some unintended outcomes in one Florida DOT district where wildlife structures were designed and permitted without adequate input from road or wildlife ecologists after the preliminary engineering phase. In response to these errors, the district initiated a biological review of wildlife crossing designs by adding the in-house environmental department to the plan reviews. But the formal review-comment-response procedure was still too cumbersome to allow for adjusting the standard designs to the unfamiliar needs of wildlife. An adjustment made by one specialty would induce a change by another specialty, which would create a different obstruction to wildlife, but the new obstruction wouldn't be detected until the next iteration of plans. Projects were approaching critical deadlines and still did not have functional accommodations.

The district finally convened an in-house Wildlife Crossing Design Team, which employed a more flexible, collaborative approach and a higher degree of involvement by road and wildlife ecologists than had usually been practiced in design. The team included not just roadway, structural, and drainage engineers, but also included an engineer from the maintenance department, an engineer from the construction department, an engineer and a biologist from preliminary engineering, and an experienced road ecologist.

Instead of passing finished draft drawings from office to office, all members of the team first met together with the design project manager to discuss early concept sketches. With all the specialized professionals in the same room, unintended consequences of design features and changes could be addressed immediately. Issues that had stalled in formal plan reviews over periods of months could be resolved in one or two meetings and some e-mail exchanges.

The Ocala National Forest case study (Section 6.3.1.3) describes how this approach was independently adopted by consultant design teams working on projects with major wildlife crossing structures; this took place as a result of commitments made in the preliminary engineering phase that mandated the involvement of wildlife experts during the design phase (see Sections 6.3.1 and 12.2.1). In spite of the differing approaches of ecologists and engineers discussed earlier, when both disciplines were engaged from the beginning to pursue the same goal (i.e., a fully functional wildlife structure), a mutual understanding and a better product resulted.

It is important to note that agency biologists must also participate in the scoping, contract writing, selection process, and staff-hours negotiations when consultant engineering firms are engaged to design a project with wildlife accommodations. Early and continued collaboration with wildlife experts must be included in the schedule and in the budget of the project.

6.7.2. Value Engineering

Transportation agencies often employ a formal process called value engineering, whereby staff from several different disciplines who have *not* been involved in developing the project examine it to determine if there might be changes in the design that would save money or improve performance. It has usually been employed near the end of the detailed design phase, but is becoming more common in preliminary engineering. It is also very common in the construction phase for the selected construction contractor to propose similar value engineering suggestions, as discussed in Section 11.2. After its examination, the value engineering team (or the construction contractor) submits a list of suggestions, which are reviewed by the original design team for acceptance, modification, or rejection.

This additional perspective often results in significant cost savings and improved designs, but it also necessitates detailed input and review by agency biologists because the need for wildlife features and their design requirements are usually the least understood

aspects of a road project. It is very important and mutually beneficial for wildlife experts and agency biologists to be a part of the review team when the project includes accommodations for wildlife.

6.8. Lessons Learned

If wildlife experts are not included in the detailed design of wildlife accommodations, roads will be built with accommodations whose functioning is impaired or completely ineffective. An unfortunate example is two major state roads in Florida, which were developed at approximately the same time for widening from two to four lanes. Both roads traversed kilometers of largely undeveloped rangeland, and extensive coordination with wildlife agencies during the projects' preliminary engineering phases had resulted in commitments to wildlife accommodations.

The preliminary engineering concept plans provided for retrofitting existing bridges and culverts at major riparian wildlife corridors with wildlife ledges and shelves. A total of seven structures were modified between the two projects. Of those seven, only two were constructed without major design flaws that seriously impaired or completely eliminated the wildlife connections. The problem was not with any single engineering firm or transportation agency project manager; the errors were distributed among three different design contractors and occurred on both roads. The problem was in the traditional process, which failed to incorporate wildlife expertise beyond the preliminary engineering phase. Examples of these problems were presented in a poster at the International Conference on Ecology and Transportation (ICOET; Tonjes 2006; Figure 6.4).

The structures were never inspected and the design flaws were never noted by any of the resource agencies—an example of the lack of understanding and lack of resources available in most agencies to ensure mitigation of this kind is effective. The problems were discovered by an environmental scientist (author, S. Tonjes) in the state DOT, not because of any explicit monitoring requirement but rather in conjunction with wildlife research he had initiated on his own to investigate the efficacy of the accommodations. Three years after construction, design was started on a project to correct the major errors. Design and permitting of

Figure 6.4. Constructed wildlife ledges with no connection to adjacent habitat. *Credit: Stephen Tonjes.*

the retrofits were much more complicated than they would have been if incorporated into a standard road project at the beginning. The project was finally completed five years later.

These mistakes could be prevented by proactive policies in state DOTs to employ wildlife experts throughout the project development process, and by permit conditions from the resource agencies that require post-construction monitoring of wildlife accommodations and adaptive management, as discussed in Chapter 12. Citizens interested in reducing wildlife impacts could make a difference by advocating for such conditions (see Chapter 5).

6.9. Key Points

- The development of most road projects proceeds in identifiable phases. The first two are preliminary planning and project delivery (preliminary engineering / conceptual design and detailed design). There are opportunities in each phase to ensure that wildlife considerations are included in road projects. The best and most effective opportunities occur earlier.
- The planning phase is when needs for improvement in, and expansion of, the regional road network are evaluated, and individual road projects are proposed. This phase is where planners need to be informed of how the road network will interact with the surrounding ecological landscape, and

where inclusion and funding of wildlife accommodations are most appropriate.

- Roads that are federally funded, approved, or built on federal lands must comply with NEPA. This act requires evaluation of alternative actions and public disclosure of their impacts on the environment. The NEPA process is usually accomplished in the preliminary engineering phase and provides another opportunity for influencing the inclusion and conceptual design of wildlife accommodations.
- Wildlife agencies must be provided the opportunity to comment during the NEPA process; this presents another chance to promote inclusion of wildlife accommodations.
- Transportation engineers, biologists working in transportation agencies, and biologists working in wildlife agencies all need technical help from wildlife and road ecology experts in order to ensure that wildlife accommodations are designed to function effectively.
- Roads in national parks can carry significant volumes of traffic. The NPS has its own procedures and manuals that provide opportunities to incorporate wildlife accommodations, in addition to those resulting from the NEPA process.
- Projects on public roads traversing public lands must comply with the procedures of the agency managing the public land, in addition to NEPA. Different agencies may need to be consulted, depending on how the land is managed; therefore, there are opportunities to work within their procedures to ensure wildlife concerns are addressed.
- Unless state and local roads have a federal nexus, they are not subject to NEPA; however, they still must comply with various environmental laws and also receive permits from environmental resource agencies. Wildlife advocates must learn the local process for road planning, design, and permitting, and can work with the environmental resource agencies to ensure wildlife concerns are addressed.
- Wildlife mitigation projects independent of planned road projects may be identified through wildlife resource agency plans or transportation

BOX 6.3. ONCE UPON A TIME

Chapter the First, the Castle of Dreams

Once upon a time, there was an overcrowded little road that crossed a pretty little stream flowing through a full little culvert. The road builders and the gamekeepers and the local citizenry were called together to decide what to do. The GIS wizards had earlier helped the gamekeepers make maps of the lands that were needed to preserve good hunting for the king, and the gamekeepers saw that the stream and its floodplain provided for the movement of many valuable animals. They also knew that many animals were being killed or turned back as they moved along the riverbanks and tried to cross the road. And if the road was made wider, it would be even worse for them.

The assembly decided that the movement of the animals under the bridge was important; therefore, they decided to replace the full little culvert with a new bigger bridge that would be long enough to both cross the stream and leave dry banks on either side for the animals to walk on. The nice new bridge was put into the Dream Book, and everybody thought they would live happily ever after.

Chapter the Second, the Castle of Drawings

But the wicked wizard Murphy had other ideas. He had worked his evil magic to undo the good intentions of the Castle of Dreams before. He knew that the Castle of Drawings had different towers—Roadway, Structures, Drainage—each of which would fashion a part of the bridge. And he knew that each had its own Book of Rules and that none of the Books had any spells for animal crossings. When Drainage designed special culverts with wildlife ledges, Murphy saw to it

that neither they nor Roadway would consider how the animals would reach the ledges. When Structures made a longer bridge for both water and animals to pass underneath, Drainage lined the extra space with impassable boulders to protect the ends of the bridge from damage by floods. And nobody noticed the willy-nilly placement of Roadway's fences that were supposed to keep the animals off the road and guide them to the openings.

The gamekeepers in the Castle of Dreams had been dismayed to see how their plans had gone awry, and went to the Castle of Drawings.

"No problem!" said the project manager. "I'll put you on the perusal of plans mailing list."

"A good start," said Wiley Cologist, the head gamekeeper.

But he remembered when Murphy had deleted the gamekeepers from the mailing list during an update. He also knew that the makers of drawings labored long hours to produce even a preliminary set of plans. They were far more willing to change concept sketches than finished drawings, and the best time for emissaries from the towers to meet together was before the plans were drawn up.

So the gamekeepers met with emissaries from each of the towers in the Castle of Drawings, and from the Castle of the Builders and the Castle of the Groundskeepers as well, and they made a treaty that whenever a project was to have a wildlife crossing, they would first convene at the Round Table and discuss concept sketches among themselves, before any plans were drawn, so that there would be no more unpleasant surprises during the perusal of plans. And *then* they lived (somewhat more) happily ever after.

---

agency adaptive management procedures. Federal road funds may be used for this purpose and various grants may be available.

- Transportation engineers are generally unfamiliar with wildlife accommodations and there are few

standards available for guidance. There are design details that are critical for effective wildlife accommodations but are not ordinarily considered in road design. Wildlife and road ecology experts must be included throughout the preliminary engineering and design phases to ensure these details are addressed. Without this involvement, projects have been built with accommodations whose functioning is impaired or completely ineffective.

LITERATURE CITED

Andrews, K. M., and J. W. Gibbons. 2005. How do highways influence snake movement? Behavioral responses to roads and vehicles. Copeia 2005:772–782.

Andrews, K. M., J. W. Gibbons, and D. M. Jochimsen. 2006. Literature synthesis of the effects of roads and vehicles on amphibians and reptiles. Federal Highway Administration (FHWA), US Department of Transportation, Washington, DC, USA.

Andrews, K. M., J. W. Gibbons, and D. M. Jochimsen. 2008. Ecological effects of roads on amphibians and reptiles: a literature review. Pages 121–143 in Urban herpetology. J. C. Mitchell, R. E. Jung Brown, and B. Bartholomew, editors. Herpetological Conservation Vol. 3. Society for the Study of Amphibians and Reptiles, Salt Lake City, Utah, USA.

Aresco, M. J. 2005. Mitigation measures to reduce highway mortality of turtles and other herpetofauna at a north Florida lake. Journal of Wildlife Management 69:549–560

Beaudry, F., P. G. deMaynadier, and M. L. Hunter. 2010. Identifying hot moments in road-mortality risk for freshwater turtles. Journal of Wildlife Management 74:152–159.

Brown, J. W. 2006. Eco-logical: an ecosystem approach to developing infrastructure projects. Report FHWA-HEP-06-011. Federal Highway Administration, Washington, DC, USA.

Carsignol, J. 2005 (translated to English, 2007). Facilities and measures for small fauna. technical guide. Ministry of Transport and Infrastructure, Technical Department for Transport, Roads and Bridges Engineering and Road Safety, Bagneux Cedex, France.

CEE (Center for Environmental Excellence). 2009. Wildlife and roads: decision guide. American Association of State Highway and Transportation Officials. http://environment .transportation.org/environmental_issues/wildlife_roads /decision_guide/manual/1_0.aspx. Accessed 11 January 2015.

Clevenger, A. P., and M. P. Huijser. 2011. Wildlife crossing structure handbook: design and evaluation in North America. Report No. FHWA-CFL/TD-11-003. Federal Highway Administration, Washington, DC, USA.

CRAFT (Comparative risk assessment framework and tools). 2012. CRAFT. http://www.fs.fed.us/psw/topics/fire _science/craft/craft/index.htm. Accessed 15 September 2014.

Fahrig, L., J. H. Pedlar, S. E. Pope, P. D. Taylor, and J. F. Wegner. 1995. Effect of road traffic on amphibian density. Biological Conservation 73:177–182.

Findlay, C. S., and J. Bourdages. 2000. Response time of wet-

land biodiversity to road construction on adjacent lands. Conservation Biology 14:86–94.

Forman, R.T.T. 2000. Estimate of the area affected ecologically by the road system in the United States. Conservation Biology 14: 31–35.

Gibbs, J. P., and W. G. Shriver. 2002. Estimating the effects of road mortality on turtle populations. Conservation Biology 16:1647–1652.

Gunson, K. E., D. Ireland, and F. Schueler. 2012. A tool to prioritize high-risk road mortality locations for wetland-forest herpetofauna in southern Ontario, Canada. North-Western Journal of Zoology 8:409–413.

Hels, T., and E. Buchwald. 2001. The effect of road kills on amphibian populations. Biological Conservation 99:331–340.

Huijser, M. P., P. McGowen, A. P. Clevenger, and R. Ament. 2008. Wildlife–vehicle collision reduction study: best practices manual. Federal Highway Administration, Office of Safety Research and Development, McLean, Virginia, USA.

Huijser, M. P., P. McGowen, J. Fuller, A. Hardy, A. Kociolek, A. P. Clevenger, D. Smith, and R. Ament. 2008. Wildlife-vehicle collision reduction study: report to congress. Federal Highway Administration, Office of Safety Research and Development, McLean, Virginia, USA.

Mifsud, D. A. 2014. Michigan amphibian and reptile best management practices. Herpetological Resource and Management Technical Publication, Chelsea, Michigan, USA. http://www.herprman.com/amphibian-reptile-management -practices-michigan. Accessed 13 September 2014.

NPS and FHWA (National Park Service and Federal Highway Administration). 2008. Park roads and parkways program handbook: guidelines for program implementation. US Department of the Interior and US Department of Transportation. http://www.nps.gov/features/dscw/88_PRPPHand book/prpp_home.htm. Accessed 15 September 2014.

Road and Delivery Performance Division. 2010. Fauna sensitive road design. Vol. 2. Department of Transport and Main Roads, Queensland, Australia. http://www.tmr.qld.gov .au/business-industry/Technical-standards-publications /Fauna-Sensitive-Road-Design-Volume-2.aspx. Accessed 15 September 2014.

Sarver, L. 2012. An analysis of box turtle injury and mortality facilitated by anthropogenic activity. Master's thesis, Southern Illinois University, Edwardsville, Illinois, USA.

Schonewald-Cox, C., and M. Buechner. 1992. Park protection and public roads. Pages 373–395 in Conservation biology: The theory and practice of nature conservation. P. L. Fiedler and S. K. Jain, editors. Chapman and Hall, New York, New York, USA.

SETRA (Service d'Etudes techniques des routes et autoroutes). 2005. Facilities and measures for small fauna, technical guide. Ministere de I'Ecologie du Developpment et de I'Amenagement durables, Chambéry, France. [In French.]

Smith, D. J., and R. F. Noss. 2011. A reconnaissance study of actual and potential wildlife crossing structures in central Florida. Final Report. University of Central Florida–Florida Department of Transportation, Contract No. BDB-10. Florida Department of Transportation, District Five, Deland, Florida, USA.

Tonjes, S. D. 2006. Ledges to nowhere: structure to habitat transitions. Page 607 *in* Proceedings of the 2005 international conference on ecology and transportation. C. L. Irwin, P. Garrett, and K. P. McDermott, editors. Center for Transportation and the Environment, North Carolina State University, Raleigh, North Carolina, USA.

White, P. A., and M. Ernst. 2004. Second nature: improving transportation without putting nature second. Defenders of Wildlife, Washington, DC. http://transact.org/wp-content /uploads/2014/04/Second_Nature.pdf. Accessed 17 September 2014.

7 — Sources of Funding

PATRICIA A. WHITE
AND KEVIN MOODY

7.1. Introduction

Once a need for a crossing structure has been identified, and questions regarding where and what kind of structure have been addressed, two other questions must be asked: How can it be funded? Where are the resources to make it a reality?

Funding for wildlife conservation measures, including passage structures, can come from several different sources, depending on the location and circumstance. Wildlife experts often find funding for their work through various academic or research channels, as well as through resource agencies. This chapter demonstrates that non-traditional sources of funding that allow for consideration of wildlife impacts can be part of the following projects: (1) transportation improvement projects; (2) transportation enhancements and alternatives; (3) transportation safety improvements; and (4) maintenance, renewal, and replacement of aged facilities, though these sources are not mutually exclusive. Funding mechanisms will continue to evolve, however, and will vary among geographic locations; thus, wildlife experts must continue to explore new sources and creative ways to leverage available resources.

Each project is unique and construction and materials costs are constantly fluctuating, thus, it is nearly impossible to develop firm cost guidelines. Generally speaking, single-passage structures appropriate for small animals, such as culverts (e.g., Donaldson 2006), are less expensive than structures typically appropriate for large animals, such as underpasses or bridges (Forman et al. 2003; Ruediger and DiGiorgio 2007); though the latter can often work for small animals also (Chapter 9). However, as small animals move shorter distances relative to large animals, inter-tunnel distances tend to be shorter and thus more passage structures may be needed in a given project or on a given stretch of road (Chapter 10). This is particularly true for terrestrial passage structures (Chapter 9). These structures range in price based on size and landscape considerations, as well as project objectives and particular species considerations. The price tag for projects involving more structures geared to small animals may actually be comparable to that for a single large structure geared toward large animals. As such, it has become apparent that customized structure configurations, particularly with respect to aquatic species and habitats, are ideal (Chapter 9).

7.2. Transportation Improvement Projects

Road building includes avoiding, minimizing, and mitigating adverse impacts to wildlife; it may even include measures that enhance wildlife conservation values relative to pre-construction conditions. While the cost of road building varies greatly depending on location and type, a portion of the cost of every transportation project can be dedicated to mitigating the negative environmental impacts of the project; the challenge is making the case that there is a negative environmental impact. Though there is no single statutory or regulatory definition of the "environment," it is interpreted within transportation agencies to encompass much more than wildlife and plants, and includes aspects of the human environment, such as cultural, archaeological, and noise factors. Everything from vegetation management and historic preservation to construction of noise-abatement walls may be considered environmental mitigation. Thus, all road construction projects will include some environmental considerations, which potentially includes consideration of effects on wildlife.

Wildlife-vehicle collisions remain a serious safety hazard on many highways simply because these roads bisect wildlife habitat. Therefore, measures that reduce the risk of accidents should be considered justified transportation expenses. However, even in the case of large wildlife, it is not typical to incorporate mitigation measures to improve human safety.

In 2006, the Federal Highway Administration (FHWA) commissioned a study of a variety of highway projects, from updating a rural highway to full replacement of an urban interchange on a busy interstate. Environmental analyses ranged from Categorical Exclusions (CE) to Environmental Impact Statements (EIS; Section 6.3). The study concluded that environmental compliance can account for 2–12% of a project budget and averaged approximately 8% (Smith 2006). However, some projects deemed to have avoided all impacts are not required to allocate any project costs toward environmental mitigation.

Any transportation agency is motivated to keep costs down and devote the maximum amount of a project budget to direct construction costs (e.g., concrete and steel). With tight budgets and limited state tax dollars, project managers want to be fiscally responsible. It is not easy for a budget-conscious agency to add anything above basic compliance. Conservation and wildlife experts will need to continue to demonstrate, in terms of cost savings and public safety considerations, that mitigation measures are justified and should be required for past, present, and future impacts of the road. Passage structures benefit wildlife by restoring the habitat connectivity (i.e., ecological infrastructure lost due to the construction of the highway). However, explanations and justifications can also be described in terms of benefits and long-term savings, including explanations and justifications based on the costs of taking no action.

7.2.1. Proactive Mitigation
Experts and advocates continue to pursue integrating wildlife passage structures as standard practice for the transportation industry. There was a time when safety

features and pedestrian facilities like guard rails and crosswalks were not considered necessary, but guard rails and crosswalks have now become standard practice in transportation projects. Similarly, stormwater management is considered a standard practice by transportation engineers. While some elements of stormwater drainage systems are implemented to comply with water quality regulations, without proper stormwater conveyance, the road or bridge cannot function properly and will be damaged or washed out. Thus, incorporation of this mitigation measure saves money over the long term. These are the examples that conservation and wildlife experts should keep in mind when promoting the inclusion of wildlife passage structures as part of a given transportation project.

7.2.2. Example: Washington State Department of Transportation's (WSDOT's) Proactive Transportation Improvement Planning
There are a lot of lessons learned from large animal passage projects, primarily when it comes to proactive planning. The WSDOT did not only include the cost of wildlife crossings in widening Interstate 90 through Snoqualmie Pass, they included the restoration of habitat connectivity in the purpose and need of the project (WSDOT 2008). While the purpose of the project, as with most transportation improvement projects, is to meet projected traffic demands, improve public safety, and meet the identified project needs for the local community, federal land management plans determined that I-90 formed a barrier to wildlife movement. The department proactively addressed ecological connectivity across the highway, which reduced fish and wildlife population isolation, but also reduced the risks to wildlife and the public from collisions between vehicles and wildlife.

7.2.3. Retroactive Mitigation
The Federal-aid Highway Program includes funding eligibility for environmental mitigation (23 USC 119 and 133). FHWA's regulation in 23 CFR 777 provides policy and procedures for the evaluation and mitigation of adverse environmental impacts to wetlands and natural habitats resulting from federal-aid projects. The regulations allow for "mitigation or restoration of historic impacts to wetlands and natural habitats caused

by past highway projects funded pursuant to title 23m US Code, even if there is no current federally funded project in the immediate vicinity" (23 CFR 777.9[d]).

In this regulation, "natural habitat" is defined as "a complex of natural, primarily native or indigenous vegetation, not currently subject to cultivation or artificial landscaping, a primary purpose of which is to provide habitat for wildlife, either terrestrial or aquatic" (23 CFR 777.2). Actions eligible for federal funding include "activities required for the planning, design, construction, monitoring and establishment of wetlands and natural habitat mitigation projects and acquisition of land or interests therein" (23 CFR 777.5[b]). These funds may not be used unless the area will be maintained in the intended state as a wetland or natural habitat.

7.2.4. Timing of Wildlife Passage and Mitigation Project Proposals

While the transportation planning process itself does not offer a source of funding for wildlife passage structures, it is important to note that early in the planning process is the best time to introduce wildlife and habitat conservation needs. Protection and enhancement of the environment is included as one of eight transportation planning factors. FHWA's regulations require that long-range transportation plans include discussions of various "potential environmental mitigation activities and potential areas to carry out these activities, including activities that may have the greatest potential to restore and maintain the environmental functions" affected by the transportation plan (23 CFR 450). Early engagement in the transportation planning process reduces the likelihood of conflicts with wildlife conservation goals (Chapter 6). One of the greatest benefits to early engagement in planning is to influence road alignment and placement (where possible) and inform the development of project alternatives that would lessen the need for retroactive wildlife mitigation. However, for many small animal species there are no mitigation options like this possible; therefore, proactive planning is the only way to reduce effects on these populations. If widening or improvement projects are planned with restoration of habitat connectivity in mind, funding to address wildlife impacts is more likely to be included in the project budget.

Early planning for wildlife and habitat conservation can also provide a good opportunity to explore mitigation options, and wildlife and habitat conservation resources, to identify the best remaining sites for acquisition and restoration (Chapters 6, 8, 9, and 10). For example, habitat mapping data in the State Wildlife Action Plans (SWAP; http://www.teaming .com/state-wildlife-action-plans-swaps) can help to identify planned transportation projects that may have a major impact on wildlife. Additionally, SWAPs can also be used to identify where priority species and conservation areas may be at risk from development and transportation threats (White et al. 2007). Often, by the time a road project develops through the planning, review, and design process, many of the opportunities for high-quality and affordable wildlife and habitat mitigation have been lost.

7.3. Transportation Enhancements and Alternatives

Each federal fiscal year, Congress passes appropriations for Transportation, Housing, and Urban Development and each bill authorizes new or consolidates existing programs. In 1991, recognizing that transportation is more than just cars, concrete, and steel, Congress created the Transportation Enhancements (TE) program with the enactment of the Intermodal Surface and Transportation Equity Act (ISTEA) to fund bicycle and pedestrian facilities, scenic or historic easements, welcome centers, and roadside beautification. The 1998 Transportation Equity Act for the 21st Century (TEA-21) included a new activity eligible for funding under the TE program that covered projects related to "environmental mitigation to address water pollution due to highway runoff or reduce vehicle-caused wildlife mortality while maintaining habitat connectivity" (PL 105–178).

In 2012, the Transportation Alternatives Program (TAP) was authorized under Section 1122 of Moving Ahead for Progress in the 21st Century Act (MAP-21). The TAP was authorized for two years and provides funding for programs and projects defined as transportation alternatives, including:

- on- and off-road pedestrian and bicycle facilities, infrastructure projects for improving non-driver access to public transportation and enhanced mobility, community improvement activities, and environmental mitigation;

- recreational trail program projects;
- safe routes to school projects; and
- projects for planning, designing, or constructing boulevards and other roads largely in the right-of-way of former interstate system routes or other divided highways.

Disbursement of TAP funding is at the discretion of state highway agencies. Depending on state priorities, it is still a potential source of funding for mitigation and habitat connectivity projects.

To qualify for consideration, TAP projects do not have to be associated with a specific highway improvement project, but they must be within the acceptable categories and must relate to surface transportation. Eligibility is limited to

- local governments;
- regional transportation authorities;
- transit agencies;
- natural resource or public land agencies;
- school districts, local education agencies, or schools;
- tribal governments; and
- any other local or regional governmental entity with responsibility for or oversight of transportation or recreational trails (other than a metropolitan planning organization or a state agency) that the state determines to be eligible.

In general, the federal award is 80% of the project cost, and the sponsor contributes the balance of 20%, or the nonfederal match. The federal share is even higher in states with a large proportion of land in the public domain. Matching monies can come from a variety of sources, including (but not limited to) the following:

- Local, county, and state agency funds (taxes; bonds; wildlife, natural heritage, or conservation programs; state environmental quality department).
- Federal land management agency funds (see below).
- Donated property, materials, services, or labor (including that from individuals, volunteers, private sources, businesses, nonprofit organizations, or local governments).
- Donated cash via project fundraisers.

Typically, federal funds cannot be used to match a federal funding award. However, because of the special nature of the TAP program, FHWA allows funds and the value of contributions received from federal land management agencies to be credited toward the nonfederal share with some limitations. Some potential federal sources for matching funding include (but are not limited to) the following:

- Federal Lands Transportation Program.
- National Coastal Wetlands Conservation Grant Program (US Fish and Wildlife Service) provides grants to acquire, restore, and enhance wetlands of coastal states.
- Five-Star Restoration Program (US Environmental Protection Agency) provides grants for restoration projects that involve five or more partners, including local government agencies, elected officials, community groups, businesses, schools, and environmental organizations.

Potential wildlife-related projects under TAP that are eligible for funds include, but are not limited to the following projects and activities:

- Wildlife-crossing structures, including the necessary project feasibility studies, planning, research, scoping, designing, engineering, and construction.
- Bridge extensions to further accommodate terrestrial crossings.
- Wildlife exclusionary fencing or other structure installations to guide wildlife toward crossings.
- Technologies, including those to deter wildlife-vehicle collisions, such as radio collars to track wildlife and identify common crossing locations, or remote-sensing devices that trigger warnings to drivers.
- Monitoring, research projects, and mapping to address impacts of habitat fragmentation due to the road and related mortality; these activities include data collection on migration patterns, habitat use, distribution, and crossing behaviors.
- Collection of wildlife-vehicle collision data.
- Identifying collision hotspots through tracking, telemetry, and wildlife cameras.
- Creating or updating state or regional habitat connectivity plans.
- Restoring aquatic passages and watersheds

to provide adequate wildlife corridors and streamflows.

- Evaluating roadside vegetation, removing invasive species, and planting native species along rights-of-way and in neighboring properties; a transition to native vegetation can provide wildlife habitat, erosion control, and stormwater management.
- Training and planning related to wildlife-vehicle collision reduction and habitat connectivity.
- Motorist education to reduce wildlife-vehicle collisions.

7.3.1. Case Study: Monkton Road's Salamanders

Patricia A. White

On a few wet nights every spring, thousands of amphibians, including rare, blue-spotted salamanders (*Ambystoma laterale*) and spotted salamanders (*A. maculatum*), attempt to cross a busy 1.29 km section of Monkton Road in western Vermont. Biologists consider the stretch one of the most important seasonal migration crossing areas for amphibians in the state. Roadkill levels in the area are so high that experts believe these populations may simply cease to exist in coming years (Gibbs and Shriver 2005).

Looking for a permanent solution to this problem, the town of Monkton applied for Transportation Enhancements funding in 2006 to plan and construct a series of culverts and retaining walls that would safely usher the native amphibians, as well as reptiles and small mammals, under the road. With a strong network of public support, community outreach, and local media attention, the Monkton Road project was granted $25,000 in 2007 to scope and plan the project, and another $150,000 grant was awarded in 2010 to complete preliminary engineering and construction on the very first transportation enhancements (TE) wildlife project in Vermont. Financing the entire project, however, was a challenge. After obtaining nearly 90% of their engineer's estimate of the costs from other sources, the group resorted to crowdfunding via the internet. As a result, they were able to meet the engineer's estimate; however, in order to put the project out for bid, additional contingency funds were required for construction. The entire project, from design through construction, cost approximately $300,000.

7.4. Transportation Safety Improvements

While small animals are less likely to cause vehicle accidents, they can still benefit from passage structures designed for larger animals. Because wildlife-vehicle collisions are now more widely recognized as a serious safety hazard for the traveling public, safety funding can be used to build wildlife crossings or to fund any other mitigation measure. Under current legislation, the definition of a highway safety improvement project includes the "addition or retrofitting of structures or other measures to eliminate or reduce crashes involving vehicles and wildlife" (23 USC 148). If the transportation or natural resource agencies that oversee the location of a potential mitigation measure have collected wildlife-vehicle collision data, those data can be used to make a case for safety improvements, such as wildlife passage structures.

Highway safety concerns are paramount among road operators and owners, and many such issues include the problem of wildlife-vehicle collisions. As such, safety-related considerations can be a win-win for wildlife and transportation professionals. The most obvious opportunities are those that reduce collisions by maintaining ecological infrastructure (e.g., safe and reliable connections among habitat patches around the road). There are less obvious opportunities, such as using wildlife habitat patch size to justify acquiring access easements along roads in undeveloped areas. The easements can serve as safety measures that prevent driveways from being inserted onto the road. The lack of driveway access makes it much more expensive to develop the surrounding land, which in turn can preserve ecological infrastructure by helping to keep the surrounding wildlife habitat relatively intact.

7.5. Maintenance, Repairs, and Renewals

Road maintenance divisions provide the ongoing, necessary services to ensure that physical infrastructure is in good working order and conditions are safe for the motoring public. Maintenance measures are also essential for protecting the significant public investment that is our surface transportation system. By prolonging the life of our existing infrastructure, they also reduce the need to continuously build more new highways that may ultimately consume and fragment remaining natural areas and essential wildlife habitat. In addition, as detailed in Chapter 11, the activities of maintenance and operations divisions can also provide opportunities to improve habitat quality through voluntary stewardship actions. Some agencies have been proactive in incorporating endangered species considerations. For example, the WSDOT collaborated with the National Marine Fisheries Service, local government agencies, and other partners to develop a set of road maintenance policies and practices that contribute to the conservation of endangered aquatic species through 10 program elements, including best maintenance practices and a workforce training program. However, because maintenance budgets within transportation agencies are frequently inadequate, it is not a recommended course of action to attempt to seek funding from your state or local maintenance divisions. That being said, small changes in maintenance practices can often make a big difference for wildlife (Chapter 11).

FHWA recognized the need for sharing information on best maintenance practices for wildlife conservation, and developed the Keeping it Simple website dedicated to going beyond compliance to identify simple techniques that protect wildlife through road maintenance (FHWA 2013). In addition, through the National Cooperative Highway Research Program, transportation officials developed a comprehensive compendium of practices for integrating environmental stewardship into construction, operations, and maintenance activities entitled "Environmental Stewardship Practices, Procedures and Policies for Highway Construction and Maintenance" (Venner et al. 2004).

7.5.1. Bridges and Culverts

As part of maintenance and upkeep of highways, road owners and operators periodically inspect bridges and culverts, keep records of their conditions, and schedule these structures for maintenance, restoration, and full reconstruction when necessary. Such activities create opportunities for wildlife experts to propose enhancements to bridges or culverts that accommodate wildlife crossing.

The American Society of Civil Engineers (ASCE) 2013 Report Card for America's Infrastructure gave America's roads a D grade, and bridges a C+ grade. According to the report card, $3.6 trillion dollars would be required to update America's infrastructure by 2020.

More specifically, one in nine bridges is structurally deficient, and would require an investment of $20.5 billion per year to address the backlog of needs by 2028 (ASCE 2013). A similar assessment of culverts would fare no better. Older culverts tend to

- block aquatic and terrestrial passage,
- degrade streambanks and habitat,
- change channel shape,
- promote stream erosion, and
- alter water temperature and chemistry (FSSWG 2008).

There are economic impacts from poorly designed or inadequate culverts to consider as well. These impacts can serve as further justification for installing enhanced culverts to better accommodate wildlife. The cost of flooding and erosion repair typically becomes the responsibility of transportation agencies and municipalities. Damage can extend to private property, and lead to road closures, limited emergency access, longer commutes, and lost business revenue.

Over the next few decades, deficient bridges and culverts will require a significant investment for restoration. Wildlife experts can add value to these investments by identifying opportunities to restore habitat connectivity and improve ecological infrastructure. Sometimes, just extending the footprint of a bridge or span by a few meters on either side can make a big difference. The FHWA has developed two documents that work well when used in concert to guide culvert and road-stream crossing design for aquatic organism passage (Hotchkiss and Frei 2007; Kilgore et al. 2010).

To help justify the need for a new structure that will function as a wildlife passage, the following actions can be helpful:

- Map and catalog the bridges and culverts in the location of interest.
- Determine if the structures are currently being used by wildlife as a means of passage across roads, or if not, how they could be modified to accommodate such passage.
- Check with the transportation agency's bridge department for a schedule of upcoming improvements and cross check these with the location of interest.

- Meet with bridge department officials to discuss opportunities to integrate measures to improve habitat connectivity, hydrologic function, and other ecological processes.

The *Massachusetts Stream Crossings Handbook* (Singler et al. 2012) states the following: "Crossings designed with rivers in mind . . . have been found to safely pass huge volumes of water, sediment, and debris stirred up by high flows, as well as maintain safe passage for emergency personnel and residents. While initial installation costs for an open arch or bridge span may be more than traditional culvert approaches, long-term costs are significantly reduced as the road crossing survives larger precipitation events and operates with limited maintenance."

7.5.2. Example: Culvert Upgrades in Worthington

Bronson Brook is a high-quality cold water tributary to the East Branch of the Westfield River in Massachusetts, where Atlantic salmon (*Salmo salar*), eastern brook trout (*Salvelinus fontinalis*), and blacknose dace (*Rhinichthys atratulus*) persist. In 2003, Hurricane Irene caused extensive damage to the concrete double box culvert on Dingle Road, clogging it with debris, eroding the fill, and damaging the stream bank (Figure 7.1a). Through a collaborative effort, the Town of Worthington worked with local conservation partners and several agencies to raise $380,000 in grants

and donations to replace the failing culvert. The new structure is a 12.19 m wide open-bottomed arch culvert spanning the channel width, allowing for a natural streambed and flows (Figure 7.1b). In addition to the improvements for wildlife, Worthington reduced maintenance costs and protected nearby municipal and private infrastructure.

7.6. Federal Lands Transportation Program

The Federal Lands Transportation Program (FLTP; formerly Federal Lands Highway Program) was created in 1982 to fund a coordinated roads program for transportation needs of federal and Indian lands, such as national parks, fish and wildlife refuges, and tribal lands. Often referred to as "the DOT for federal lands," FLTP's purpose is stated specifically as follows:

1. To ensure effective and efficient funding and administration for a coordinated program of public roads and bridges serving federal and tribal lands.
2. To provide needed transportation access for Native Americans.
3. To protect and enhance our nation's resources.

Wildlife mitigation measures are eligible for FLTP funds if they are on highways within or serving our public lands system. Specifically, funds can apply to projects that "improve public safety and reduce vehicle-caused wildlife mortality while maintaining

Figure 7.1. Double box culvert bridge in Worthington, Massachusetts, United States (a), damaged in 2003 following Hurricane Irene. Replacement with an open-bottomed arch culvert bridge (b) allowed for restoration of the natural streambed and water flow. *Credit: Massachusetts Division of Ecological Restoration.*

habitat connectivity; and to mitigate the damage to wildlife, aquatic organism passage, habitat, and ecosystem connectivity, including the costs of constructing, maintaining, replacing, or removing culverts and bridges, as appropriate" (23 USC 202).

When a project to potentially benefit wildlife is planned on federal land, it is appropriate to contact the FLTP regional office to discuss the inclusion of a mitigation measure as a specific item on the project list, or else to identify opportunities to incorporate wildlife passage structures into other pending projects. For example, in 2004, wildlife conservation experts partnered with Colorado Department of Transportation (CDOT) to explore a potential wildlife crossing bridge just west of Vail Pass on I-70, a location identified as high priority habitat linkage for a variety of species and one of the last remaining forested connections for wildlife moving north-south through the heart of the Rocky Mountains. In 2005, Congress appropriated $500,000 through FLTP's Public Lands Highway Discretionary Program to conduct preliminary studies and planning for the Vail Pass wildlife crossing.

7.7. State Ballot Measures

According to the Initiative and Referendum Institute (2013), in nearly half of the United States, citizens may introduce state ballot initiatives by collecting a threshold number of signatures on a petition to demonstrate public support. After the signature threshold has been met, the initiative usually is certified by the state for the election and then presented on a ballot for public vote; in some cases it goes through the state's legislature first (Initiative & Referendum Institute 2013). Conservation advocates often use ballot initiatives to protect open space and bring much-needed funding for habitat acquisition. Now, ballot initiatives are being used to raise money for wildlife passages.

As an example, Pima County, Arizona, voted to pass a half-cent sales tax increase to fund their Regional Transportation Authority's $2.1 billion regional transportation plan in 2006. The plan included several highway and transit projects, but also set aside $45 million for a "Critical Landscape Linkages" category to fund wildlife passages critical to accomplishing the vision of the Sonoran Desert Conservation Plan's Multi-species Conservation Plan (Pima County 2012).

7.8. The Cost of Doing Nothing

Wildlife passages can be expensive, especially when they (a) are done carefully and correctly, (b) are supported by the necessary predesign research, (c) are of adequate size and number, (d) connect protected and high-quality habitat on either side, and (e) are maintained and monitored for the most effective use. However, measures to protect, monitor, manage, and recover threatened and endangered species can be far more expensive. Certain wildlife species, such as herpetofauna, are particularly vulnerable to impacts from roads and highways. Early engagement in planning as well as pursuing opportunities to integrate passage structures as part of a transportation project can offer cost savings, and prevent the increased costs of addressing wildlife impacts later (Chapter 6).

Wildlife experts have known about the negative impacts of roads on wildlife and habitat for decades now and continue to improve their understanding of these impacts and how to begin to address them. The science and technology of road ecology are only now catching up with the knowledge base, providing solutions to help restore wildlife habitat connectivity and maintain or improve ecological infrastructure. Eventually, policy and funding should follow suit, making wildlife passages standard practice. Until then, wildlife experts can explore the above-mentioned funding sources and continue to find creative, innovative ways to make wildlife passages a reality.

7.9. Key Points

- As transportation agencies must be budget-conscious, recommendations to add wildlife passage measures should be supported by demonstrating cost savings and public safety considerations.
- Continue to collect data and engage the public about the importance and value of wildlife mitigation measures and passage structures. Ultimately, the goal is to make mitigation measures for wildlife a standard practice in the transportation industry, justified in part by cost savings and public safety considerations.
- Include restoration of habitat connectivity in the stated purpose and need of transportation projects, as appropriate.

- Engage with transportation professionals early in the planning process for the mutual benefit of avoiding sensitive wildlife areas, thus reducing the need, and associated costs, of adding passage structures during or after construction.
- Apply for federal TAP funds for habitat connectivity projects.
- Apply wildlife-vehicle collision data to support the use of safety funds for wildlife passage structures.
- Meet with transportation agency maintenance department officials to discuss opportunities to incorporate considerations for wildlife in maintenance practices.
- Meet with transportation agency bridge department officials to discuss maintenance schedules.
- Contact the FLTP to discuss inclusion of mitigation measures on their project lists.
- Introduce a state ballot initiative to protect habitat or fund wildlife passages.
- Pursue opportunities to integrate passage structures into a transportation project. While these can be expensive, measures to protect, monitor, manage, and recover threatened and endangered species can be far more costly.

LITERATURE CITED

ASCE (American Society of Civil Engineers). 2013. Report Card for America's Infrastructure. http://www.infrastructurereportcard.org/. Accessed 15 September 2014.

Donaldson B. M. 2006. Use of highway underpasses by large mammals and other wildlife in Virginia and factors influencing their effectiveness. Pages 433–441 in Proceedings of the 2005 international conference on ecology and transportation. C. L. Irwin, P. Garrett, and K. P. McDermott, editors. Center for Transportation and the Environment, North Carolina State University, Raleigh, North Carolina, USA.

FHWA (Federal Highway Administration). 2013. Wildlife Protection: Keeping It Simple. http://www.fhwa.dot.gov/environment/wildlife_protection/index.cfm. Accessed 15 September 2014.

Forman, R.T.T., D. Sperling, J. A. Bissonette, A. P. Clevenger, C. D. Cutshall, V. H. Dale, L. Fahrig, et al. 2003. Road ecology. Science and solutions. Island Press, Washington, DC, USA.

FSSWG (Forest Service Stream-Simulation Working Group). 2008. Stream simulation: an ecological approach to providing passage for aquatic organisms at road-stream crossings.
US Forest Service, National Technology and Development Program Report No. 0877 1801-SDTDC. FSSWG, National Technology and Development Program, San Dimas, California, USA.

Gibbs, J., and W. Shriver. 2005. Can road mortality limit populations of pool-breeding amphibians? Wetlands Ecology and Management 13:281–289.

Hotchkiss, R. H., and C. M. Frei. 2007. Design for fish passage at road-stream crossings: synthesis report. Publication No. FHWA-HIF-07-033. Federal Highway Administration, Washington, DC, USA.

Initiative & Referendum Institute (IRI). 2013. State I&R. http://www.iandrinstitute.org/statewide_i&r.htm. Accessed 16 September 2014.

Kilgore, R. T., B. S. Bergendahl, and R. H. Hotchkiss. 2010. Aquatic organism passage design guidelines for culverts. Hydraulic Engineering Circular, No. 26. First edition. Publication No. FHWA-HIF-11-0008. Federal Highway Administration, Washington, DC, USA.

Pima County. 2012. Multi-species conservation plan for Pima County, Arizona: public draft. Submitted to the Arizona Ecological Services Office of the US Fish and Wildlife Service. Pima County, Tucson, Arizona, USA.

Ruediger, B., and M. DiGiorgio. 2007. Safe passage: a user's guide to developing effective highway crossings for carnivores and other wildlife. Southern Rockies Ecosystem Project, Denver, Colorado, USA.

Singler, A., B. Graber, and C. Banks, editors. 2012. Massachusetts stream crossings handbook. Second edition. Massachusetts Department of Fish and Game, Division of Ecological Restoration, Boston, Massachusetts, USA.

Smith, L. S. 2006. Report to congress on costs associated with the environmental process: impacts of federal environmental requirements on federal-aid highway project costs. Federal Highway Administration, Washington, DC, USA. https://collaboration.fhwa.dot.gov/dot/fhwa/ReNepa/Lists/aReferences/DispForm.aspx?ID=327. Accessed 16 September 2014.

Venner, M., and P. Brinckerhoff. 2004. Environmental stewardship practices, procedures, and policies for highway construction and maintenance. Transportation Research Board, Washington, DC, USA.

White, P., J. Michalak, and J. Lerner. 2007. Linking conservation and transportation: using the State Wildlife Action Plans to protect wildlife from road impacts. Defenders of Wildlife, Washington, DC, USA. http://www.defenders.org/publications/linking_conservation_and_transportation.pdf?ht. Accessed 16 September 2014.

WSDOT (Washington State Department of Transportation). 2008. I-90 Snoqualmie Pass East final environmental impact statement and section 4(f) evaluation. Federal Highway Administration Report No. FHWA-WA-EIS-05-01-F. FHWA, Washington, DC, USA.

Lessons from Terrapin Mortality and Management on the Jekyll Island Causeway, Georgia, USA

Introduction

In order to improve our ability to estimate, prioritize, and reduce road threats to wildlife while satisfying stakeholders, we used four years of intensive monitoring of a declining species, the diamond-backed terrapin (*Malaclemys terrapin*), at Jekyll Island, Georgia, United States. Coastal roads cause direct mortality throughout the range of terrapins as adult females are killed during summer months while attempting to nest on roadsides (Wood and Herlands 1997). As seen in other species, causeways can indirectly affect terrapins when subsidized predators, such as raccoons (*Procyon lotor*), depredate nests while patrolling roadsides. Terrapins exhibit environmental sex determination where the proportion of female hatchlings increases with temperature (Ewert et al. 1994), and roadside vegetation causes a range of ground temperatures that could alter survival and sex ratios of nests. While the effects of road mortality, nest predation, and roadside vegetation remain undocumented across the majority of the terrapin's range, mortality in specific areas, such as along causeways to barrier islands, is suspected to contribute to population declines (Wood and Herlands 1997; Grosse et al. 2011).

The Jekyll Island Causeway (JIC) is a state highway leading to Jekyll Island that bisects 8.7 km of salt marsh inhabited by terrapins; this relatively small stretch of road presents a broad area for mortality and management. Typical of roads leading to summer tourist destinations, traffic on the JIC peaks from May through July (Georgia Department of Transpor-

tation 2011; "GDOT Traffic Counts Portal," http://trafficserver.transmetric.com/gdot-prod/gdot_report.html), coinciding with the terrapin nesting season. Since 2009, researchers at the University of Georgia (UGA) have worked with the Georgia Sea Turtle Center (GSTC) to assess and reduce the impacts of the JIC on nearby terrapin populations. There were three equally important issues that warranted research. First and most obvious, the GSTC had documented 100–400 adult terrapins struck and killed on the JIC each year since 2007 in addition to substantial nest predation. Second, Jekyll Island is a state park with a publicized mission for the protection and conservation of wildlife. With every turtle carcass, or every "managed" raccoon, that accumulates on the JIC, the public perception of Jekyll may be negatively influenced. Third, even small animals like turtles present a potential hazard to human safety when drivers swerve to miss or hit wildlife, or walk on the road to remove an animal. To address these issues on the JIC and serve as a case study to inform management of regional hotspots of road mortality, UGA and GSTC researchers adopted a three-stage conservation program (Table PE 1.1) placing equal emphases on wildlife ecology, stakeholder values, and public education.

Stage 1. Threats and Consequences on the Jekyll Island Causeway

We started with a question: Are 100 dead terrapins per summer enough to extirpate a population? In Stage 1, we initiated our project by estimating the

Table PE 1.1. An integrated progression for developing research questions and management approaches where the information from one stage formulates the question and objective of the next stage

Questions	Objectives	Methodologies
Stage 1—Are roads threatening the diamond-backed terrapin (*Malaclemys terrapin*) population?	Estimate rates of mortality attributed to the Jekyll Island Causeway. Determine the impacts on population growth.	Intensive road and nest monitoring over multiple years, mark-recapture, population modeling, and monitoring.
Stage 2—Where and when is management needed?	Prioritize hotspots and hot moments of mortality for management.	Geographic Information Systems (moving window) and regression analyses.
Stage 3—What management options will most satisfy stakeholders?	Prioritize management options based on stakeholder attitudes.	On-site and online surveying of visitors, residents, and employees of Jekyll Island.
Stage 4—Is management successful?	Implement management, monitor, and revise.	Integrative network of stakeholders for research, monitoring, and education.

rates of road mortality and nest predation for terrapins and used these to model the consequences on population growth. Additionally, we measured the effects of roadside vegetation on sex ratios of surviving hatchlings.

Beginning in 2009, our research team as well as citizen volunteers conducted a mark-recapture study of terrapins attempting to cross the JIC. We used intensive driving surveys during summer nesting seasons, which allowed us to derive annual rates of road mortality for each study year. Next, during the summer of 2011, we monitored nest predation of eggs seen laid on roadsides. Lastly, we measured the hatchling sex ratios produced from nests in three habitat types on JIC roadsides: open areas, hedges, and artificial nest mounds constructed as an alternative solution to increase egg survival (Buhlmann and Osborn 2011).

Given our mortality rates, we constructed a population model in readily available, user-friendly spreadsheet software (Microsoft Office Excel 2007) to determine the consequences of threats as well as mitigation strategies to terrapin population growth near the JIC. Individuals in the population model were grouped by life stages (e.g., hatchlings, juveniles, adults)—each with stage-specific rates of survival and reproduction. This approach allowed us to identify the stage and parameter with the largest impact on population growth and play "what if" scenarios to simulate the expected benefits of actions targeting specific life stages (e.g., protecting eggs from predators, "head starting" hatchlings

in captivity, or reducing adult road mortality). We first simulated a stable population under no threats using survival and fecundity rates from the literature. However, we knew these projections did not reflect conditions on the JIC. Next, we incorporated causeway threats, and we iteratively reduced both adult survival by current rates of road mortality and egg survival by the rate of nest predation. Lastly, we simulated the effects of producing female-biased sex ratios to augment population growth (λ) as a potential management solution.

Overall, we observed substantial road mortality and nest predation of terrapins on the JIC between 2009 and 2012, and the population was predicted to decline in the absence of conservation actions (Figure PE 1.1). Over 622 adult terrapins were struck and researchers intervened to save an additional 254 individuals. Adult road mortality had the largest negative impact to the population growth rate relative to mortality of any other stage. Thus, adults should be the priority for future mitigation, a notion that is supported by most studies examining turtle species (e.g., Doak et al. 1994; Heppell 1998; Enneson and Litzgus 2008). Population models predicted declines when more than 3% of adults are killed by vehicles each year; our estimates showed that road mortality of adult terrapins varied annually from 4 to 16%. These results indicate the threat of road mortality was significant even in low-mortality years.

In 2011, we estimated that the majority of nests (62%) were depredated by raccoons, most within the first 24 hours after egg deposition. The effects

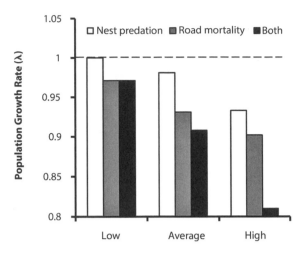

Figure PE 1.1. Expected population responses in growth rates (λ) to nest predation, road mortality, or both threats occurring at varying rates on the Jekyll Island Causeway, Georgia, United States. The dashed line represents stable population growth. Any scenario falling below this line represents a population predicted to decline. *Credit: Brian A. Crawford.*

of nest predation on population growth were smaller relative to road mortality, but our models showed that the current rate of nest predation could exacerbate declines if not managed. We observed substantial variation in hatchling sex ratios from nests laid in different roadside habitats. Specifically, nests on open roadsides and artificial nest mounds produced 100% females while only 16% of hatchlings from nests in hedgerows were female. By altering the hatchling sex ratio in our model, we could stabilize the population using a combined approach for management that moderately reduced road mortality and nest predation while increasing the percentage of females produced. Employing habitat modification (e.g., clear-cutting of roadside hedges to create more open nesting habitat) or creating more artificial nesting mounds on the JIC could expand warmer nesting areas that increase the number of female hatchlings entering the population and assist in its recovery.

In addition to modeling the population's fate, we directly sampled terrapins in two creeks adjacent to the JIC with the help of citizen scientists. This effort continued a long-term monitoring program, initiated in 2007 (Grosse et al. 2011), that will enable us to measure terrapin population trends given the

threats of the JIC and any management we apply. We captured, sexed, measured, individually marked, and released terrapins each year prior to the onset of nesting season. To date (2014), we have caught 254 individuals that composed an extremely male-biased population where males outnumbered females 4 to 1. We also rarely captured old, large adult females, consistent with other studies that show road mortality can alter population structures and eventually cause declines (Steen and Gibbs 2004).

Stage 2. Hotspots and Hot Moments of Road Mortality

Given that Jekyll's terrapin population is predictably declining and management is needed primarily to reduce road mortality, we identified hotspots and "hot moments" (spatial and temporal peaks, respectively) of terrapin crossing activity to determine where efforts would be most effective. We used data from road surveys (Stage 1), where we recorded the following for each live and dead terrapin observed on the road: (1) the date to measure seasonal activity patterns; (2) the time to the nearest scheduled high tide to measure daily patterns; and (3) the location, recorded using a handheld Global Positioning System (GPS: Garmin International, Olathe, Kansas, USA), to determine spatial aggregations of crossing activity. We used an approach similar to the moving-window analyses available in Geographic Information Systems like ArcGIS (ESRI, Redlands, California) to identify localized hotspots on the JIC. Next, we identified daily hot moments of activity by grouping terrapin encounters into 30-minute intervals preceding, during, and following the scheduled high tide when their activity is the highest. For broader seasonal patterns of activity, we analyzed our dataset using an approach similar to regression analyses.

Overall, we found consistent hotspots and hot moments of crossing activity on the JIC. The moving-window analysis identified three hotspots; collectively, these zones only spanned 800 m (less than 10% of the entire causeway's length), but 30% of terrapins crossed within them. We observed terrapins on the JIC between late April and mid-July with a seasonal peak in late May. Although this seasonal peak only provides a broad target for management, we also identified finer-scale daily peaks of activity.

The majority (55%) of terrapins emerged on the road within a three-hour window during the diurnal high tide. Furthermore, the consistency of hotspots and hot moments between years allows for the potential integration of permanent mitigation measures and management regimes.

Consistency of activity peaks may be attributed to terrapin life history and behavioral adaptations. For example, we found evidence of nest site fidelity, as in other terrapin studies (e.g., Szerlag-Egger and McRobert 2007) where 50% of recaptured females were observed crossing within 50 m of their original capture location. This behavior suggests that hotspots may remain constant across annual crossing activity peaks as terrapins return to the same locations—assuming they survive. Ultimately, we plan to use these local peaks of road mortality to develop targeted mitigation measures such as location- and time-specific warning systems for motorists. For a more detailed description of Stages 1 and 2, see Crawford, Maerz, Nibbelink, Buhlmann, and Norton (2014) and Crawford, Maerz, Nibbelink, Buhlmann, Norton, and Albeke (2014).

Stage 3. Integrating Human Perspectives into Management

Before finalizing and implementing mitigation and management plans, we surveyed visitors, residents, and employees of Jekyll Island who also have a stake in the JIC. The aim was to assess their attitudes toward potential management strategies. For example, we asked these island patrons to rate their level of agreement with the following statements: "we should protect terrapins from being struck by vehicles on the causeway," "we should protect terrapin nests from raccoon predators," and "we should maintain the current speed limit regardless of the impact on terrapins." Lastly, we asked patrons to gauge their level of concern if they were to strike various animals with their vehicles, including deer, snakes, and turtles.

We found strong support among Jekyll Island patrons for mitigation options that did not affect causeway aesthetics or speed limits, such as warning signage, short fencing, and additional artificial nesting mounds (Figure PE 1.2; Crawford et al. 2015). However appealing, we caution against using fencing as the primary measure to mitigate road mortality for reasons put forward in Table PE 1.2 and Chapter 9. Although mesopredator removal is often acceptable to wildlife managers, patrons had polarized, but mostly negative, opinions of both lethal and nonlethal tactics. This raises a difficult issue given that previous results supported reduction of nest predation. The results of the survey of public attitudes suggest a lack of understanding of the role humans play in causing overabundance of subsidized predators, and many patrons expressed the need for better communication between researchers and the public. Perhaps not surprisingly, patrons expressed a high level of concern for hitting turtles and the lowest level of concern for hitting snakes (Figure PE 1.3; Crawford and Andrews, submitted manuscript). Although we will not use these results solely to guide

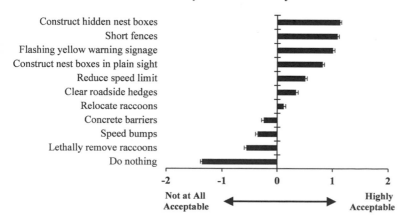

How acceptable is it for the following management actions to occur on the Jekyll Island Causeway?

Figure PE 1.2. Average (± standard error, or SE) attitudes of Jekyll Island patrons (*n* = 1,238) based on surveys assessing potential management actions on the Jekyll Island Causeway, Georgia, United States. *Credit: Brian A. Crawford.*

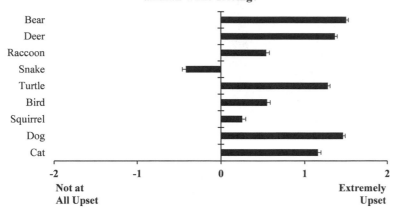

Figure PE 1.3. Average (± SE) level of concern of Jekyll Island patrons (*n* = 1,238) based on surveys assessing wildlife-vehicle collisions involving different taxa. *Credit: Brian A. Crawford.*

Table PE 1.2. Reasons why fences are not a favorable option as the sole technique for mitigating road effects on nesting turtles

Groups negatively affected by fencing	Reason
Turtle adults	Potentially creates hotspots of road mortality at edges of fence; limits turtle access to portions of road shoulders that could be used for nesting safely.
Turtle eggs	Nesting along the fence provides predators (e.g., raccoons [*Procyon lotor*]) a transect on which to more efficiently find and destroy eggs.
Other wildlife	Creates barrier to movement and gene flow among populations of small vertebrates.
Human drivers	Limits vehicle access to road shoulders in case of emergency; reduces road aesthetics and may obstruct scenic views of surrounding habitat.

management on the JIC, evaluating public attitudes toward at-risk groups of species like snakes and turtles can further improve our ability to educate stakeholders about conservation issues.

Future Directions for Turtles, Drivers, and Causeways

In light of these results, we began to initiate an adaptive management program in 2013 to conserve terrapins on Jekyll Island. We plan to implement mitigation measures that are most acceptable to the public, to assess the effects on rates of road mortality and nest predation, and to adjust our efforts to form a cost-effective conservation plan. The program will also seek involvement from all stakeholders in order to foster conservation stewardship via educational and volunteer programs facilitated by the GSTC. With the help of the Georgia Department of Transportation and Jekyll Island Authority, we have already installed two terrapin crossing signs with

flashing yellow lights (analogous to school zones) that are deployed during daily hot moments at high tides in summer nesting seasons. Ultimately, our approach will serve as a model for developing solutions to the broader issue of human-influenced road mortality of turtles and other at-risk species while promoting human safety and stewardship.

LITERATURE CITED

Buhlmann, K. A., and C. P. Osborn. 2011. Use of an artificial nesting mound by wood turtles (*Glyptemys insculpta*): a tool for turtle conservation. Northeastern Naturalist 18:315–334.

Crawford, B. A., and K. M. Andrews. Drivers' attitudes toward wildlife-vehicle collisions with respect to reptiles and other wildlife taxa (submitted manuscript).

Crawford, B. A., J. C. Maerz, N. P. Nibbelink, K. A. Buhlmann, and T. M. Norton. 2014. Estimating the consequences of multiple threats and management strategies for semi-aquatic turtles. Journal of Applied Ecology 51:359–366.

Crawford, B. A., J. C. Maerz, N. P. Nibbelink, K. A. Buhlmann, T. M. Norton, and S. E. Albeke. 2014. Hot spots and hot moments of diamondback terrapin road-crossing activity. Journal of Applied Ecology 51:367–375.

Crawford, B. A., N. C. Poudyal, and J. C. Maerz. 2015. When drivers and terrapins collide: assessing stakeholder attitudes toward wildlife management on the Jekyll Island Causeway. Human Dimensions of Wildlife.

Doak, D., P. Kareiva, and B. Kleptetka. 1994. Modeling population viability for the desert tortoise in the western Mojave Desert. Ecological Applications 4:446–460.

Enneson, J. J., and J. D. Litzgus. 2008. Using long-term data and a stage-classified matrix to assess conservation strategies for an endangered turtle (Clemmys guttata). Biological Conservation 141:1560–1568.

Ewert, M. A., D. R. Jackson, and C. E. Nelson. 1994. Patterns of temperature dependent sex determination in turtles. Journal of Experimental Zoology 270:3–15.

Georgia Department of Transportation (GDOT). 2011. "Georgia's State Traffic and Report Statistics (STARS)." https://data.georgiaspatial.org/index.asp?body=download&dataID=41250. Accessed 29 August 2014.

Grosse, A. M., J. C. Maerz, J. Hepinstall-Cymerman, and M. E. Dorcas. 2011. Effects of roads and crabbing pressures on diamondback terrapin populations in coastal Georgia. Journal of Wildlife Management 75:762–770.

Heppell, S. S. 1998. Application of life-history theory and population model analysis to turtle conservation. Copeia 1998:367–375.

Steen, D. A., and J. P. Gibbs. 2004. Effects of roads on the structure of freshwater turtle populations. Conservation Biology 18:1143–1148.

Szerlag-Egger, S., and S. P. McRobert. 2007. Northern diamondback terrapin occurrence, movement, and nesting activity along a salt marsh access road. Chelonian Conservation and Biology 6:295–301.

Wood, R., and R. Herlands. 1997. Turtles and tires: the impact of roadkills on northern diamondback terrapin, Malaclemys terrapin terrapin, populations on the Cape May Peninsula, southern New Jersey, USA. Pages 46–53 in Proceedings: conservation, restoration and management of tortoises and turtles-an international conference. J. Van Abbema, editor. New York Turtle and Tortoise Society, New York, USA.

8

Tom A. Langen, Kari E.
Gunson, Scott D. Jackson,
Daniel J. Smith, and
William Ruediger

Planning and Designing Mitigation of Road Effects on Small Animals

8.1. Introduction

In Chapters 2, 3, and 4, we reviewed *why* small animals are negatively affected by the presence of roads and the effects of road vehicle traffic. But what can be done to avoid or reduce the impacts of roads and road traffic? In this chapter, we will focus on the first two planning steps toward making roads and road traffic less harmful to small animals: (1) deciding what the management objectives are, and (2) deciding where to focus management measures to be most effective at meeting these objectives. In the subsequent chapters, we detail management practices, design solutions, and technologies that can be implemented to meet the management objectives.

Mitigation of road impacts is often ad hoc and post hoc. For example, when a segment of road is found to have unusually high seasonal roadkill, and the public and conservation agencies recognize this as a problem, sometimes actions are taken such as installation of barrier fencing. Or, the permit-issuing agency for a local road project requires that a specific mitigation measure such as wildlife fencing be incorporated into the work plan. While ad hoc mitigation measures may be effective at the immediate site in question, solutions are often constrained by the pre-existing public infrastructure. As such, the mitigation measures may not address the root problems they are intended to solve, and it is possible that mitigation at road segments elsewhere in the road network would be more effective at meeting conservation goals, and less expensive.

So why plan? Planning allows for more effective and more efficient accomplishment of management objectives by identifying and implementing the most promising mitigation solutions at the optimum locations. In other words, planning can increase the chance of success at conserving small animals, and at a lower cost, than retrospective action. It is important that ecological infrastructure is part of the discussion early in the planning process (Chapters 6 and 7) so that it is considered equitably with the other major planning considerations (e.g., safety, public's expectations of service, cost, social and economic impacts, aesthetics, political interests).

8.1.1. What Are the Goals?

What are the goals of a planning process? At least three questions must be answered to define the goals:

1. What are the focal species of concern?
2. What impacts of roads and road traffic need to be addressed for the focal species?
3. What spatial extent or scale must the plan encompass to accomplish the goal?

What are the focal species of concern? In some cases, the planning process may be focused on a single species, perhaps one of great conservation concern that is known to be negatively affected by the barrier effect of the road or is at unacceptable risk of road mortality. Alternatively, the planning process may be focused on a whole assemblage of species—for example, wetland-associated small animals, or all forest-interior animals, big and small. It also is possible for the planning process to encompass a hybrid of the two: A single

Table 8.1. Attributes of planning strategies and methods for a local population-, metapopulation-, and landscape- or regional-scale conservation of small animals

Spatial scale	Goals	Data needed	Temporal planning scale	Implementation
Population	Reduce road mortality Maintain habitat accessibility Reduce roadside environmental degradation	Locations of populations Seasonal habitat use	Immediate	Easy
Metapopulation	Maintain connectivity among local populations	Locations of populations Distribution of available habitat Interpopulation movement behavior	Medium term	Moderate
Regional/Landscape	Maintain connectivity among regional populations Provide movement corridors to accommodate current and future environmental conditions	Locations of populations Distribution of available habitat Interpopulation movement behavior Projected regional environmental change	Long term	Difficult

"umbrella species" may be a focus. An umbrella species is one that will also benefit many other species of management concern, if its management objectives are met. For example, the planning process could focus on spotted salamanders (*Ambystoma maculatum*) as the umbrella species, but with the greater management goal of conserving all amphibians associated with vernal pools.

What effects of roads and associated traffic need to be addressed for the focal species? It may be that excessive road mortality is the effect that most needs to be addressed; roadkill not only harms individual animals, but may pose public safety or population decline risks (Chapter 3). Alternatively, it may be the barrier effect of the road and traffic, either because the road prevents habitat access (i.e., habitat required for survival or reproduction; Chapter 2) or because it prevents dispersal and population interchange between populations separated by the road (Chapter 3). Finally, other environmental impacts of roads such as contamination, noise, microclimate alteration of bordering habitat, and human access may be most in need of mitigation (Chapter 4).

It is not uncommon that more than one impact must be addressed—for example, both road mortality and disrupted habitat connectivity could be considered in one project. In such instances, the potential solutions may require an analysis of costs and benefits to the focal species. As an example, both barrier and directional fencing can reduce road mortality, but may re-

sult in reduced habitat connectivity. For certain target populations, loss of some habitat connectivity may not be a significant problem, but for others it may be severe enough that passage structures or other solutions (e.g., road closures instead of fencing) must be incorporated into the plan. Moreover, solutions for focal species may create new problems for others; as an example, fencing that is successful at reducing road mortality for a target species as intended may also create a movement barrier for a nontarget species (e.g., Jaeger and Fahrig 2004).

What spatial extent or scale must the plan encompass to accomplish the goal? This depends on whether the reason for the plan is to conserve a local population (or assemblage of species at a particular locality), to conserve a network of populations that is maintained by dispersal among them, or to conserve connectivity of habitat across a landscape or region. The planning process is often driven by the scale or extent of the problem to be addressed. The next four Sections (8.2–8.5) provide a review of planning strategies and methods for a local population, metapopulation, and for landscape- or regional-scale conservation of small animals (see also Table 8.1).

8.1.2. What Are the Potential Solutions?

The earlier that planning can occur within the development of a road network, the more effective and the less expensive the solutions will be for mitigating the effects on small animals. It is better to plan routes that proactively minimize impacts on wildlife and incorporate appropriate design features during the road construction process than it is to implement mitigation retroactively on existing roads. The most direct way to reduce the effects of roads on small animals is to route new roads away from sensitive habitat and wildlife movement corridors. Where roads are planned, incorporate specific design features that reduce road mortality, preserve habitat connectivity, and prevent habitat degradation adjacent to the road.

When plans are likely to increase the negative effects of existing roads, such as widening to add new lanes, it is better to anticipate these effects and implement mitigation measures before the changes occur. Similarly, incorporating design features beneficial to animals into otherwise necessary road reconstruction and maintenance projects (e.g., culvert replacement or bridge reconstruction) is much less expensive, less contentious, and can offer a wider set of solutions than advocating for a new major road project strictly for the benefit of wildlife (Chapters 7, 10, and 11).

A promising example of this is in New York State, where the state Department of Transportation (DOT) partnered with The Nature Conservancy to identify, via Geographic Information System (GIS) habitat models, where existing culverts could potentially be upgraded for wildlife passage across the entire state highway system. Designs that are more suitable for wildlife passage are installed during scheduled periodic maintenance and replacement of the structures. After flooding caused by Hurricane Irene resulted in severe road infrastructure damage in the Adirondack Mountains, the New York State DOT incorporated animal passage in the reconstruction plan for culverts and bridges (Levine 2013).

In many regions, the road network is already in place and the opportunity for such proactive planning has been lost. However, in the developing world (see Case Study 4.4.3.1), in rapidly growing suburban belts around cities, and in areas of rapid energy production and infrastructure development (see Case Study 4.5.1.1), new extensive road networks are being constructed and offer proactive opportunities for route planning. Although relevant for all scales of planning, we will discuss some particular strategies for road network planning in Section 8.5.1. Mitigation options for existing roads are numerous and can include installation of wildlife barriers and passage structures,

measures to alter driver behavior, management of the vegetation and other features of the road verge, and measures to prevent degradation of roadside habitat (Chapter 4).

8.2. Planning at Local or Population Scales

The objective of local-scale planning is to conserve populations that occur along existing roads or in the vicinity of planned new roads. The extent of the planning activity may be restricted to one segment of road, or may encompass a whole road network. The focus of local-scale planning is on avoiding or mitigating the impacts of roads and traffic on individual populations of a target species. The planning steps include locating populations that may be negatively affected by roads, prioritizing populations when resources are limited, and finding solutions that are appropriate for avoiding or reducing negative impacts. Criteria for prioritization might include population size (bigger is better), degree of protection of the population and its habitat (e.g., populations within parks and protected areas vs. those in areas subject to residential development), and likelihood that the population will respond positively to mitigation. Local-scale challenges may include excessive mortality due to roadkill, habitat fragmentation, and other environmental consequences (e.g., noise, contamination, human trespass); often it is a combination of two or all three of these that must be addressed.

How are locations detected where roads are having immediate negative effects, or will have in the future? Some options might include:

1. surveying planned road routes, or existing roads and roadside habitat, to directly locate populations that are likely to be affected;
2. using experts or mapped habitat data to indicate where at-risk populations are likely to be; or
3. using computer models to predict where individual animals are likely to encounter planned or existing roads.

However, it is important to keep in mind that each option has its benefits and limitations.

The limitations of any single option can be reduced by employing multiple methods (e.g., combining information from habitat and mapping of known populations, direct field surveys, expert opinion, computer modeling of habitat connectivity) to more reliably infer where attention should be focused. Different techniques can provide complementary data. For example, camera traps or track stations can provide good information about how frequently a potential crossing location or passage structure is used. Radio tracking can provide information about animals' general movements in relation to a road and rates of road crossing by individual animals, but without the spatial precision of camera traps and track stations.

8.2.1. Direct Methods of Inferring Critical Locations

8.2.1.1. Distribution Data

Particularly for species of great conservation concern, data may already exist on the locations of populations. State natural heritage programs or state fish and wildlife agencies often collect such information for their management obligations, for conservation, or for environmental impact studies prior to development. To infer where these species may be affected by roads, species distribution data can be plotted with the proposed or existing road network, and relevant land use or land cover data, in GIS. Then, by applying what is known about the natural history of each species (e.g., daily and seasonal movements, habitat requirements), at-risk populations may be identified. This may enable prioritization of the populations, species, or locations that merit additional attention. The major limitation of this option is that few accurate and precise distributional data are available for small animal species.

Many forms of distributional data (e.g., atlas programs) are too coarse for road planning at the fine spatial scale necessary to identify specific points for mitigation. However, such atlas data can be useful to indicate areas to survey for species of concern. Species occurrence data from state natural heritage programs may provide locations that are sufficiently precise for road-planning purposes, but such data are rarely complete. Lack of a species occurrence record for a location is not adequate evidence of absence.

If species of conservation concern are suspected to occur in the vicinity of a road project, it may be worthwhile to contact regional experts on the species; such experts often have access to distributional records that are otherwise publicly unavailable, and can also provide knowledgeable judgments about the quality of existing distributional data. Similarly, biologists with either unpublished data or data that differ from state natural heritage program or state fish and wildlife agency datasets should consider sharing these with state data managers to help keep information current and readily accessible. Transportation agencies typically contact federal and state fish and wildlife agencies to obtain occurrence data because doing exhaustive literature searches or trying to contact all potential data sources is prohibitively time consuming.

8.2.1.2. Presence Indicators

In lieu of existing distribution data for small animal species of concern, direct surveys may be necessary. Such surveys are typically time consuming, and therefore most practical if restricted to a limited set of road segments with the highest likelihood of species occurrence (and thus the highest chance of wildlife-road conflicts). This approach requires some understanding of the natural history of the species of concern. For example, determining the maximum distance from a road at which a population is likely to be negatively affected is important for establishing the appropriate scale for surveys.

In some cases, visual searches for animals or their traces (e.g., scat, burrows, tracks), or call surveys for vocalizing animals (e.g., frogs), can reliably indicate presence. Using baited live traps for small mammals or pitfall traps along drift fences for amphibians and reptiles can be practical and effective (Gibbs and Shriver 2005; Rytwinski and Fahrig 2011). Several compilations and reviews of valid survey methods for amphibians, reptiles, small mammals, and birds include Corn and Bury (1990), Heyer et al. (1994), Wilson et al. (1996), Bibby et al. (2000), Long et al. (2008), and Graeter et al. (2013).

Track stations, baited or placed where animals are likely to pass if present, can be used to detect presence based on the track imprints (Smith 2006a). Substrate used for collecting tracks includes snow, sand, silt, gypsum, and soot (Smith 2006a; Clevenger and Huijser 2011; Rytwinski and Fahrig 2011). This method is low cost but labor intensive, and is affected by weather conditions, especially wind and rain. Where rain is frequent, designs that protect the track station substrate may be needed (e.g., Rytwinski and Fahrig 2011).

Motion-detecting and infrared-detecting (through body heat) camera traps provide another way to survey small animals, although this technique generally works best for medium-sized or larger mammals and birds (Manley et al. 2006; Smith et al. 2006; Long et al. 2008; O'Connell et al. 2011). Like track stations, these can be baited or placed where animals are likely to pass if present. For a few species, photographic images may reveal distinct characteristics or markings that can be used to identify specific individuals (Mendoza et al. 2011; O'Connell et al. 2011). Cameras, unlike track stations, can indicate what time of day animals pass, and whether they travel singly or in groups. Camera traps are limited by the trip sensor's capability to detect smaller animals and by the sensor's range (e.g., typical daytime range is 20 m, while nighttime flash range is 12 m; see www.trailcampro.com). Cameras are also expensive and highly subject to theft and vandalism.

Track stations and camera traps can be placed adjacent to a road to detect crossing attempts. They can also be placed along fencing, along features within the road verge (e.g., drainage ditches, cover structures, different vegetation types), or within potential crossing structures to document use (Gompper et al. 2006). Visual searches, traps, track stations, and camera traps arrayed at a set of specified distances along transects perpendicular to a road can be used to detect the spatial scale of any behavioral road deterrence or attraction (Smith 2006a).

8.2.1.3. Movement Data

On a fine scale, movement data can indicate how animals behave as they approach roads (deterred, attracted, or unaffected), how they use features in the road verge (vegetation, cover objects, rock riprap, drainage ditches), how they react to potential road obstructions (fencing, Jersey barriers, gutters, wingwalls), or how they behave upon arrival at passage structures. On a coarse scale, animal movement data can indicate where movement corridors occur, where animals interact with roads, and whether animals attempt to cross roads, use passage structures, or are deterred by the road or road traffic.

While radio tracking is the most popular and a generally useful way to directly document animal movements (discussed below), there are other methods that permit limited inferences of small animal movements.

These movements can be directly observed as small animals encounter roads (e.g., Bouchard et al. 2009). Snow tracking can be used to infer the behavior of some small to medium-sized mammals along winter roads in cold regions (Zielinski and Kucera 1995; Clevenger et al. 2003). Capture-mark-release-recapture (i.e., mark-recapture) studies may be useful in certain cases, although perhaps too labor intensive for most road-planning purposes. Genetic (DNA) markers for identifying individuals can be done using tissue, hair, or scat, but again may be too expensive and labor intensive for most planning purposes (Gompper et al. 2006; Long et al. 2008).

Still-frame camera studies and tracking stations provide limited information on movement behavior, but can be useful for detecting direction of movement and, when placed on both ends of a crossing structure, whether animals move through the entire structure as intended. Movement-triggered video cameras can be used to document behavior of animals approaching roads, at features within the road verge, at wildlife fences, or at crossing structures (Clevenger and Huijser 2011). Passive Integrated Transponder (PIT) tags can be affixed onto or surgically inserted into animals. PIT tag readers installed on each end of a passage structure can be used to record patterns of use by individual small animals.

Lightweight spools or bobbins with thread have been used to track movements of reptiles and small mammals over short distances (e.g., Boonstra and Crane 1986; Dodd 2002; Dorcas and Wilson 2009). This technique must be used with care, because there is increased risk of injury or predation due to the trailing thread becoming caught on vegetation or debris, and the additional bulk of the spool can impede movement of very small species. Fluorescent or iridescent powders can be coated on small animals to track their movements up to about 200 m (McDonald and St. Clair 2004; Graeter and Rothermel 2007; Furman et al. 2011; Brehme et al. 2013); however, this technique may be inappropriate for amphibians sensitive to desiccation.

Radiotelemetry can typically provide information on home range size (basic area requirements); use of and preferences for different habitats; and movement paths and distances traveled for foraging, dispersing to and from breeding sites, seasonal migration, and dispersal between populations (Manley et al. 2006; Silvy 2012). Radiotelemetry is useful for determining single or recurring movement patterns by individuals, including the number of road crossings and the frequency of use of roadside habitats (Smith 2006a; Rouse et al. 2011). It can also be used to evaluate age- or sex-related differences in spatial requirements, habitat use, and movement patterns (Silvy 2012). Radio-tracking data can be very useful for ensuring that planned roads avoid movement corridors, or if the intersection is unavoidable, the data can help determine where mitigation measures should be implemented.

Radiotelemetry data are limited primarily by the functionality and precision of the equipment, frequency of locational fixes, and user proficiency (e.g., Silvy 2012). Radiotelemetry uses standard VHF technology and is most applicable for smaller vertebrate applications, but is typically labor intensive. Newer GPS technology has reduced labor requirements and allows for remote collection of data at shorter time intervals (as frequently as every 15 minutes). With GPS transmitters, it is much more feasible to make precise determinations of actual road-crossing locations (Guthrie 2012). Unfortunately, the miniaturization of this technology is either unaffordable, or unavailable (those available are not appropriate), for use on small animals (Wikelski et al. 2006), but frequent technological advances may soon make GPS technology available for these animals (Soutullo et al. 2007; Kays et al. 2011).

8.2.1.4. Roadkill and Road Encounter Surveys

Spatial patterns of roadkill or road crossings can be used to locate segments of road that need attention. Data on collisions between large mammals and vehicles are routinely recorded as a consequence of property damages and injuries, but data on small vertebrate roadkill along extensive road networks is not systematically collected. Some citizen science efforts engage the public to record data on small vertebrate roadkill they encounter (e.g., California Roadkill Observation System, http://www.wildlifecrossing.net/california/doc/about_cros; Chapter 5), but these data are typically collected too haphazardly to be of much use for planning that is relevant to small animals.

Systematic roadkill surveys are a popular way to locate segments of roads that may be a problem for wildlife. When a species is common and easily detected when present, these surveys can help to determine where animals most frequently encounter and attempt to cross roads (e.g., Ashley and Robinson 1996). Spatial patterns of roadkill reflect the spatial patterns of road-crossing attempts (where and how many attempt to cross), traffic volume (risk of mortality when crossing a road), and persistence of roadkill remains on a road. Periodically patrolling a network of roads and recording the identity and location of dead and live animals adjacent to or crossing roads can help to identify hotspots (clusters) of road-crossing attempts and road mortality. Such surveys, though labor intensive, provide data that can be used to develop predictive wildlife road encounter models such as those described in Section 8.2.2. Road surveys are ineffective where road mortality has already resulted in severe population declines in the adjacent habitat (Fahrig and Rytwinski 2009) and for species that avoid roads (Andrews and Gibbons 2005).

One must be cautious in interpreting spatial patterns of road mortality as detected by roadkill surveys. Apparent spatial patterns may merely reflect detectability or persistence of roadkill; well-designed methodology can control for this bias (Slater 2002). More problematic, a road segment at which roadkill appears to be relatively infrequent may be so for various reasons: (1) the road segment is not close to habitat for the species nor does it lie along a movement corridor, (2) many animals cross at the segment but traffic volume is low at the time animals routinely cross, (3) many animals encounter the road segment but are deterred from attempting to cross at it, or (4) a past history of heavy road mortality has caused a population decline in the vicinity of the road segment. High levels of roadkill at a given location may indicate a potential long-term population decline, or it may indicate a population that is large, healthy, and able to sustain this level of roadkill, at least in the short term. Each of these alternatives requires different management solutions. Langen (2010) discusses further challenges and potential solutions to using roadkill data for planning mitigation.

Roadkill surveys are frequently conducted using a standardized methodology and by either driving a motor vehicle, walking, or bicycling. Driving provides an advantage in terms of the length of road that can be surveyed, but detectability of small animals can be very poor using this method (Slater 2002; Langen et al.

2007). Langen et al. (2007) provides a comparison of driving, walking, and hybrid surveys for amphibians and reptiles.

Roadkill surveys have a number of methodological challenges that affect what proportion of roadkill animals is detected (Slater 2002). The timing of surveys is important; poorly timed surveys may provide little data, or worse, may provide a misleading picture of where animals most frequently encounter roads. Small animal movements are often seasonal, and road surveys should be timed for these peak movements (Chapter 2). The number of carcasses found on a stretch of road is also a function of the length of the interval between surveys and how long carcasses persist; a number of studies have devised ways to control for survey interval and duration of carcass persistence when estimating roadkill numbers (e.g., Barthelmess and Brooks 2010; Santos et al. 2011). The length of time that small vertebrate carcasses persist on roads varies depending on weather conditions, abundance of scavengers, and traffic volume, but the remains often disappear quite rapidly (Slater 2002). For some nocturnal amphibians whose movements are triggered by rainfall, surveys must be conducted at night before morning traffic and rain obliterate the delicate carcass remains.

The skill of surveyors, the number of surveyors (e.g., driver plus observer vs. solo driver), and the viewing conditions (day vs. night, volume of vehicle traffic encountered during the survey, weather conditions) all affect detectability. Well-designed studies take all of these methodological challenges into account so that despite many roadkill animals being undetected, the observed spatial pattern of detected roadkill correlates with the actual spatial pattern of road mortality.

8.2.2. Predictive Modeling of Critical Locations

The goal of predictive modeling of critical locations is to develop a valid model that uses available data to predict with accuracy and appropriate precision within a planned or existing road network where attention is needed (e.g., potential connectivity blockages or road mortality hotspots). Ideally, such models can also provide a priority rank indicating the likely relative severity of road effects. Potentially, such models can greatly simplify the task of locating critical sites along

extensive road networks. There are three approaches to predictive modeling that are currently being used for planning. The first uses behavior-based models, which use data on how animals move through landscapes, and data on landscape features around a road, in order to predict where road encounters most frequently occur. The second involves road-encounter models, which use data on actual detected road encounters (especially roadkill) to detect spatial aggregations. Road and landscape features associated with the aggregations are used to predict other such critical road segments along road networks. The third approach consists of habitat models, which use data on habitat patterns near roadways to predict where animals are likely to encounter roads.

8.2.2.1. Behavior-Based Models

Behavior-based models use data on how and why individual animals move through the landscape and how they behave when encountering roads; the models then are able to predict where other animals are likely to encounter roads. They further assess how those roads are likely to affect animals' movements and risk of death while crossing. Radiotelemetry is the primary method used to acquire the high-resolution movement data needed to model movement trajectories (see Section 8.2.1). The locations of the primary habitat patches among which animals move must be accurately mapped in relation to roads using GIS. Depending on the taxon, the relevant resource patches may include breeding sites, foraging habitat, and hibernacula (e.g., for modeled movements between upland forest and vernal pools by pond-breeding salamanders, see Compton et al. 2007). By simulating movements of many individuals among habitat patches, it is possible to predict the segments of road where animals are most likely to attempt to cross, or segments where they would cross if not deterred by the road or road traffic (e.g., Litvaitis and Tash 2008). These simulations can also help to elucidate what habitat is accessible to animals when roads provide a barrier to movement (e.g., Eigenbrod et al. 2008).

The simplest models are gravity models, which predict trajectories that are of reduced risk (least cost) in terms of distance traveled between patches, with some cutoff for distances that are considered too long to be feasible for the modeled species (Sen and Smith 1995). By ranking the quality of resource patches, trajectories can be simulated that both minimize the cost of travel while maximizing the quality of the resource patch to which an animal moves (e.g., turtles, Beaudry et al. 2008). Traffic volume and the target species' speed of movement can be incorporated to estimate the probability of successful road crossings, and can be combined with model predictions of hotspots.

A more sophisticated approach is to incorporate landscape resistance, in terms of travel cost coefficients (also known as resistance coefficients), for traversing areas of different land use or for crossing roads. For the purposes of modeling movement trajectories, these cost coefficients are based on how they affect animal behavior—how likely individuals would be to traverse habitat of each resistance class. Although ideally the resistance coefficients are generated directly from behavioral data, they may also be inferred via expert opinion or expert consultation (e.g., Compton et al. 2007). Similar to the simpler gravity models, modeled trajectories are least cost in terms of the product of distance and the resistance coefficients of the landscape traversed (e.g., Joly et al. 2003).

A principal limitation of the behavioral modeling approach is the challenge of acquiring sufficient data about a focal species' local population distributions, its behavior, and ecologically relevant landscape features at a scale relevant for making an accurate and spatially precise prediction of critical locations. Good examples of this approach applied to small animals include Compton et al. (2007), Beaudry et al. (2008), and Patrick et al. (2012). Priorities for mitigation are locations where animals are predicted to frequently traverse but, due to traffic, the probability of a successful crossing is low; or else, they are locations that animals would normally traverse but are deterred by the road and traffic (Joly et al. 2003; Roe et al. 2006; Jaarsma et al. 2007).

8.2.2.2. Road-Encounter Models

Road-encounter modeling, also referred to as hotspot or spatial wildlife-vehicle collision modeling, is a predictive modeling approach that uses data on spatial patterns of road mortality and detected road crossings to identify spatial hotspots of road-crossing attempts. Databases of geo-referenced small vertebrate roadkill

rarely exist; thus, it is usually necessary to do a survey, using methods like those discussed in Section 8.2.1. Survey data for spatial wildlife-vehicle collision models are laborious to collect, and models only apply to the surveyed roads. Moreover, the spatial patterns have the uncertainties of interpretation discussed in Section 8.2.1. Good examples of this approach include Clevenger et al. (2003) and Ramp et al. (2005).

Two standard methods used to determine where hotspots occur are the kernel density estimator and Ripley's K clustering technique. Kernel density estimation indicates where these hotspots are located (e.g., turtles, Langen et al. 2012). Ripley's K clustering technique measures the spatial scale at which roadkills are significantly clustered—that is, it measures how long a stretch of road a hotspot encompasses (e.g., small mammals, Clevenger et al. 2003; turtles, Langen et al. 2012). The information combined from both techniques can be used to locate and prioritize where road mitigation measures should be sited for populations affected by roads. Some computer applications used to do these spatial analyses include SANET (Spatial Analysis along Networks, http://sanet.csis.u-tokyo.ac.jp/) and the R package Spatstat (http://www.spatstat.org/); both sites include useful introductory tutorials on spatial statistical analysis.

A road-encounter model can be extended to predict locations for monitoring and mitigation on roads that have not been surveyed by identifying the local landscape and road attributes that correlate with hotspots. Presuming the model is properly validated, the informative features of the landscape and road can then be used to predict hotspots on other roads, as long as there are data available on the relevant features. Examples of this approach are Langen et al. (2010, 2012).

GIS technology and the growing availability of high-resolution geospatial data will increase the use of predictive models as tools for conservation planning along road networks in coming years. Reviews of spatially explicit predictive models applied to small vertebrate encounters with roads include Langen (2010) and Gunson et al. (2011). A recent extension of the approach incorporates temporal clustering (hot moments, Practical Example 1) as well as spatial clustering (hotspots, Mountrakis and Gunson 2009; Beaudry et al. 2010; Cureton and Deaton 2012).

8.2.2.3. Habitat Models

Spatially explicit habitat modeling is used to predict where road encounters are most likely to occur by integrating GIS data (e.g., topography, hydrology, land cover, and land use), animal occurrences on roads, density and distribution of populations, and habitat use data. The habitat use data may be derived from expert opinion, radio tracking data, or records of roadkill or road crossings. Data needed to create habitat-based models may be less onerous to collect than those needed for the other classes of models, given the wide availability of high-resolution environmental data layers for GIS. It is important to verify that the spatial resolution of these data is adequate for the scale of small vertebrate movements. Standard GIS applications permit the selection of a complex set of attributes (e.g., habit patch minimum size, maximum distance of a patch from the road, maximum distance of two habitat patches divided by the road). Clevenger et al. (2002) provide an example of this approach.

8.2.3. Indirect Methods of Inferring Critical Locations

8.2.3.1. Habitat Maps

Habitat maps are available via remote sensing data (e.g., high resolution orthoimagery) or existing maps of landscape features. For example, data layers available for the United States include National Land Cover Data, National Wetlands Inventory, Digital Elevation Models, and National Hydrology Data (see http://nationalmap.gov/viewer.html). Also, most states have clearinghouses of other kinds of relevant habitat data (e.g., the Massachusetts Natural Heritage and Endangered Species Program, http://www.mass.gov/dfwele/dfw/nhesp/gis_resources.htm).

Using GIS, these data can be overlaid on planned or existing roads to infer where populations of small animals are most likely to be at risk of being negatively affected by roads. For example, in the absence of species occupancy data, high-risk road segments may be inferred from those dividing large patches of habitat where a population of a target species of concern may occur. High-risk road segments may also be inferred from locations passing between two habitat types that are both required for population viability (e.g., vernal pools and upland forest for pond-breeding amphib-

ians). Other high-risk segments are those bisecting habitat that is likely used for movement corridors.

8.2.3.2. Local Informants and Expert Opinion

Local natural historians, conservation or transportation agency personnel, zoologists, and other environmentally aware community members may be able to provide accurate and detailed information on the locations of populations of target species within their region; they also may know where and how populations of the species are affected by roads. Local informants can greatly expedite the planning process by indicating places to avoid when building roads and where to implement mitigation on existing roads. Experts on the target species could be asked to inspect habitat maps, species presence data, and maps of planned or existing roads to provide their opinions. Formal structured processes exist in which teams of experts can evaluate and prioritize potential critical locations; evidence suggests that expert opinions can be superior to predictive models for identifying the best candidate locations for mitigation (Hurley et al. 2009).

8.2.4. Validation

It is essential to validate any model used to predict critical locations for mitigation. First, conduct field surveys at the predicted priority sites. Next, validate the model by applying it to roads that have not been surveyed to evaluate how well it identifies the most promising sites for mitigation. Finally, conduct field surveys to verify the predicted critical locations. These validation steps may reveal either additional factors that should be incorporated to more accurately and efficiently identify critical locations along roads, or else that the model has only limited application to other roads.

Even after a model has been well validated, it is essential that field surveys are conducted in the priority locations identified in order to confirm that these are indeed appropriate for mitigation. As small animals exhibit high degrees of interannual variability and detectability, this validation process ideally would occur over more than one wildlife activity season. When site surveys do not detect the target species, or one or more of these species is observed at an unusually low density, this may indicate either a need for model parameter refinement, or a naturally low year in activity and the need for more data. Alternatively, such low detection could suggest that the road has already negatively affected the species. It is a great challenge to interpret these survey and validation data when collected under limited time frames; this conundrum further reinforces the need to plan ahead when determining mitigation priorities.

8.2.5. Case Study: Population-Level Planning for Blanding's Turtle Populations in Massachusetts

Scott D. Jackson

In Massachusetts, biologists became concerned when a large number of Blanding's turtles (*Emydoidea blandingii*, a state threatened species) were found dead on a highway traversing a conservation area (MA Route 2 at Oxbow National Wildlife Refuge). The majority of Blanding's turtle habitat was south of the highway, but every year females attempted to nest along the highway, and some individuals tried to cross the road to access a small area of wetland to the north. Although creating a more viable connection between habitat north and south of the highway was desirable, it would have been expensive. Further, the amount of habitat available north of the highway was quite small, and it appeared that population connectivity already existed at a bridge crossing over a river; this was based on expert opinion of a local biologist who had studied the population for many years. The more immediate concern was road mortality, which prompted a mitigation project to reduce turtle loss on the highway.

Over 2 km of turtle-proof barrier fencing was installed by the Massachusetts DOT on each side of the highway. The placement of the fencing was based on roadkill surveys (weekly during May and June) conducted by the Massachusetts Division of Fisheries and Wildlife for Blanding's turtles and other species. Previously installed barrier fences contained many gaps between the ground and fence bottom, providing ample opportunity for turtles to continue to access the highway. In 2007, prior to the installation of the turtle fence, 43 turtles were killed along this road segment, 5 of which were Blanding's turtles. The year after the fencing was installed (2008), the number of dead turtles found along the same stretch of highway dropped to four, and none of these were Blanding's

turtles (Lori Erb, Massachusetts Division of Fisheries and Wildlife, personal communication).

It should be noted that the solution to the Blanding's turtle problem could result in serious consequences for nontarget species. The fencing that protected turtles from road mortality may also function as a fragmenting feature in the landscape preventing other species (e.g., cottontail rabbits [*Sylvilagus* spp.] and snowshoe hares [*Lepus americanus*]) from accessing habitat on the other side of the highway. In this specific case, the traffic volume was so high that few species would be able to cross the road in numbers large enough to maintain population continuity. However, it is possible to envision a scenario where mitigation is planned on a small road that might solve a problem for one species while creating problems for other, nontarget species. This illustrates the potential for conflicting trade-offs in mitigation planning. While many mitigation projects may be based on the needs of a particular target species or group of species, maintaining a healthy ecosystem bordering a road project should take into account all species that need to cross the section of road.

8.3. Planning at the Metapopulation Scale

Metapopulation-scale planning focuses on locations where roads intersect and affect animal movements among populations; these sites are often termed habitat linkages (Austin et al. 2006; Beier et al. 2008). A metapopulation consists of a group of spatially distinct populations that are within dispersal distances of each other, thereby allowing both colonization of vacant habitat patches and gene flow between populations (Levins 1969; Hanski 1999). Metapopulation dynamics can be important when populations of a species are small and individually at risk of extinction, but collectively can be maintained by occasional population supplementation and genetic exchange via (1) movements between populations, and (2) recolonization of vacant habitat by members of persisting populations after extirpation of some local populations by chance or after habitat change has resulted in unoccupied suitable habitat (see also Section 3.3.1). Many small animals are likely to be distributed in some form of metapopulation structure (e.g., Gibbs 2000; Semlitsch 2000).

Similar to local-scale planning, metapopulation

planning may target one species or an assemblage of ecologically or taxonomically similar species. However, metapopulation-scale planning focuses on long-range movements that typically do not occur as often, are made by only a small fraction of a population, and therefore are less predictable (Semlitsch 2008). In order to preserve long-term regional persistence of a species, planning for habitat connectivity is essential at both the population and metapopulation scales (Semlitsch 2000, 2008).

The primary goal for metapopulation-scale planning is to maintain long-term regional persistence of a target species by planning road routes that minimize disruption of ecological infrastructure among populations; also, for existing roads, it is important to place mitigation measures, such as passage structures, where they are likely to be effective at maintaining metapopulation connectivity (Hanski 1999; Marsh and Trenham 2001). To fully understand the metapopulation dynamics of a species' local populations, detailed information on population sizes, habitat requirements and distribution, population dynamics, and movement behavior are needed. However, such detailed data are rarely available for metapopulation-scale planning of roads for small animals.

Road mitigation planning at a metapopulation scale is typically focused on habitat. Minimally, a clear understanding of the species' habitat requirements (often termed habitat suitability modeling) and movement behavior (trajectories, movement corridor habitat selection, behavior when encountering roads) are required for metapopulation planning. Collectively, information on habitat requirements, movement corridor preferences, and dispersal distances are termed an "ecological profile," and can be ascertained through the methods described in Section 8.2.1 or from published literature (Vos and Chardon 1998; van der Grift and Pouwels 2006; Eigenbrod et al. 2008; Opdam et al. 2008; Case Study 8.4). Habitat mapping can be done for multispecies assemblages using the mean parameter values for different species (e.g., Vos et al. 2001; Patrick et al. 2012). Other parameters, such as the minimum size of a core habitat needed to maintain breeding populations, are important for studies that focus on population viability (van der Grift and Pouwels 2006).

The information relevant to the ecological profiles of the target animals is then matched with data describing the landscape in GIS (see Section 8.2.2). Once core habitat areas are mapped, they can be connected into a habitat network based on critical thresholds of dispersal distances for the target species (e.g., D'eon et al. 2002; Case Study 8.4). The maximum dispersal distance can be based on field data such as telemetry studies (Macarney et al. 1988; Bartelt et al. 2004; Beaudry et al. 2009), local distribution studies, or other methods (see Section 8.2.1). For example, Vos and Chardon (1998) showed that the mean distance between occupied ponds for three anurans in Europe was less than 1 km, indicating that dispersal distance was likely less than 1 km for these species in that landscape.

Movements among populations (or core habitat patches) within a metapopulation depend on (1) the size, shape, configuration, and spatial dispersion of the core habitat patches; (2) the extent, shape, configuration, and spatial dispersion of habitat suitable for movement corridors; (3) the sizes of populations and their population trajectories (growing, stable, declining); and (4) potential dispersal distances of individual animals (Opdam et al. 1995; Verboom et al. 2001). Increasingly, least-cost or friction methods are used to determine the important linkages between populations (Ray et al. 2002; Joly et al. 2003, Patrick et al. 2012). As described in Section 8.2, several methods exist to help determine places where roads intersect these important connections, both among populations (habitat linkages, metapopulation-scale planning) and within a population's core habitat (hotspots, local-scale planning); these places may be prioritized for mitigation.

In principle, it is important to know how many individuals must disperse among populations to maintain sufficient gene flow and to sustain the metapopulation. Successful exchange of a small number of individuals per generation is usually enough to sustain genetic connectivity (Mills and Allendorf 1996; Taylor and Goldingay 2009). Population viability analysis (PVA) is a powerful modeling approach for assessing whether a metapopulation is likely to persist under current conditions and how it will respond to anticipated changes. Such changes include improved connectivity due to addition of passage structures on roads, or loss of connectivity, core habitat, or individuals from roadkill due to creation of new roads (Jaeger et al. 2005; van der Grift and Pouwels 2006). There are several PVA applications that can be used with GIS for spatially explicit population viability modeling (e.g., RAMAS GIS, http://www.vortex9.org/vortex.html); such technologies are developing rapidly. Spatially explicit PVA can be used to prospectively evaluate alternative scenarios of different combinations of mitigation measures and locations (e.g., Haines et al. 2006).

One additional aspect of metapopulation dynamics of concern to road planners is the creation of "sink" habitat that results in an "ecological trap." Sink habitat is poor quality habitat in which a population cannot be sustained (i.e., survivorship or reproduction is low), and therefore the population can only persist through immigration from elsewhere; populations from which these immigrants originate are called "sources" (Pulliam 1988). An ecological trap is habitat that is attractive to immigrants but is actually a sink (Schlaepfer et al. 2002). Roadside drainage ditches and settling ponds, or road verge vegetation may be sink habitat for some small vertebrate species. If so, from a metapopulation perspective, it may be worthwhile to create barriers that prevent members of viable local populations from colonizing sink habitat. A PVA could be done to evaluate how harmful roadside sink habitat is for metapopulation persistence, and if deemed unacceptably harmful, some mitigation could be put in place (e.g., habitat alteration or barriers to discourage or prevent access).

8.4. Case Study: Road Mitigation Planning for Amphibians at a Metapopulation Scale in Ontario

Namrata Shrestha and Kari E. Gunson

The Region of Peel in the Greater Toronto Area, Ontario, Canada, has developed a framework for urban road mitigation planning for amphibians across its regional jurisdiction (1,253 km²). The region, not including urban areas, has a road density of approximately 0.6 km/km² with a north-south gradient of increasing road density (Figures 8.1 to 8.3). Spring peepers (*Pseudacris crucifer*) and wood frogs (*Lithobates sylvaticus*) were selected as focal species because (1) they were representative of the common amphibian species in this urban landscape and their presence indicates suitable wetland and upland habitat; (2) they were likely to experience conflicts with roads because they move between wetland and upland habitat; and (3) occu-

pancy data existed for these species, as did wetland and upland forest habitat dispersion data suitable for analysis in GIS. The overarching goal was to prioritize areas and road segments where mitigation measures would be most effective for enhancing population-level (local-scale) and metapopulation-level connectivity. At the population scale, the goal was to reduce road mortality between upland forest and wetland habitat areas, and at the metapopulation scale the goal was to improve connectivity among populations.

To map the preferred or core habitat for the target species, two years of occupancy data for amphibians at 20 selected ponds in the study area were used. A logistic regression analysis indicated that both spring peepers and wood frogs preferred wetlands with at least 40% forest cover within 400 m of the wetland. The preferred habitat was mapped in GIS, and the preferred habitat patches were then connected to create habitat networks using 1 km as the critical maximum dispersal distance between the edges of habitat patches (D'eon et al. 2002). This distance was based on the capacity of some anurans to disperse up to 1 km (e.g., Vos and Chardon 1998), and because the mean distance between occupied ponds for some species is less than 1 km (e.g., Helferty 2002). The habitat network map (Figure 8.1) reflects the baseline layer of preferred habitat in the landscape that is available to maintain amphibian population and metapopulation processes.

For population-scale movements, a GIS-based query tool using a gravity model (i.e., least-cost path approach, Section 8.2.2.1) was applied to locate where roads may be barriers to movements between patches of preferred wetland and upland forest habitat. The tool incorporated the following three parameters at locations where a road intersected a wetland-forest connection: (1) size of the wetland, (2) percentage of total forest area across the road (within a distance of 400 m from wetland), and (3) species presence data (reflecting validated existing functionality of preferred habitat). In this study, the road segment was identified as a priority for road mitigation wherever more than 75% of the forest habitat within 400 m of a preferred wetland was on the other side of a road.

A circuit theory model using Circuitscape (McRae et al. 2008) identified likely paths for movements at the metapopulation scale (i.e., dispersal and recolonization movements within the habitat network). In a

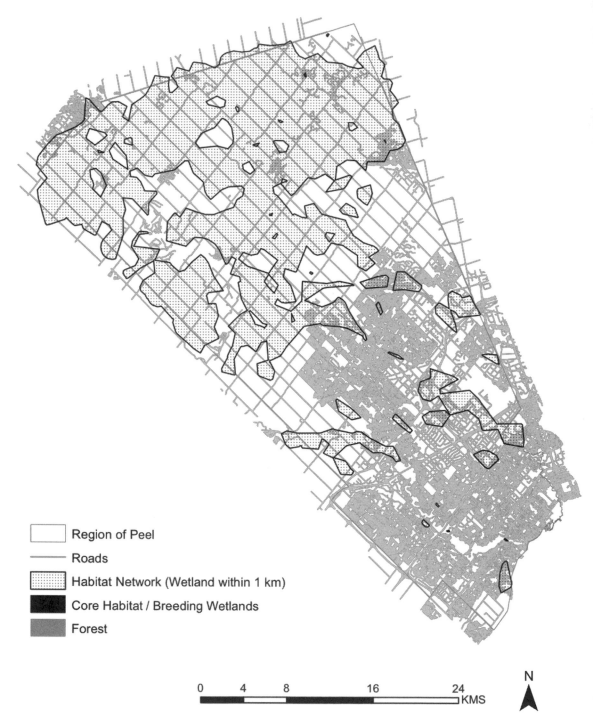

Region of Peel

Roads

Habitat Network (Wetland within 1 km)

Core Habitat / Breeding Wetlands

Forest

0   4   8   16   24
$\Box$ KMS

N

Figure 8.1. Region of Peel, Ontario, Canada, as an example of priority areas for improving amphibian habitat connectivity. This habitat network map reflects most likely areas for amphibian movements among preferred habitats within 1 km of each other. *Credit: Namrata Shrestha.*

circuit theory model, land use, and land cover classes in GIS are assigned scores that weight them as resistance or conductance surfaces. One of the measures generated by this model is "current density," which reflects the probability of the target species moving from one preferred habitat to another and is directly proportional to its weighted conductance surface (Figure 8.2). High current densities indicate the areas that are most permeable, assuming no barriers, and therefore that are likely the critical dispersal pathways for wildlife movement given the state of the surrounding landscape. The road segments that occur in these high current density areas are identified as the priority locations for mitigation to enhance regional connectivity. Both population- and metapopulation-scale priorities were combined to identify the most effective priority locations for mitigation of road impacts for amphibians in the Region of Peel (Figure 8.3). These proposed priority locations were then reviewed by local experts, the Toronto Region Conservation Authority, Credit Valley Conservation, and the Region of Peel.

8.5. Planning at the Landscape or Regional Scale

Road planning at the landscape scale focuses on the big picture and long term. Environmental conditions are always changing; ecological succession, habitat-altering ecological disturbances (e.g., fire, flood, hurricanes, shifting river courses), human alteration of land cover and land use, and climate change all contribute to dynamic changes in landscapes that cause shifts in species distributions. Occupied habitat becomes unsuitable, and new suitable habitat becomes available for colonization. Fundamentally, landscape-scale road planning is intended to avoid having roads be the reason that new suitable habitat fails to be colonized, and it can also ensure genetic interchange among populations, which is important for long-term population viability and evolutionary responsiveness. The task is to identify the habitat linkages that are key to maintaining connectivity across landscapes or regions (geographically contiguous areas that contain multiple landscapes). Good reviews of planning at this scale include Beckman et al. (2010) and Jaeger et al. (2011).

The geographic extent of landscape-scale planning usually depends on the agencies' or organizations' missions that are undertaking the task. It may be a watershed, a state, or a "bioregion" (e.g., the Yukon to Yellowstone northern Rocky Mountain corridor, Chester 2003). Planning at this scale can use all of the tools of the smaller planning scales (local, metapopulation) but it also requires a greater focus on large-scale habitat connectivity and some forecasting of regional environmental change. Landscape- or regional-scale planning of roads almost always includes multiple species as its targets. Most efforts at landscape-scale planning for road mitigation have focused on large animals or on ecosystem-scale processes and landscape connectivity that encompass a wide range of species, both large and small (Hoctor et al. 2000; Ng et al. 2004; Dickson et al. 2005; Beier et al. 2008; Brost and Beier 2012). This type of application is sometimes referred to as an "umbrella" approach, where planning for certain indicator species serves to benefit many species, but whether the needs of all or certain key species are addressed is dependent on selecting an appropriate umbrella or surrogate species (Fleishman et al. 2000; Section 10.3.2).

8.5.1. Landscape-Scale Planning Based on Core Habitat and Corridors

Landscape-scale transportation planning uses methodologies that delineate networks of core habitat areas and movement corridors for wildlife that connect the core habitat. Core habitat and corridor networks (also referred to as green or reserve networks) are well established in Europe (e.g., the Netherlands, Germany; Jongman 1995), the United States (e.g., the states of Arizona, Florida, California; Hoctor et al. 2000; Beier et al. 2006), and in Ontario, Canada (Toronto and Region Conservation Authority 2007; Ontario Ministry of Natural Resources 2010). In connectivity analyses, roads are overlaid on the core habitat and corridor network maps (see Austin et al. 2006), and prioritized locations for mitigation are selected. Priorities include (1) roads that are incursions into, or, worse, bisect, large core habitat areas; and (2) roads that bisect major movement corridors between core habitat areas. These types of analyses are well established at a state scale in the United States (e.g., Maine's Beginning with Habitat program [http://www.beginningwithhabitat.org/], "Arizona's Wildlife Linkages Assessment" [https://www.azdot.gov/business/environmental-services-and-planning/programs/wildlife-linkages]).

Figure 8.2. Region of Peel, Ontario, Canada, as an example of priority areas for improving amphibian habitat connectivity. This cumulative current density map was developed using a circuit theory approach to reflect the likely areas for amphibian movement within the habitat networks. *Credit: Namrata Shrestha.*

Roads
▢ **Region of Peel**
▭ **Habitat Network (Wetland within 1 km)**

Metapopulation level connectivity priority

Low
High

≡≡≡ **Population level connectivity priority**

```
0    4    8        16        24
                              ⌐KMS
```

N
▲

Figure 8.3. Region of Peel, Ontario, Canada, as an example of priority areas for improving amphibian habitat connectivity. This map is an overlay of a cumulative current density map and the population-level priority road segments where 75% more forests are across the preferred habitat. *Credit: Namrata Shrestha.*

Effective landscape-scale planning requires long-term, multi-agency collaboration and commitment, and incorporates other smaller-scale road-planning efforts. Coordination between transportation planners and wildlife professionals is essential. Integration of long-term planning for transportation alongside natural resource protection has only recently been implemented in transportation policy in the United States (e.g., Safe Accountable Flexible Efficient Transportation Equity Act: A Legacy for Users [SAFETEA-LU]; Thorne et al. 2009) and in Europe (Jaeger et al. 2011).

Methodologies for regional planning borrow from those used for population- or metapopulation-scale planning. However, the quality of data available for planning at the scale of regions or landscapes is not typically feasible due to cost and time required to complete detailed surveys across a road network. Thus, planning at this scale relies on the less precise data on population distributions and movement patterns.

8.5.1.1. Core Habitat and Corridor Mapping

Core habitat areas are large areas of protected natural habitat able to sustain a complete community assemblage of wildlife that is characteristic of the landscape. For example, the Yellowstone to Yukon initiative in western North America classifies national parks, national forests, and wildlife refuges as core habitat and delineates a network of movement corridors among them spanning the northern Rocky Mountains from Canada to the northern United States (Clevenger 2012). Core habitat may be designated by state conservation plans that identify biodiversity hotspots or areas with significant or rare ecosystems and species (National Gap Analysis Program; Scott et al. 1993).

In some cases, corridor mapping is based on the relationship between core habitat areas and the composition of the matrix surrounding them (Forman 1995; Ray et al. 2002; Toronto and Region Conservation Authority 2007). In other cases, corridor mapping uses computer models to predict the movements of large animals, such as black bears (*Ursus americanus*) and grizzly bears (*Ursus arctos horribilis*), because they move at a scale relevant to the planning area (Boone and Hunter 1996; Schippers et al. 1996; Singleton and Lehmkuhl 1999; Clevenger et al. 2002; Larkin et al. 2004). Two frequently used classes of models are least-cost paths and circuit theory (see summary in Clevenger and Ford 2010; Section 8.2.2.1).

Corridor-mapping efforts that focus on small animals as the target species are not as common as those that focus on wider-ranging dispersers, because it is difficult to obtain spatial data measured at a resolution that is relevant to small animal movements. Smaller animals also tend to move shorter distances, so it can be harder to conceptualize their movements across large landscapes. However, it remains important to maintain landscape connectivity for smaller animals, and their habitat requirements should be incorporated into plans for landscape linkages. As an example, for the Santa Monica Mountains to Sierra Madre linkage plan as part of the Missing Linkages Project in southern California, small animals, such as desert woodrats (*Neotoma lepida*), western toads (*Anaxyrus boreas*), California kingsnakes (*Lampropeltis californiae*), and tiger whiptail lizards (*Aspidoscelis tigris*), were included in corridor mapping in addition to the more usual large, wide-ranging animals, such as mountain lions (*Puma concolor*) and mule deer (*Odocoileus hemionus*; Beier et al. 2006).

Clevenger and Ford (2010) present the types of geographic data that are useful for landscape-scale planning and can be readily acquired from local authorities. These include land cover and land use, soils, geology, hydrology, topography, land ownership and zoning regulations, linear infrastructure (e.g., roads, transmission line corridors), and rare, localized ecological communities. Quality core habitat and corridor maps, combined with a good knowledge of the behavior and ecology of the animals that are the planning focus, can be user-friendly and effective tools for planning. Movement corridors for many species assemblages typically overlap and follow recognizable features, such as riparian vegetation and mountain ranges.

8.5.1.2. Habitat Linkage Modeling

Habitat linkage modeling is an approach that relies on a computer algorithm to identify core areas and major habitat linkages. Algorithms can be used to both locate and prioritize or rank habitat linkages; this modeling is usually done using GIS. It is possible to focus on locating important linkages in an existing landscape, or to investigate the likely consequences of changes in infrastructure arrangement (e.g., the road network) and land use that increase or decrease habitat connectivity and fragmentation of core habitat areas (McRae et al. 2012). An outstanding source of information and tools for habitat linkage modeling is CorridorDesign.org, developed by Paul Beier and colleagues, which includes the essential introductory resource "Conceptual steps for designing wildlife corridors" (http://corridordesign .org/designing_corridors/).

There are many approaches to creating habitat linkage models; the methodology and tools are developing rapidly in parallel with more sophisticated GIS applications and more powerful computers. Modeling approaches include individual-based movement models, least-cost or habitat resistance corridor modeling, circuit theory, graph theory, and centrality analyses (McRae et al. 2012). One popular habitat linkage modeling software application, which uses a least-cost or habitat resistance corridor modeling approach, is LinkageMapper (http://code.google.com/p/linkage -mapper/). An advantage of habitat linkage modeling is that it makes it possible to objectively analyze spatial patterns across a large and complex region, integrate many different kinds of information and decision criteria, and produce a map that indicates where habitat linkages are likely to be most worth preserving or enhancing. However, the configuration of the habitat linkage map will be sensitive to the specific algorithm used to map linkages and the quality (spatial resolution, precision, and accuracy) of the data used. One must be appropriately cautious about the output of a habitat modeling project and take measures to evaluate the validity of the modeling approach and its resulting map (Beier et al. 2008, 2009; Sawyer et al. 2011; Brost and Beier 2012).

As one example, Thorne et al. (2009) integrated regional conservation designs, also known as "greenprints," with early multi-project mitigation assessment in California (see also Girvetz et al. 2008). They used GIS to identify the footprint, or areal extent, of planned road projects. The footprints were overlaid on habitat maps to quantify potential habitat impacts of road projects in two regions; this generated a summary analysis of aggregate regional impacts. Marxan, a reserve-selection algorithm (http://www.uq.edu.au /marxan/), was used to identify a regional greenprint

for each site and to also identify parcels for acquisition that would mitigate or offset the impacts of the road project and contribute to the greenprint. The two regions differed by the types and amount of data available, their respective conservation objectives, and land management; these regional differences are typical of the range of conditions that conservation practitioners experience. This type of GIS-based assessment can contribute to collaboration among transportation planners and environmental managers by informing systematic mitigation planning. Other good recent case studies of applied habitat linkage modeling include van der Grift and Pouwels (2006), Gurrutxaga et al. (2011), and Beier et al. (2012).

8.5.1.3. Case Study: The Conservation Assessment and Prioritization System

Scott D. Jackson

An example of how GIS can be used to locate road mitigation projects builds upon the landscape modeling and assessment capabilities of the Conservation Assessment and Prioritization System (CAPS, www.umasscaps .org). Developed at the University of Massachusetts Amherst (UMass), CAPS is a computer software program that uses an ecosystem-based approach to assess the ecological integrity of land and water across a landscape and subsequently to identify and prioritize land for habitat and biodiversity conservation.

The CAPS approach begins with the characterization of both the developed and undeveloped elements of the landscape. With a computer base map depicting various classes of developed and undeveloped land, a variety of landscape-based variables ("metrics") are evaluated for every point in the landscape. A metric may, for example, take into account the microclimatic alterations associated with edge effects, intensity of road traffic in the vicinity, nutrient loading in aquatic ecosystems, or the effects of human development on landscape connectivity.

Various metrics are applied to the landscape and then integrated in weighted linear combinations as models for predicting ecological integrity. This process results in a final index of ecological integrity (IEI) for each point in the landscape, based on models constructed separately for each ecological community (e.g., forest, shrub swamp, headwater stream) within an area. Results for the individual metrics are saved to facilitate analysis—thus, one can examine not only a map of the final indices of ecological integrity, but also maps of road traffic intensity, connectedness, microclimate alterations, and so on.

In 2006, The Nature Conservancy approached UMass about using CAPS to complete a comprehensive analysis of Massachusetts to identify areas where connections must be protected and restored to support the state's wildlife and biodiversity. The Critical Linkages project (http://umasscaps.org/applications/critical-linkages.html) was initiated in 2008 to develop spatially explicit tools for use in mitigating the impacts of roads and railroads on the environment, including models, maps, and scenario-testing software.

The Critical Linkages project employed a "coarse-filter" approach to assess connectivity (i.e., one that does not involve any particular focal species but instead holistically considers ecological systems). In order to address biodiversity conservation in its broadest sense, two scales were distinguished for assessing connectivity. Local connectivity refers to the spatial scale at which the dominant organisms interact directly with the landscape via demographic processes such as dispersal and home range movements. Regional connectivity refers to the spatial scale exceeding that in which organisms directly interact with the landscape. This is the scale at which long-term ecological processes such as species range expansion or contraction and gene flow occur.

Because CAPS provides a quantitative assessment for IEI as well as for each metric used in the ecological integrity models, it can be used for comparing various scenarios. Scenario analysis involves running CAPS separately for each scenario, and comparing results to determine the loss (or gain) in IEI or in the index for a specific metric. At the local scale, the Critical Linkages project used the scenario-testing capabilities of CAPS to assess changes in the connectedness and aquatic connectedness metrics for dam removal, culvert or bridge replacement, and construction of wildlife passage structures on roads. The connectedness metric is a measure of the degree to which a focal cell is interconnected with other cells in the landscape that are a potential source of individuals or materials contributing to long-term ecological integrity. Connectedness is based on a "resistant kernel" model (Figure 8.4), introduced

by Compton et al. (2007), which is a hybrid between two existing approaches: the standard kernel estimator and least-cost paths based on resistant surfaces.

An assessment of connectedness (Figure 8.5) and aquatic connectedness (Figure 8.6) provided a statewide baseline scenario to use for comparison of restoration options. Scenario-testing software was developed to efficiently assess restoration potential for large numbers of possible restoration projects and then applied statewide to dams, road-stream crossings (Figure 8.7), and road segments (Figure 8.8). Results of these analyses indicate that a relatively small proportion of culvert replacements or dam removals would result in substantial improvements in aquatic connectivity.

A hybrid of the resistant kernel estimator approach (used at the local scale) and a graph theoretic approach (where the landscape is represented graphically via nodes and the linear connections among them) was used to assess connectivity at a regional scale. This hybrid system maintains the spatial realism of the resistant kernel estimator approach and capitalizes on the computational efficiencies of the graph matrix representation.

8.5.1.4. Expert Opinion

Consultation with wildlife professionals (e.g., wildlife biologists, landscape ecologists, hydrologists, environmental engineers) can be valuable for both identifying core habitat and critical movement corridors and for validating existing core habitat and corridor maps (e.g., Clevenger and Ford 2010). Facilitated workshops and discussions among experts, stakeholders, decision makers, and transportation planners can be conducted to evaluate, negotiate, and prioritize where mitigation should be implemented (e.g., the Rapid Assessment Process; Ruediger and Lloyd 2003; Box 6.1). In addition to professional experts who are active in the relevant biological fields, experts may also include landowners, highway maintenance crews, and concerned citizens familiar with the local study area. Stakeholders and decision makers generally include transportation and land-use planners, engineers, and officials in land regulatory agencies. To be effective, such workshops must provide experts with relevant data (e.g., core habitat and corridor maps and aerial imagery, land-use planning and zoning maps).

A helpful final product of a workshop can be a table

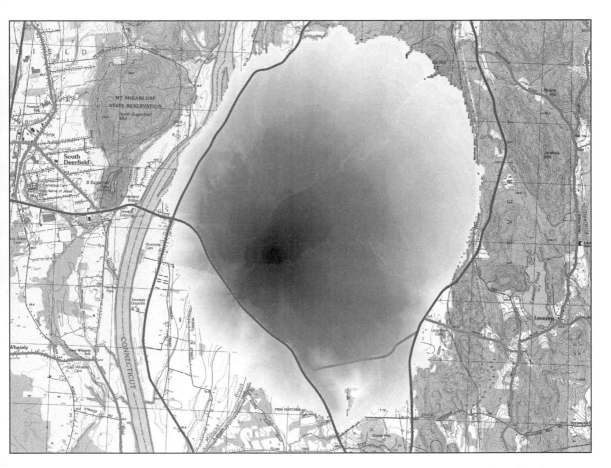

Figure 8.4. An example of a resistant kernel. Resistant kernels are calculated for each cell of the undeveloped landscape using a resistant surface that is unique for that cell. Kernel values are highest at the focal cell (the darker the color, the greater the interconnectedness and the greater the ease with which an organism can move through the landscape). The value is lost as the kernel spreads through the resistant landscape (with the color becoming lighter as the case with which an organism can move decreases). *Credit: Bradley W. Compton.*

that includes all candidate sites for mitigation and relevant information about each, including geo-referenced locations. Such a table can help justify and encourage implementation, and it can help guide both short-term and long-term mitigation efforts. Lastly, workshop products should be ground truthed by biologists, engineers, and highway construction officials prior to finalization. This is to ensure that the mitigation plans can be implemented and maintained effectively and over the long term within the landscape and its associated topography and hydrology.

8.5.2. Mitigation Options

One outcome of the planning process must include strategic determination of *where* to prioritize mitiga-

tion efforts. A second outcome should be how extensive and expensive a mitigation option is justified at each priority site. As budgets are always limited for environmental mitigation, there is an optimization challenge; in other words, the more money spent on mitigation at each site, less money is available for mitigation at other sites. For example, where a major highway bisects two high priority core habitat areas, a mitigation measure as technically complex, politically sensitive, and expensive as a wildlife overpass crossing structure may be justified. At other sites, less expensive mitigation measures such as culverts, fencing, or signage may be sufficient. It takes multidisciplinary teams of planners, engineers, landscape architects, wildlife biologists, transportation operations personnel, and

Figure 8.5. Connectedness metric for an area on the north shore of Massachusetts, United States. Areas in darker colors are more interconnected with similar areas nearby than those depicted in lighter colors. White areas are developed land. Aquatic connectedness functions much like terrestrial connectedness but is constrained to move only along the center lines of streams and rivers or through water bodies and wetlands. *Credit: Scott D. Jackson.*

economists to determine an optimal mitigation plan for an entire landscape. Further details of specific mitigation measures and options are discussed in Chapters 9 and 10.

8.5.2.1. Maintaining Roadless Areas

In an ideal world, the best way to avoid or reduce the negative impact of roads is by preventing new roads from being routed through priority core habitat or across key linkage corridors. An important planning objective at the landscape or regional scale would be to identify large roadless areas and to avoid or minimize road-associated fragmentation (Selva et al. 2011). Although politically contentious, road closures are a way to remove existing linkage blockages and to temporarily defragment core habitat. Havlick (2002) provides a

good review of the challenges and benefits of road closures on national forests and other public lands within the United States.

Mesh analysis provides a tool to analyze how much existing roads fragment a landscape, to what degree proposed roads will further fragment the landscape, and how much can be gained by alternatives (e.g., removing roads or changing planned locations of roads) to defragment or minimize fragmentation where possible (Jaeger 2000; Jaeger et al. 2006; Jaeger 2007; Jaeger et al. 2007; Jaeger et al. 2011; Chapter 3). In this analysis, the road network is considered a mesh, and the spaces between the mesh are habitat patches. Planning strategies that result in larger habitat patches include "road bundling," which is locating (or relocating) roads in a road network in such a way that parallel

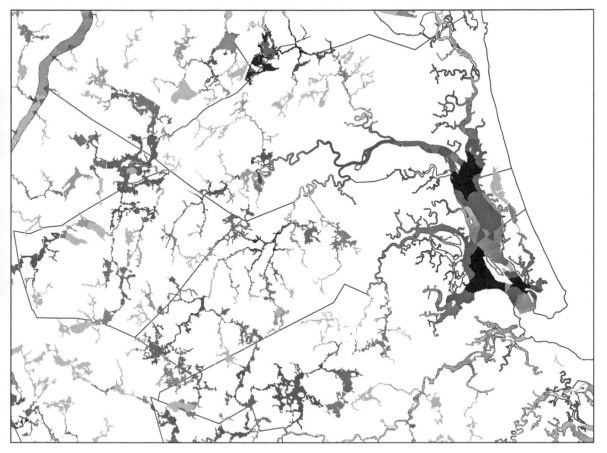

Figure 8.6. Aquatic connectedness metric for an area on the north shore of Massachusetts, United States. This metric is applied only to wetland and aquatic communities. Darker areas are more interconnected with similar areas nearby than those depicted in lighter color. *Credit: Scott D. Jackson.*

roads tend to be clustered. Road bundling may have environmental benefits, but may also result in less convenient travel routes and other social and economic consequences.

Another way to avoid or reduce negative road effects is to require a no-development buffer around key habitat or habitat features. For example, when planning new roads, a planner can stipulate that a road must be located a minimum distance away, or buffer distance, from water bodies. This distance should be set by biological considerations (Semlitsch 1998; Semlitsch and Bodie 2003), such as the distance that aquatic turtles move onto land to nest (Steen et al. 2012). Studies have shown that some regulatory buffer distances may be based on limited scientific evidence, and thus may be inadequate to minimize negative impacts (Houlahan and Findlay 2004).

8.5.2.2. Strategic Placement and Design of Passage Structures

Effective landscape-scale planning will indicate where important disruptions in ecological infrastructure are likely to occur along the road network. One of the most important outcomes of planning is identifying where mitigation efforts that improve habitat connectivity will result in the greatest conservation gains. However, prioritizing where to focus mitigation efforts should also consider projected conditions in the future. Changes in several factors, such as patterns of human development, land use, traffic patterns, and climate conditions, may result in shifts in movement corridors for both people and animals. Thus, these alterations result in shifts in the locations of important habitat linkages and in the location and severity of disruptions in habitat connectivity (Nuñez et al. 2013). A strategic

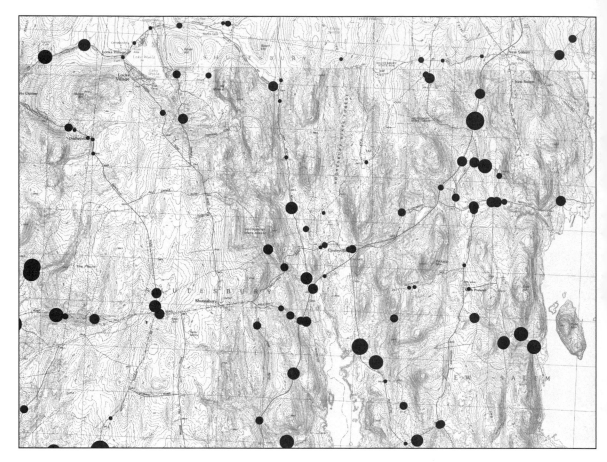

Figure 8.7. Results of scenario analyses for culvert or bridge replacements for a portion of Massachusetts, United States. Size of the circles is proportional to the change in "aquatic connectedness" that would be achieved by replacing a crossing structure. The larger the circles, the greater the improvement in aquatic connectedness. *Credit: Scott D. Jackson.*

plan for placement of passage structures should incorporate landscape change considerations, because this change can happen on a time scale that is relevant to road design decisions. Landscape change is also relevant to selecting what mitigation measures are appropriate. That is, will a passage structure be adequate for future needs, or will it be designed in a way that can encompass later upgrades or alterations? A good discussion of why and how to develop plans that result in strategic placement and design of mitigation measures is provided in Brown (2006).

8.5.3. Prioritization and Planning for the Long Term

When a large region with a vast road network is the focus of planning, many locations along roads may be identified as worthy of attention, but it may not be fea-

sible for transportation agencies to mitigate all road effects at all locations. Therefore, the challenge lies in determining where a transportation agency should invest mitigation funds to best meet the conservation needs of the target species, the objectives of the planning project, and the policy obligations of the agencies. The goals of the responsible agencies involved in a landscape-scale assessment are numerous; hence, criteria should be established to prioritize locations for mitigation. These criteria should be discussed through expert consultation or the use of multicriteria decision tools involving transportation and conservation planners and other stakeholders (Opdam et al. 2008; Clevenger 2012). Other tools that use a set of user-defined criteria to prioritize locations include computer applications such as Marxan (e.g., Thorne et al. 2009, see also Section 8.5.1.2) and CAPS (Conservation

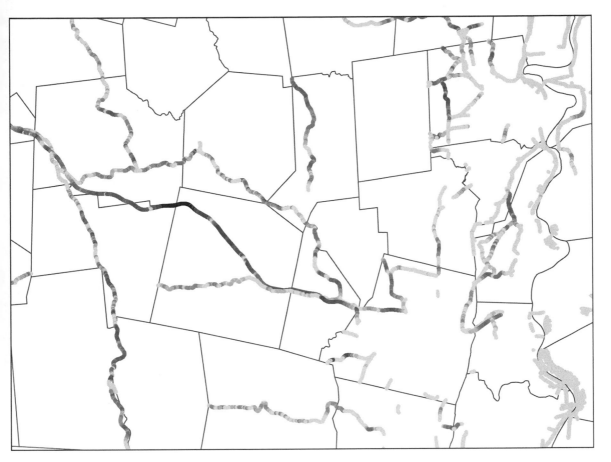

Figure 8.8. Results of scenario analyses for potential wildlife passage structures for a portion of Massachusetts, United States. The color of the lines is proportional to the change in "connectedness" that would be achieved by the construction of a wildlife passage structure. The darker the color, the greater the benefit of using a passage structure at that location. *Credit: Scott D. Jackson.*

Assessment and Prioritization System, www.umasscaps .org; Case Study 8.5.1.3). Examples of relevant factors in the decision-making process include land use, land ownership, landowner and community attitudes, conservation significance, known road impacts on wildlife, degree of habitat fragmentation, and mitigation options (Thorne et al. 2009; Clevenger 2012). An excellent recent review on how to prioritize and plan for the long term is provided in Beier et al. (2012).

For example, to integrate road-project planning into statewide conservation objectives, Smith (1999, 2006b) created a rule-based GIS model to identify and prioritize ecological hotspots on state roads for the Florida DOT (Case Study 10.6.3). The aim was to target locations at the statewide level where construction of new underpasses or culverts (or modification

to existing structures) should be considered to restore landscape connectivity and ecological processes. The economics of an effort of such a large scope dictated the need for a method to identify and prioritize such projects. Priorities might include chronic roadkill sites, locations of listed species or species of conservation interest, focal species concentrations, known or predicted movement or migration routes, land cover, riparian corridors, greenway linkages, strategic habitat conservation areas, existing and proposed conservation lands, traffic volume, and existing and future road projects. Each of the criteria was grouped into the following six categories of information: biological features, landscape features, chronic roadkill sites, conservation planning, public ownership, and traffic and other infrastructure. A weighting algorithm was applied to these categories to create a scale of importance or ranking. Weightings were assigned based on expert opinion surveys. A database of both upcoming programmed road projects and existing structures suitable for wildlife passage upgrades was developed. Several road projects and suitable existing structures were identified at locations with high rankings, and gradually mitigation measures have been implemented to maintain and improve permeability for wildlife across Florida.

Comprehensive prioritization and planning efforts advance the development of wildlife mitigation solutions while addressing economic, political, and social constraints on conservation and transportation infrastructure needs. Lastly, a more comprehensive and organized planning process involving representatives of relevant stakeholder groups saves both time and money while minimizing impacts on wildlife and ecological infrastructure.

8.5. Key Points

- Planning and proactive mitigation is more effective and less expensive than reactive mitigation of road impacts after problems have become evident.
- Effectively determining priority species and locations for mitigation and management is the first and critical step in the assessment of road effects on wildlife and ecological infrastructure.
- Priority mitigation locations are determined by identifying spatial peaks (hotspots) and times (hot moments) where wildlife-road conflicts occur.
- An effective plan requires a multidisciplinary team of land planners, civil engineers, transportation agency operations personnel, wildlife biologists, landscape architects, and others.
- Planning for avoidance or mitigation of road impacts on small animals may be targeted at the local (population) scale, metapopulation scale, and landscape scale. The scope or scale of planning will be determined primarily by the goals of the stakeholders, but also by the budget, target species, and data availability. While scales may be assessed independently, multiple scales can be incorporated in a given project.
- From a local-scale planning perspective, reducing road mortality, local negative environmental

impacts of roads, and barrier effects (habitat fragmentation) may all be objectives.
- From the metapopulation- and landscape-scale planning perspectives, reduction of habitat fragmentation and increased habitat connectivity are the primary objectives. Metapopulation scales tend to focus on habitat whereas landscape scales investigate wildlife habitat, movement corridors, and human land use configurations. Both of these scales often require the development of technology-based models to assess large-scale effects.
- There are usually fewer data available on small animals than large mammals. Large mammals are the typical focus of wildlife-road conflict reduction and planning projects. Planners must incorporate and synthesize a variety of sources of data to locate critical areas for avoiding road construction or for mitigation projects benefitting small animals.
- During the planning process, transportation officials often request species occurrence data from federal and state natural resource agencies as their primary sources of species distribution information; as such, it is important for biologists and other species experts to provide and maintain current findings with these agencies. This reporting will allow a better assessment by transportation officials of where road conflicts may occur for a given species and which locations may be priorities for pre-construction surveys and mitigation.
- Methodologies for collecting field data and modeling road effects are well established for large animals, and they can often be used for small animals as well. However, there are unique challenges in both directly studying small animal movements in relation to roads and using computer simulations of their movements. Thus, innovative techniques and approaches may be required to plan and implement effective mitigation measures that minimize negative road impacts.
- Field data should be collected whenever possible before a plan is finalized. These data serve multiple purposes for direct impact assessments, model parameterization (before model development), and model validation (after model assessment).

- While predictive modeling, inferential habitat mapping, and expert opinion are useful tools when direct field assessments are not possible or are insufficient, these models or inferences must be ground truthed with site visits and monitoring to confirm whether identified road segments do warrant management action. When critical locations for mitigation are identified by models and inferences, collecting field data (ground truthing) is again an essential step to validate the accuracy of the identification method.

- When site surveys are conducted to ground truth model results and do not detect the target species, or one or more of these species is observed at an unusually low density, either model parameter refinement or additional field survey data are needed; alternatively, this could suggest that the road has already negatively affected the species' local population.

- While road development is necessary for all functioning aspects of society, proactively determining and then avoiding or minimizing development in roadless areas, particularly in those known or determined to be core habitat or corridor linkage areas for wildlife, can minimize the negative impacts of transportation infrastructure.

- A primary consideration in planning is projecting how the landscape is likely to change in the future, and whether this will be at a time scale relevant to the project's outcomes. In many regions, landscapes are changing rapidly, and therefore planning recommendations must anticipate and incorporate these changes.

LITERATURE CITED

Andrews, K. M., and J. W. Gibbons. 2005. How do highways influence snake movement? Behavioral responses to roads and vehicles. Copeia 2005:772–782.

Ashley, E. P., and J. T. Robinson. 1996. Road mortality of amphibians, and other wildlife on the Long Point Causeway, Lake Erie, Ontario. Canadian Field Naturalist 110:403–412.

Austin J. M., K. Viani, F. Hammond, and C. Slesar. 2006. A GIS-based identification of potentially significant wildlife habitats associated with roads in Vermont. Pages 185–196 in Proceedings of the 2005 international conference on ecology and transportation. C. L. Irwin, P. Garrett, and K. P. McDermott, editors. Center for Transportation and the Environment, North Carolina State University, Raleigh, North Carolina, USA.

Bartelt, P. E., C. R. Peterson, and R. W. Klaver. 2004. Sexual differences in the post-breeding movements and habitats selected by Western toads (Bufo boreas) in southeastern Idaho. Herpetologica 60:455–467.

Barthelmess, E. L., and M. S. Brooks. 2010. The influence of body-size and diet on road-kill trends in mammals. Biodiversity and Conservation 19:1611–1629.

Beaudry, F., P. G. deMaynadier, and M. L. Hunter. 2008. Identifying road mortality threat at multiple spatial scales for semi-aquatic turtles. Biological Conservation 141:2550–2563.

Beaudry, F., P. G. deMaynadier, and M. L. Hunter. 2009. Seasonally dynamic habitat use by spotted (Clemmys guttata) and Blanding's turtles (Emdoidea blandingii) in Maine. Journal of Herpetology 43:636–645.

Beaudry, F., P. G. deMaynadier, and M. L. Hunter. 2010. Identifying hot moments in road-mortality risk for freshwater turtles. Journal of Wildlife Management 74:152–159.

Beckmann, J. P., A. P. Clevenger, M. P. Huijser, and J. A. Hilty. 2010. Safe passages: Highways, wildlife and habitat connectivity. Island Press, Washington, DC, USA.

Beier P., D. Majka, and S. Newell. 2009. Uncertainty analysis of least-cost modeling for designing wildlife linkages. Ecological Applications 19:2067–2077.

Beier, P., D. Majka, and W. Spencer. 2008. Forks in the road: choices in procedures for designing wildlife linkages. Conservation Biology 22:836–851.

Beier, P., K. L. Penrod, C. Luke, W. D. Spencer, and C. Cabanero. 2006. South Coast missing linkages: Restoring connectivity to wildlands in the largest metropolitan area in the USA. Pages 555–585 in Connectivity conservation. K. R. Crooks and M. Sanjayan, editors. Cambridge University Press, New York, New York, USA.

Beier, P., W. Spencer, R. F. Baldwin, and B. H. McRae. 2012. Toward best practices for developing regional connectivity maps. Conservation Biology 25:879–892.

Bibby C. J., N. D. Burgess, D. A. Hill, and S. Mustoe. 2000. Bird census techniques. Second edition. Academic Press, San Diego, California, USA.

Boone, R. B., and M. L. Hunter. 1996. Using diffusion models to simulate the effects of land use on grizzly bear dispersal in the Rocky Mountains. Landscape Ecology 11:51–64.

Boonstra, R., and I.T.M. Crane. 1986. Natal nest location and small mammal tracking with a spool and line technique. Canadian Journal of Zoology 64:1034–1036.

Bouchard, J., A. T. Ford, F. E. Eigenbrod, and L. Fahrig. 2009. Behavioral responses of northern leopard frogs (Rana pipiens) to roads and traffic: implications for population persistence. Ecology and Society 14:23. http://www.ecologyandsociety.org/vol14/iss2/art23/. Accessed 23 August 2014.

Brehme, C. S., J. A. Tracey, L. R. McClenaghan, and R. N. Fisher. 2013. Permeability of roads to the movement of scrubland lizards and small mammals. Conservation Biology 27:710–720.

Brost, B. M., and P. Beier. 2012. Comparing linkage designs based on land facets to linkage designs based on focal species. PLoS ONE 7(11):e48965.

Brown, J. W. 2006. Eco-logical: an ecosystem approach to developing infrastructure projects. Federal Highway Administration Report FHWA-HEP-06-011. Federal Highway Administration, Washington, DC, USA.

Chester, C. C. 2003. Responding to the idea of transboundary conservation: an overview of the public's reaction to the Yellowstone to Yukon (Y2Y) Conservation Initiative. Journal of Sustainable Forestry 17:103–125.

Clevenger, A. P. 2012. Mitigating continental-scale bottlenecks: how small-scale highway mitigation has large-scale impacts. Ecological Restoration 30:300–307.

Clevenger, A. P., B. Chruszcz, and K. Gunson. 2003. Spatial patterns and factors influencing small vertebrate fauna road kill aggregations. Biological Conservation 109:15–26.

Clevenger, A. P., and A. T. Ford. 2010. Wildlife crossing structures, fencing and other highway design considerations. Pages 17–49 in Safe passages: Highways, wildlife and habitat connectivity. J. P. Beckmann, A. P. Clevenger, M. P. Huijser, and J. A. Hilty, editors. Island Press, Washington, DC, USA.

Clevenger, A. P., and M. P. Huijser. 2011. Wildlife crossing structure handbook: design and evaluation in North America. Report No. FHWA-CFL/TD-11-003. Federal Highway Administration, Washington, DC, USA.

Clevenger, A. P., J. Wierzchowski, B. Chruszcz, and K. Gunnison. 2002. GIS-generated, expert based models for wildlife habitat linkages and mitigation planning. Conservation Biology 16:503–514.

Compton, B. W., K. McGarigal, S. A. Cushman, and L. R. Gamble. 2007. A resistant kernel model of connectivity for vernal pool amphibians. Conservation Biology 21:788–799.

Corn, P. S., and R. B. Bury. 1990. Sampling methods for terrestrial amphibians and reptiles. General Technical Report PNW-GTR-256. Pacific Northwest Field Station, US Forest Service, Portland, Oregon, USA.

Cureton, J. C., and R. Deaton. 2012. Hot moments and hot spots: identifying factors explaining temporal and spatial variation in turtle road mortality. Journal of Wildlife Management 76:1047–1052.

D'eon, R. G., S. M. Glenn, I. Parfitt, and M. Fortin. 2002. Landscape connectivity as a function of scale and organism vagility in a real forested landscape. Conservation Ecology 6:10. http://www.consecol.org/vol6/iss2/art10/. Accessed 23 August 2014.

Dickson, B. G., J. S. Jenness, and P. Beier. 2005. Influence of vegetation, topography, and roads on cougar movement in southern California. Journal of Wildlife Management 69:264–276.

Dodd, C. K., Jr. 2002. North American box turtles: A natural history. University of Oklahoma Press, Norman, Oklahoma, USA.

Dorcas, M. E., and J. D. Wilson. 2009. Innovative methods for studies of snake ecology and conservation. Pages 5–37 in Snakes: Ecology and conservation. S. J. Mullin and R. A. Seigel, editors. Cornell University Press, Ithaca, New York, USA.

Eigenbrod, F. S., J. Hecnar, and L. Fahrig. 2008. Accessible habitat: an improved measure of the effects of habitat loss and roads on wildlife populations. Landscape Ecology 23:159–168.

Fahrig, L., and T. Rytwinski. 2009. Effects of roads on animal abundance: an empirical review and synthesis. Ecology and Society 14:21. http://www.ecologyandsociety.org/vol14/iss1/art21/. Accessed 24 August 2014.

Fleishman, E., D. D. Murphy, and P. F. Brussard. 2000. A new method for selection of umbrella species for conservation planning. Ecological Applications 10:569–579.

Forman, R.T.T. 1995. Land mosaics: The ecology of landscapes and regions. Cambridge University Press, Cambridge, UK.

Furman, B.L.S., B. R. Scheffers, and C. A. Paszkowski. 2011. Use of fluorescent powdered pigments as a tracking technique for snakes. Herpetological Conservation and Biology 6:473–478.

Gibbs, J. P. 2000. Wetland loss and biodiversity conservation. Conservation Biology 14:314–317.

Gibbs, J., and W. Shriver. 2005. Can road mortality limit populations of pool-breeding amphibians? Wetlands Ecology and Management 13:281–289.

Girvetz, E. H., J. H. Thorne, A. M. Berry, and J.A.G. Jaeger. 2008. Integration of landscape fragmentation analysis into regional planning: a statewide multi-scale case study from California, USA. Landscape and Urban Planning 86:205–218.

Gompper, M. E., R. W. Kays, J. C. Ray, S. D. Lapoint, D. A. Bogan, and J. R. Cryan. 2006. A comparison of noninvasive techniques to survey carnivore communities in northeastern North America. Wildlife Society Bulletin 34:1142–1151.

Graeter, G. J., K. A. Buhlmann, L. R. Wilkinson, and J. W. Gibbons, editors. 2013. Inventory and monitoring: recommended techniques for reptiles and amphibians. Technical Publication IM-1. Partners in Amphibian and Reptile Conservation, Birmingham, Alabama, USA.

Graeter, G. J., and B. B. Rothermel. 2007. The effectiveness of fluorescent powdered pigments as a tracking technique for amphibians. Herpetological Review 38:162–166.

Gunson, K. E., D. Ireland, and F. W. Schueler. 2012. A tool to prioritize high-risk road mortality locations for wetland-forest herpetofauna in southern Ontario, Canada. North-Western Journal of Zoology 8:409–413.

Gunson K. E., G. Mountrakis, and L. J. Quackenbush. 2011. Spatial wildlife-vehicle collision models: a review of current work and its application to transportation mitigation projects. Journal of Environmental Management 92:1074–1082.

Gurrutxaga, M., L. Rubio, and S. Saura. 2011. Key connectors in protected forest area networks and the impact of highways: a transnational case study from the Cantabrian Range to the Western Alps (SW Europe). Landscape and Urban Planning 101:310–320.

Guthrie, J. M. 2012. Modeling movement behavior and road crossing in the black bear of south central Florida. Master's thesis, University of Kentucky, Lexington, Kentucky, USA.

Haines, A. M., M. E. Tewes, L. L. Laack, W. E. Grant, and J. Young. 2006. Evaluating recovery strategies for an ocelot (*Leopardus pardalis*) population in the United States. Biological Conservation 26:512–522.

Hanski, I. 1999. Metapopulation ecology. Oxford University Press, Oxford, UK.

Havlick, D. G. 2002. No place distant: Roads and motorized recreation on America's public lands. Island Press, Washington, DC, USA.

Helferty, N. 2002. Natural heritage planning for amphibians and their habitats. Supplementary Report for Oak Ridges Moraine Richmond Hill Ontario Municipal Board Hearing. Save the Rouge Valley System, Inc. and the City of Toronto, Toronto, Ontario, Canada.

Heyer, W. R., M. A. Donnelly, R. W. McDiarmid, L. C. Hayek, and M. S. Foster. 1994. Measuring and monitoring biological diversity: Standard methods for amphibians. Smithsonian Institute Press, Washington, DC, USA.

Hoctor, T. S., M. H. Carr, and P. D. Zwick. 2000. Identifying a linked reserve system using a regional landscape approach: the Florida ecological network. Conservation Biology 14:984–1000.

Houlahan, J. E., and C. S. Findlay. 2004. Estimating the "critical" distance at which adjacent land-use degrades wetland water and sediment quality. Landscape Ecology 19:677–690.

Huber, P. R., J. H. Thorne, and N. R. Siepel. 2013. Convergence of green- and blueprints: integrating long-range transportation planning and landscape connectivity. *In* Proceedings of the 2013 international conference on ecology and transportation. http://www.icoet.net/ICOET_2013/documents/papers/ICOET2013_Paper205B_Huber_at_al.pdf. Accessed 24 August 2014.

Hurley, M. V., E. K. Rapaport, and C. J. Johnson. 2009. Utility of expert-based knowledge for predicting wildlife-vehicle collisions. Journal of Wildlife Management 73:278–286.

Jaarsma, C. F., F. van Langevelde, J. M. Baveco, M. Van Eupen, and J. Arisz. 2007. Model for rural transportation planning considering simulating mobility and traffic kills in the badger *Meles meles*. Ecological Informatics 2:73–82.

Jaeger, J. 2000. Landscape division, splitting index, and effective mesh size: new measures of landscape fragmentation. Landscape Ecology 15:115–130.

Jaeger, J. 2007. Effects of the configuration of road networks on landscape connectivity. Pages 267–280 *in* Proceedings of the 2007 international conference on ecology and transportation. C. L. Irwin, D. Nelson, and K. P. McDermott, edi-

tors. Center for Transportation and the Environment, North Carolina State University, Raleigh, North Carolina, USA.

Jaeger, J., R. Bertiller, and C. Schwick. 2007. Degree of landscape fragmentation in Switzerland: quantitative analysis 1885–2002 and implications for traffic planning and regional planning. Condensed version. Federal Statistical Office, Neuchâtel, Switzerland.

Jaeger, J.A.G., J. Bowman, J. Brennan, L. Fahrig, D. Bert, J. Bouchard, N. Charbonneau, et al. 2005. Predicting when animal populations are at risk from roads: an interactive model of road avoidance behavior. Ecological Modeling 185:329–348.

Jaeger J. A., and L. Fahrig. 2004. Effects of road fencing on population persistence. Conservation Biology 18:1651–1657.

Jaeger J. A., L. Fahrig, and K. C. Ewald. 2006. Does the configuration of road networks influence the degree to which roads affect wildlife populations? Pages 151–163 *in* Proceedings of the 2005 international conference on ecology and transportation. C. L. Irwin, P. Garrett, and K. P. McDermott, editors. Center for Transportation and the Environment, North Carolina State University, Raleigh, North Carolina, USA.

Jaeger, J.A.G., T. Soukup, L. F. Madriñán, C. Schwick, and F. Kienast. 2011. Landscape fragmentation in Europe. Report No. 2/2011. European Environmental Agency, Copenhagen, Denmark.

Joly, P., C. Morand, and A. Cohas. 2003. Habitat fragmentation and amphibian conservation: building a toolbox for assessing habitat connectivity. Comptes Rendus Biologies 326:S123–S139.

Jongman, R.H.G. 1995. Nature conservation planning in Europe: developing ecological networks. Landscape and Urban Planning 32:169–183.

Kays, R., S. Tilak, M. Crofoot, T. Fountain, D. Obando, A. Ortega, F. Kuemmeth, et al. 2011. Tracking animal location and activity with an automated radio telemetry system in a tropical rainforest. The Computer Journal 54:1931–1948.

Langen, T. A. 2010. Predictive models of herpetofauna road mortality hotspots in extensive road networks: three approaches and a general procedure for creating hotspot models that are useful for environmental managers. Pages 475–486 *in* Proceedings of the 2009 international conference on ecology and transportation. P. J. Wagner, D. Nelson, and E. Murray, editors. Center for Transportation and the Environment, North Carolina State University, Raleigh, North Carolina, USA.

Langen T. A., K. Gunson, C. Scheiner, and J. Boulerice. 2012. Road mortality in freshwater turtles: identifying causes of spatial patterns to optimize road planning and mitigation. Biodiversity and Conservation 21:3017–3034.

Langen, T. A., A. Machniak, E. Crowe, C. Mangan, D. Marker, N. Liddle, and B. Roden. 2007. Methodologies for surveying herpetofauna mortality on rural highways. Journal of Wildlife Management 71:1361–1368.

Langen, T. A., K. Ogden, and L. Schwarting. 2009. Predicting hotspots of herpetofauna road mortality along highway road networks: model creation and experimental validation. Journal of Wildlife Management 73:104–114.

Larkin, J. L., D. S. Maehr, T. S. Hoctor, M. A. Orlando, and K. Whitney. 2004. Landscape linkages and conservation planning for the black bear in west-central Florida. Animal Conservation 7:23–34.

Levine, J. 2013. An economic analysis of improved road-stream crossings. The Nature Conservancy, Keene Valley, New York, USA. http://www.nature.org/ourinitiatives/regions /northamerica/road-stream-crossing-economic-analysis .pdf. Accessed 24 August 2014.

Levins, R. 1969. Some demographic and genetic consequences of environmental heterogeneity for biological control. Bulletin of the Entomological Society of America 15:237–240.

Litvaitis, J. A., and J. P. Tash. 2008. An approach toward understanding wildlife-vehicle collisions. Environmental Management 42:688–697.

Long, R. A., P. MacKay, J. Ray, and W. Zielinski. 2008. Noninvasive survey methods for carnivores. Island Press, Washington, DC, USA.

Macarney, J., P. T. Gregory, and K. W. Larsen. 1988. A tabular survey of data on movements and home ranges of snakes. Journal of Herpetology 22:61–73.

Manley, P. N., B. Van Horne, J. K. Roth, W. J. Zielinski, M. M. McKenzie, T. J. Weller, F. W. Weckerly, and C. Vojta. 2006. Multiple species inventory and monitoring technical guide. General Technical Report WO-73. US Forest Service, Washington, DC, USA.

Marsh, D. M., and P. C. Trenham. 2001. Metapopulation dynamics and amphibian conservation. Conservation Biology 15:40–49.

McDonald, W., and C. C. St. Clair. 2004. The effects of artificial and natural barriers on the movements of small animals in Banff National Park, Canada. Oikos 105:397–407.

McRae, B. H., B. G. Dickson, T. H. Keitt, and V. B. Shah. 2008. Using circuit theory to model connectivity in ecology, evolution, and conservation. Ecology 89:2712–2724.

McRae B. H., S. A. Hall, P. Beier, and D. M. Theobald. 2012. Where to restore ecological connectivity? Detecting barriers and quantifying restoration benefits. PLoS ONE 7(12):e52604.

Mendoza, E., P. R. Martineau, E. Brenner, and R. Dirzo. 2011. A novel method to improve individual animal identification based on camera-trapping data. Journal of Wildlife Management 75:973–979.

Mills, L. S., and F. W. Allendorf. 1996. The one-migrant-per-generation rule in conservation and management. Conservation Biology 10:1509–1518.

Mountrakis, G., and K. Gunson. 2009. Multi-scale spatiotemporal analyses of moose-vehicle collisions: a case study in northern Vermont. International Journal of Geographical Information Science 23:1389–1412.

Ng, S. J., J. W. Dole, R. M. Sauvajot, S. P. Riley, and T. J. Valone.

2004. Use of highway undercrossings by wildlife in southern California. Biological Conservation 115:499–507.

Nuñez, T. A., J. J. Lawler, B. H. McRae, D. J. Pierce, M. B. Krosby, D. M. Kavanagh, P. H. Singleton, and J. J. Tewksbury. 2013. Connectivity planning to address climate change conservation biology. Conservation Biology 27:407–416.

O'Connell, A. F., J. D. Nichols, and K. U. Karanth. 2011. Camera traps in animal ecology: Methods and analysis. Springer, New York, New York, USA.

Ontario Ministry of Natural Resources. 2010. Natural heritage reference manual for natural heritage policies of the provincial policy statement, 2005. Second edition. Queens Printer of Ontario, Toronto, Canada. http://docs.files.ontario.ca /documents/3270/natural-heritage-reference-manual-for -natural.pdf. Accessed 24 August 2014.

Opdam, P., R. Foppen, and R. Reijnen. 1995. The landscape ecological approach in bird conservation: integrating the metapopulation concept into spatial planning. Ibis 137:139–146.

Opdam, P., R. Pouwels, S. van Rooij, E. Steingröver, and C. Vos. 2008. Setting biodiversity targets in participatory regional planning: introducing ecoprofiles. Ecology and Society 13:20. http://www.ecologyandsociety.org/vol13/iss1/art20/. Accessed 24 August 2014.

Patrick, D. A., J. P. Gibbs, V. D. Popescu, and D. A. Nelson. 2012. Multi-scale habitat-resistance models for predicting road mortality hotspots for reptiles and amphibians. Herpetological Conservation and Biology 7:407–426.

Pulliam, H. R. 1988. Sources, sinks, and population regulation. American Naturalist 132:652–661.

Ramp, D., J.K.A. Caldwell, D. Edwards, D. Warton, and D. B. Croft. 2005. Modeling of wildlife fatality hotspots along the Snowy Mountain Highway in New South Wales, Australia. Biological Conservation 126:474–490.

Ray, N., A. Lehmann, and P. Joly. 2002. Modeling spatial distribution of amphibian populations: a GIS approach based on habitat matrix permeability. Biodiversity and Conservation 11:2143–2165.

Roe, J. H., J. Gibson, and B. A. Kingsbury. 2006. Beyond the wetland border: estimating the impact of roads for two species of water snakes. Biological Conservation 130:161–168.

Rouse, J. D., R. J. Wilson, R. Black, and R. J. Brooks. 2011. Movement and spatial dispersion of *Sistrurus catenatus* and *Heterodon platirhinos*: implications for interactions with roads. Copeia 2011:443–456.

Ruediger, B., and J. Lloyd. 2003. A rapid assessment process for determining potential wildlife, fish and plant linkages for highways. Pages 206–225 in Proceedings of the 2003 international conference on ecology and transportation. C. L. Irwin, P. Garrett, and K. P. McDermott, editors. Center for Transportation and the Environment, North Carolina State University, Raleigh, North Carolina, USA.

Rytwinski, T., and L. Fahrig. 2011. Reproductive rate and body size predict road impacts on mammal abundance. Ecological Applications 21:589–600.

Santos S. M., F. Carvalho, and A. Mira. 2011. How long do the dead survive on the road? Carcass persistence probability and implications for road-kill monitoring surveys. PLoS ONE 6:e25383.

Sawyer. S. C., C. W. Epps, and J. S. Brashares. 2011. Placing linkages among fragmented habitats: do least-cost models reflect how animals use landscapes? Journal of Applied Ecology 48:668–678.

Schippers, P., J. Verbroom, J. P. Knaapen, and R.C.V. Apeldoorn. 1996. Dispersal and habitat connectivity in complex heterogeneous landscapes: an analysis with a GIS-based random walk model. Ecography 19:97–106.

Schlaepfer, M. A., M. C. Runge, and P. W. Sherman. 2002. Ecological and evolutionary traps. Trends in Ecology and Evolution 17:474–480.

Scott, J. M., F. Davis, B. Csuti, R. F. Noss, B. Butterfield, C. Groves, J. Anderson, et al. 1993. Gap analysis: a geographic approach to protection of biological diversity. Wildlife Monographs 123:1–41.

Selva, N., S. Kreft, V. Kati, M. Schluck, B. G. Jonsson, B. Mihok, H. Okarma, and P. L. Ibisch. 2011. Roadless and low-traffic areas as conservation targets in Europe. Environmental Management 48:865–877.

Semlitsch, R. D. 1998. Biological delineation of terrestrial buffer zones for pond-breeding salamanders. Conservation Biology 12:1113–1119.

Semlitsch, R. D. 2000. Principles for management of aquatic-breeding amphibians. Journal of Wildlife Management 64:615–631.

Semlitsch, R. D. 2008. Differentiating migrations and dispersal processes for pond-breeding amphibians. Journal of Wildlife Management 72:260–267.

Semlitsch, R. D., and J. R. Bodie. 2003. Biological criteria for buffer zones around wetlands and riparian habitats for amphibians and reptiles. Conservation Biology 17:1219–1228.

Sen, A., and T. E. Smith. 1995. Gravity models of spatial interaction behavior. Springer-Verlag, Heidelberg, Germany.

Silvy, N. 2012. The wildlife techniques manual. Seventh edition. John Hopkins University Press, Baltimore, Maryland, USA.

Singleton, P. H., and J. F. Lehmkuhl. 1999. Assessing wildlife habitat connectivity in the Interstate 90 Snoqualmie Pass corridor, Washington. Pages 75–84 in Proceedings of the third international conference on wildlife ecology and transportation. G. L. Evink, P. Garrett, and D. Ziegler, editors. Florida Department of Transportation, FL-ER-73-99, Tallahassee, Florida, USA.

Slater, F. M. 2002. An assessment of wildlife road casualties—the potential discrepancy between numbers counted and numbers killed. Web Ecology 3:33–42.

Smith, D. J. 1999. Identification and prioritization of ecological interface zones on state highways in Florida. Pages 209–229 in Proceedings of the third international conference on wildlife ecology and transportation. G.L. Evink, P. Garrett, and D. Zeigler, editors. Florida Department of Transportation, Tallahassee, Florida, USA.

Smith, D. J. 2006a. Ecological impacts of SR 200 on the Ross Prairie ecosystem. Pages 380–396 in Proceedings of the 2005 international conference on ecology and transportation. C. L. Irwin, P. Garrett, and K. P. McDermott, editors. Center for Transportation and the Environment, North Carolina State University, Raleigh, North Carolina, USA.

Smith, D. J. 2006b. Incorporating results from the prioritized "ecological hotspots" model into the Efficient Transportation Decision Making (ETDM) process in Florida. Pages 127–137 in Proceedings of the 2005 international conference on ecology and transportation. C. L. Irwin, P. Garrett, and K. P. McDermott, editors. Center for Transportation and the Environment, North Carolina State University, Raleigh, North Carolina, USA.

Smith, D. J., R. F. Noss, and M. B. Main. 2006. East Collier County wildlife movement study: SR 29, CR 846, and CR 858 wildlife crossing project. University of Central Florida and University of Florida IFAS. Southwest Florida REC Research Report, SWFREC-IMM 2007-01. Immokalee, Florida, USA.

Soutullo, A., L. Cadahia, V. Urios, M. Ferrer, and J. J. Negro. 2007. Accuracy of lightweight satellite telemetry: a case study in the Iberian Peninsula. Journal of Wildlife Management 71:1010–1015.

Steen, D. A., J. P. Gibbs, K. A. Buhlmann, J. L. Carr, B. W. Compton, J. D. Congdon, J. S. Doody, et al. 2012. Terrestrial habitat requirements of nesting freshwater turtles. Biological Conservation 150:121–128.

Steen, D. A., and L. L. Smith. 2006. Road surveys for turtles: consideration of possible sampling biases. Herpetological Conservation and Biology 1:9–15.

Taylor, B. D., and R. Goldingay. 2009. Can road crossing structures improve population viability of an urban gliding mammal? Ecology and Society 14:13. http://www.ecologyandsociety.org/vol14/iss2/art13/. Accessed 24 August 2014.

Thorne, J. H., P. R. Huber, E. H. Girvetz, J. Quinn, and M. C. McCoy. 2009. Integration of regional mitigation assessment and conservation planning. Ecology and Society 14:47. http://www.ecologyandsociety.org/vol14/iss1/art47/. Accessed 24 August 2014.

Toronto and Region Conservation Authority. 2007. "Terrestrial Natural Heritage System strategy," Downsview, Ontario, Canada. http://trca.on.ca/the-living-city/land/terrestrial-natural-heritage/. Accessed 24 August 2014.

van der Grift, E., and F. Pouwels. 2006. Restoring habitat connectivity across transport corridors: identifying high-priority locations for de-fragmentation with the use of an expert-based model. Pages 205–231 in The ecology of transportation: Managing mobility for the environment.

J. Davenport and J. L. Davenport, editors. Springer, Dordrecht, The Netherlands.

Verboom, J., R. Foppen, P. Chardon, P. Opdam, and P. Luttikhuizen. 2001. Standards for persistent habitat networks for vertebrate populations: the key patch approach. An example for marshland bird populations. Biological Conservation 100:89–101.

Vos C. C., and J. P. Chardon. 1998. Effects of habitat fragmentation and road density on the distribution pattern of the moor frog *Rana arvalis*. Journal of Applied Ecology 35:44–56.

Vos, C. C., J. Verboom, P.F.M. Opdam, and C.J.F. Ter Braak. 2001. Toward ecologically scaled landscape indices. American Naturalist 157:24–41.

Wikelski, M., R. W. Kays, N. J. Kasdin, K. Thorup, J. A. Smith, and G. W. Swenson Jr. 2006. Going wild: what a global small-animal tracking system could do for experimental biologists. Experimental Biology 210:181–186.

Wilson, D. E., F. R. Cole, J. D. Nichols, R. Rudran, and M. S. Foster. 1996. Measuring and monitoring biological diversity: Standard methods for mammals. Smithsonian Institution Press, Washington, DC, USA.

Zielinski, W. J., and T. E. Kucera. 1995. American marten, fisher, lynx, and wolverine: survey methods for their detection. General Technical Report No. PSW-GTR-157. Pacific Southwest Research Station, US Forest Service, Albany, California, USA.

Mitigating Road Effects on Small Animals

Scott D. Jackson,
Daniel J. Smith,
and Kari E. Gunson

9.1. Introduction

Non-passage techniques and passage structures can be implemented in both aquatic and terrestrial habitats for small vertebrate animals whose habitats are disrupted by roads. Non-passage techniques are designed to reduce road mortality without any structural accommodations to facilitate movement across roads. Aquatic passage structures include bridges and culverts at stream crossings as well as pond-to-pond, lake-to-lake, and wetland-to-wetland crossings, where the passage of water is also an important (and often the primary) consideration in the design of the structure. Structures designed primarily for terrestrial species include tunnels, underpasses, overpasses (known in Europe as ecoducts, e.g., Langton 2002), as well as poles and rope structures designed specifically for arboreal species (Case Study 9.8). Sometimes these structures are designed specifically for small vertebrates (e.g., salamander tunnels); sometimes they are designed to accommodate a broad range of species including small vertebrates; and sometimes they are designed for large animals but can be designed or adapted to also serve the needs of amphibians, reptiles, and small mammals.

9.2. Non-Passage Techniques

There are many approaches or strategies for mitigation other than using animal passage structures to reduce the negative impacts of roads on small vertebrates. These approaches include altering vehicle traffic patterns or driving behavior, placing structures (barrier fences, walls) on the roadside, and physically carrying animals over the road (see references throughout this volume). Such measures rely on public community support and flexibility, but are comparatively less expensive than wildlife crossing systems. These strategies are largely untested and their effectiveness is difficult to measure (e.g., Huijser et al. 2007; Huijser et al. 2008). Nonetheless they can be useful for gathering data about the impact of roads on wildlife populations, determining whether a more permanent solution is warranted, raising awareness of the problem, and generating community support for more long-term solutions. In some cases these techniques can be the most effective solution available; in many cases they are the most affordable, and where wildlife populations are being negatively affected by roads, they are better than doing nothing.

9.2.1. Signs

An inexpensive way to encourage motorist awareness and speed reduction on roads is the use of wildlife warning signs. This approach has been used most widely for large animals, such as deer and moose. Among small vertebrates, signs have been used most commonly for amphibians and reptiles. Signage is only effective if it prompts motorists to change their driving behavior (e.g., reduce speed), and its effectiveness at reducing animal road mortality over the long term is not known. New wildlife signs, as novel features in the road landscape, may initially result in some reduction in roadkill. However, there is concern that when signs remain along the roadside for extended periods of time motorists become increasingly desensitized to their message (for a summary of the use of warning signs to reduce wildlife vehicle collisions, see Clevenger et al. 2008). To increase the effectiveness of wildlife warning signs consider the following:

- Novel designs: Even the effect of novel signs are likely to wear off over time; consider periodically changing signs, rotating through a variety of designs. Unfortunately, novel signs that succeed in getting noticed are also at high risk of being stolen.
- Selective placement: Restrict the use of signs to the places where they will do the most good. Too many signs will reduce their effectiveness.
- Specific wording: Include distances (e.g., Turtle Xing next 0.5 km) and information on seasonality (e.g., Watch for Turtles Crossing, June–July).
- Seasonal placement: Erect signs at the specific time of year when they are most needed and then remove them outside of the movement season to reduce desensitization.

Warning signs are not recommended for species that are vulnerable to collection (e.g., certain species of turtles or snakes) or taxa that are at risk of intentional roadkill (e.g., snakes).

9.2.2. Reduction of Vehicle Speeds

Higher speeds are often associated with increased numbers of animal-vehicle collisions (Jones 2000; Gunson et al. 2011), particularly when it comes to small animals (e.g., Cristoffer 1991). Small animals in the road can be difficult to detect in time to avoid collisions. Lowering the traffic speed can increase the detection of animals in the road, the likelihood that animals will be able to take evasive action, and the ability of drivers to adjust their course in time to avoid collisions. Vehicular speed can be reduced by lowering speed limits or using speed bumps. In some cases, these reductions may be better received when combined with a public awareness about the need for them. Speed limit reductions often require enforcement to be effective. Speed bumps used in association with signs may be an effective technique for reducing roadkill on low-volume (e.g., subdivision) roads.

9.2.3. Bucket Brigades

In the late winter and early spring across eastern North America people monitor weather conditions, looking for the first warm, rainy night of the season to bring forth breeding amphibian migrations. For example, large salamanders, such as spotted salamanders (*Ambystoma maculatum*; up to 23 cm in length) often make synchronous migrations from forested habitats to vernal pool breeding sites. Where these migration routes cross roads, the road mortality can be high and obvious. In addition to the people interested in watching the annual spectacle, there are groups of people who coordinate efforts to help these slow-moving animals across roads. These efforts are sometimes referred to as "bucket brigades," and they are also used, in more or less formal efforts, to assist other species of amphibians as well as turtles in their movements across roads (see also Chapter 5). Whether or not buckets are used, this technique involves physically transporting animals across the road to prevent them from getting hit. These efforts have been reported from communities in Massachusetts, Mississippi, Pennsylvania, and Vermont.

There are a number of concerns about the effectiveness and desirability of bucket brigades. Bucket brigades focus on concentrated migrations that involve relatively large numbers of animals. In the case of spotted salamanders, the outward migration later in the year of newly transformed juveniles tends to be more spread out in time, and mortality of small salamanders is more prolonged and harder to notice. Efforts to protect a population may need to include more than one life stage, and bucket brigades may not be well suited for all life stages that cross roads. Bucket brigades are labor intensive because people must be physically present for as long as the migration lasts (or for as long as the risk of significant road mortality lasts), which frequently involves patrolling roads on dark, rainy nights when animals are moving. There are significant concerns about the safety of volunteers that spend hours moving back and forth across roads, particularly at night in these weather conditions. Roads that lack significant traffic volumes (where risks to volunteers are low) probably also represent relatively low risks to the focal species. The higher the traffic volume, the greater the need for mitigation, but also the greater the risk is for volunteers. The potential liability involved in organizing such an effort is a significant consideration, not only for bucket brigades but for any instance where people are moving animals from roads to prevent mortality or are collecting roadkill data.

9.2.4. Temporary Road Closures

Temporary road closures and detours are strategies that focus on changing traffic flow patterns on specific roads (SETRA 2005). For example, a road may be closed during a known amphibian migration or snake emergence from an overwintering den to avoid seasonal mass road mortality. This strategy is usually implemented on low volume roads with heightened community support and is not necessarily feasible for larger, high volume roads. Like bucket brigades, this approach tends to focus only on life stages that engage in concentrated migrations. Animals making more sporadic movements (e.g., juvenile spotted salamanders) generally do not benefit from road closures.

Examples of road closures include efforts in Alabama (Birmingham), New Jersey (East Brunswick), and Ontario (Burlington) for salamanders; in California (Los Padres National Forest) for endangered California red-legged frogs (*Rana draytonii*) and Arroyo toads (*Anaxyrus californicus*); in Pennsylvania and New Jersey (Delaware Water Gap National Recreation Area) for frogs and salamanders; and in Illinois (Shawnee National Forest) for snakes. The California project was triggered by recent surveys that discovered the endangered species. At Delaware Water Gap, there is a 10-year ongoing policy of seasonal road closures. In Illinois (LaRue Pine Hills Bluff Area of the Shawnee National Forest), the project is focused on migrating snakes, but the closure also serves many other species of reptiles and amphibians in the spring and fall months.

9.2.5. Barrier Fencing

Historically, fencing has been used along roadsides to separate domesticated animals found on farms and rangeland from roads, and recently this idea has been used to exclude wildlife from accessing a road (Chapter 1; see also Langton 2002). Barrier fencing is more common for larger wildlife where an animal-vehicle collision can pose a serious traffic safety hazard for motorists. For smaller animals, barrier fencing may be an adequate measure if the primary objective is to reduce road mortality, such as when animals are not able to successfully cross the road due to obstacles or high traffic volume (Figure 9.1).

Although proven successful for reducing road mortality (e.g., Aresco 2005), barrier fencing has its limitations. Careful attention to design specifications

Figure 9.1. Highway fencing in Iowa along US Highway 63 in Bremer County and US Highway 51 Monticello Bypass in Jones County, United States, modified by additional fencing material to eliminate gaps and prevent the movement of snakes, turtles, and other small animals onto the highway. *Credit: Iowa Department of Transportation.*

and material types is necessary for fencing to remain an effective barrier to small animals (SETRA 2005; Andrews et al. 2015; see also Table 10.2). Ongoing maintenance is essential in order to prevent breaches in the barrier fence; if breached, it dramatically loses its effectiveness. Barrier walls differ from barrier fences in that they present an open face in only one direction (much like a retaining wall). Made of wood, concrete, granite, marble or other materials, barrier walls are generally more durable than fencing, but are also more expensive. For more information about barrier fencing and walls, see Section 9.7.2.

If used without passage structures, fencing creates a barrier for animals that presumably need to access resources and habitat on the other side of the road. This barrier may be acceptable if the traffic intensity on the road is high enough that there is little prospect of animals successfully crossing. However, on lower volume roads, there is the potential that a fencing project designed to help one species will end up fragmenting the habitat and adversely affecting other nontarget species.

Other limitations include concerns about the creation of road mortality hotspots at barrier fence ends and the possibility that wildlife will become trapped on the road side of the fence. Wing walls or extensions at fence ends can be used to divert animals away from the road; the length of these should be determined by the spatial

Figure 9.2. One-way gate for European badgers (*Meles meles*). *Credit: Scott D. Jackson.*

extent of the road mortality as well as the relative spatial scale of the target species' movements (Chapter 8). Constructing a "dog-leg" (i.e., sharp angle; Figure PE 2.8) at the end of the fence away from the road can also help direct animals toward the right-of-way and possibly back toward the crossing structure. One-way gates or escape ramps for turtles and other small vertebrates can be used to allow animals to move through or over a fence in one direction from the right-of-way to safe habitat (SETRA 2005). An example of a one-way gate is shown in Figure 9.2.

9.2.6. Habitat Modification

Roads are not only barriers and high-risk areas for small animals, but they also provide habitat for a variety of these species (e.g., Andrews et al. 2008; Chapter 4; Practical Example 1). Female turtles often seek out sunny, sparsely vegetated road shoulders for nesting. Roadside ditches may attract pool-breeding amphibians. Snakes and small mammals can be attracted to shoulders for habitat edge foraging opportunities. If road mortality is high enough or recruitment from roadside nests is low enough, then the negative effect of the road may exceed any benefit gained by using these roadside habitats, resulting in an ecological trap (i.e., poor quality habitats that attract individuals but are not productive for reproduction or survival). One strategy for reducing road mortality is to modify habitat conditions along roadsides to eliminate features that attract wildlife. Habitat modification (e.g., creation of nesting mounds for turtles or vernal pools for amphibians) strategically placed on one side of a road near other suitable habitat can provide opportunities for

wildlife to complete their life cycle without approaching or crossing a road (Podloucky 1989).

9.2.7. Passage-Friendly Curbs

Curb designs offer a frequently overlooked opportunity to mitigate the impacts of roads on small vertebrates (see Figure 6.2). Even roads with low traffic volumes and risks of wildlife mortality can have significant effects on small vertebrates if curbing blocks their movement or funnels them into catch basins. One of this chapter's authors (Jackson) rescued 22 amphibians of 4 species out of a catch basin, presumably the result of movements that were inadvertently blocked and redirected by traditional curbing. Traditional curbs and railroad tracks can also block turtle movement (Kornlev et al. 2006); juvenile turtles are likely to be particularly vulnerable due to their smaller size. Building roads with "country drainage" (no curbs) is one way to avoid the problem. Use of gently sloping curbs (sometimes referred to as "mountable curbs" or "Cape Cod berms") instead of more traditional asphalt or granite curbing is another option (Figure 9.3).

9.2.8. Benefits and Limitations of Non-Passage Techniques

Although these techniques discussed in Section 9.2 may be appropriate for some limited circumstances and can be a first step in planning and preparing for long-term solutions, in most cases the ultimate goal is to prevent roads from fragmenting ecosystems. Wildlife crossing structures are often seen as the ultimate solution short of eliminating the road. That said, there are some circumstances where passage structures just might not be able to achieve this goal. Divided highways may require passage structures that are so long that small vertebrates may not successfully or readily pass through them. For example, in colder or drier regions, amphibians might freeze or die of dehydration if weather conditions change during a migration. In these cases, habitat modification and efforts to reduce road mortality may be the only practical measures available to protect small vertebrate populations. For purposes of this discussion, we refer to aquatic crossing structures as those that contain standing water or permanent water, and terrestrial structures as those that are generally dry or hold water temporarily. It should further be noted that these structure descriptions do not

Figure 9.3. Examples of traditional curbing (a) and sloped curbing (b), which presents less of a barrier to small vertebrates. *Credits: Scott D. Jackson.*

necessarily refer to the types of habitat they connect (e.g., a terrestrial structure may connect two wetlands).

9.3. Aquatic Road Crossings

The best opportunity for aquatic amphibians, reptiles, and small mammals to cross roads is generally through culverts or under bridges. These include stream crossings as well as areas where roads pass through lakes, ponds, wetlands, and other aquatic features. In the vast majority of cases, the culverts have been sized and designed primarily to pass water from one side of the road to the other and to prevent scour that might degrade the road or crossing structure. The common result is culverts that pass water but that are largely inadequate for fish and wildlife passage.

9.3.1. Stream Crossings

In recent years, more attention has been paid to fish passage at stream crossings, especially where migratory

fish are involved. Initially, these fish-friendly structures were planned using a hydraulic design approach. This design is essentially an engineering approach that ensures the hydraulic conditions (water depth and velocity) necessary for passage of target fish species are maintained in the culvert at particular times of the year. Hydraulic design is often successful at producing culverts that can accommodate these fish species, but these culverts typically lack the features that would be necessary for many amphibians, reptiles, and small mammals. Specifically, these structures usually lack continuous substrate (e.g., soil, leaf litter, and other base material) as well as bank edge conditions that provide low-velocity water and dry passage along the banks.

A newer approach to designing fish passage, known as stream simulation, offers far more opportunities for amphibians, reptiles, and small mammals to pass under a road. Rather than just fish passage, stream simulation seeks to achieve passage for fish and other aquatic vertebrates, aquatic invertebrates (e.g., crayfish), and semi-aquatic wildlife (e.g., stream salamanders, snakes, turtles, small mammals). Stream simulation attempts to create conditions inside a structure that mimic conditions found in the natural stream, such as water depth, water velocity, and substrate and channel characteristics. The US Forest Service Stream Simulation Working Group has published an excellent manual that provides much valuable information about the stream simulation design approach (USFSSSWG 2008).

Stream simulation is not the only approach for designing stream crossings for amphibians, reptiles, and small mammals, but it is the preferred option. The benefit of this approach is that it is not necessary to know much about the design requirements of particular species. The assumption is that if the hydraulic and benthic characteristics within the structure essentially mimic the characteristics of the stream, then those species that are indigenous to the stream should be able to pass through the structure. Another advantage is that stream simulation allows the design to encompass both target and nontarget species at the same time.

Although stream simulation may be the preferred design method, it may not always be possible to use this approach because of cost or site-specific constraints (as are often present during culvert replacement or retrofit projects). When designing a crossing for amphibians, reptiles, and small mammals, focus on the characteristics discussed below.

Water Velocity

Although this is a critical feature for fish and other aquatic organisms that typically swim in the water column, it may be less important than some other features for species that crawl rather than swim upstream or that typically stay in close association with substrate or bank edge areas. Ideally, water velocities in the structure should be comparable to those found in the natural stream channel.

Water Depth

For some large-bodied aquatic or semi-aquatic species (e.g., softshell turtles, *Apalone* spp.), water depth may be important. However, many of these species are probably able to crawl through shallow water that would be impassable for fish. Of particular importance is that the streambed within the structure be properly constructed to avoid the water flowing underground during low-flow periods. Ideally, water depths in the structure should match those present in the natural stream.

Substrate

Substrate may be the most crucial characteristic of stream crossings for amphibians, reptiles, and small mammals. Many of these species are likely to crawl or swim only short distances between cover objects as they make their way upstream. Cover is particularly important because these species are generally vulnerable to predation if they spend any sustained amount of time out in the open. Natural substrate should be continuous throughout the structure and should be comparable to what is found in the natural stream channel or otherwise appropriate for the target species.

Bank Edge Areas

Bank edges are areas of shallow water that abut land (banks) and are not inundated at least for part of the year. The shallow water against the bank generally has lower velocity and offers some refuge from fish predators that may be using the main channel of the stream. These areas are important for small animals and species that are weak swimmers or have low vagility. The banks themselves offer dry passage for terrestrial and semi-aquatic animals such as stream-dwelling salamanders, water shrews (*Sorex palustris*), and star-nosed moles (*Condylura cristata*). Cover objects (rocks, logs) on the banks are likely to enhance passage opportunities for small animals.

Light and Openness

Observations of turtle roadkill directly above stream crossings that otherwise appeared suitable for passage suggested that light and openness (size of the structure opening relative to structure length) are important in determining whether wildlife will use crossing structures. Research at the University of Massachusetts Amherst (UMass) confirmed that light is an important characteristic affecting passage success for some freshwater turtle species (Case Study 9.5.3). Unfortunately, we lack information on the light or openness requirements for just about all amphibians, reptiles, and small mammals. In general, bigger is better and lighter is better than darker. However, some species of snakes have been observed to cross more readily under more closed canopy, indicating that too much exposure can deter road crossing and culvert use for these animals (Case Study 9.6).

Fencing

Aquatic animals are confined to rivers and streams and directional fencing toward structures is therefore not needed. Semi-aquatic species are also likely to move along stream corridors and will find crossing structures without the need for fencing. However, if terrestrial species are targeted for crossing structures or there are concerns about whether the amount of light or openness will be sufficient, then wing walls or directional fencing may be appropriate to deter animals from trying to cross the road.

9.3.2. Crossings Between Water Bodies

Creating an effective crossing structure on roads that bisect ponds and lakes can be much simpler than stream crossings. Water connections are the preferred option because they can accommodate aquatic and semi-aquatic organisms. If it's possible to create a water connection beneath the road, it's not necessary to deal with water velocity and all the complications that come with it. However, efforts to perforate an existing causeway through a lake or pond might come with constraints that require building a crossing structure above the water elevation (see Section 9.4). Therefore, the following hydrological components must be considered in the initial stages of assessing mitigation options and designs.

Water Velocity

Unless there is an elevation difference between the water level on one side of the road and the water level on the other side, velocity should not be a critical issue since ponds and lakes have little or no flow.

Water Depth

To provide effective year-round passage, the design of the structure should take into account the annual variation in water levels for the target water body. Expert opinion from biologists knowledgeable about species likely to use such a structure should be used to determine what depths are necessary to pass both target and nontarget species. Ideally, the structure should be designed to provide a range of depths, including shallow areas with cover objects for smaller animals that are vulnerable to predation.

Substrate

Assuming that pond-to-pond and lake-to-lake crossings will involve minimal amounts of flow, substrate should be relatively easy to install and maintain. In general, it is best to maintain continuity of substrate from one side of the water body through the structure and into the water body on the other side. Cover objects should be used to make at least a portion of the structure suitable for use by small animals that may be reluctant to expose themselves to predation in open water.

Bank Edge Areas

Bank edges are not as important for pond-to-pond or lake-to-lake crossings because these are low velocity water bodies, but they may still be useful for facilitating movement of semi-aquatic or terrestrial wildlife that prefer dry passage over substrate that is not inundated.

Light and Openness

These issues are likely to be just as relevant in pond-to-pond and lake-to-lake passages as in stream passages. Little information is available to inform structural design. In general, the bigger and lighter the structure, the better. However, it is important to recognize that budgets are usually limited for mitigation projects, and larger structures are more expensive than small ones. It is also important to consider whether project goals are likely to be achieved with one (or a few) large structures, or a greater number of smaller structures. Open-top structures are one mechanism for increasing the amount of light in a tunnel without dramatically increasing its size.

Location

Unlike stream passages under roads where the location is obvious, crossing structures on pond, lake, and marsh causeways should be strategically located to best achieve the project's goals and objectives. Using multiple structures may address the needs of multiple species. Multiple smaller structures also allow for more flexibility in where animals can cross than a few larger structures would; but, given that budgets are typically limited, more structures may also mean reducing design features that would otherwise enhance the use of the structure. However, concentrating resources to install one large passage structure can result in having a single, well-designed structure at an ineffective location for the intended species. Road crossing or roadkill data, habitat assessment, and site constraints should all be considered when designing and placing crossing structures (Chapter 8).

Fencing

If all target species (and nontarget species to be accommodated) are aquatic, directional fencing is not necessary. However, if the structures are designed for passage of semi-aquatic or terrestrial species then directional fencing will be necessary to deter animals from trying to cross the road surface and instead funnel them toward the structure.

9.3.3. Wetland-to-Wetland Crossings

Wetlands are areas of saturated soil or shallow surface water that may be present year-round or for only part of the year and that typically support well-developed plant communities. Many wetland-to-wetland crossings are similar to crossings between water bodies. Because wetlands often experience periods without standing water, moisture and streamflow are two issues that may need careful attention when designing these types of crossings. Directional fencing also may be necessary to prevent animals from crossing over the road and thereby avoiding the structure (see Section 9.7.2).

Moisture

Amphibians are vulnerable to dehydration when exposed to dry conditions. Juveniles and other small amphibians have higher surface area to volume ratios compared to larger animals (see Chapter 2). Furthermore, rainfall is often a cue that amphibians use for migration, and it is possible that frogs and salamanders might abort their migrations if they encounter dry conditions. If passages are designed to hold standing water year-round, moisture within the structure should not be an issue. However, if the passages are designed to be above the elevation of surface water for a significant portion of the year, moisture is an important consideration. One solution is to design a passage structure so that the elevation of the structure bottom relative to the water table allows the substrate to remain wet for much of the year due to the influence of groundwater (or surface water wicking into the structure substrate). Another option is to use open-top structures that allow rain to enter and dampen the substrate.

For some sites there may be significant amounts of standing water in the wetland for some or all of the year. If the wetland-to-wetland crossing is inundated during high water periods, it may function well for aquatic and semi-aquatic species but not for terrestrial wildlife. On a stretch of Highway 93 between Lolo and Hamilton, Montana, wetland-to-wetland wildlife passage structures were modified with ramps and 63.5 cm wide shelves to allow passage of small animals during periods of high water. These ramps were used by a variety of small vertebrates including deer mice (*Peromyscus maniculatus*), short-tailed weasels (*Mustela erminea*), striped skunks (*Mephitis mephitis*), and raccoons (*Procyon lotor*; Foresman 2002, 2003).

Stream Flow

Where wetland-to-wetland crossings occur in stream and river floodplains there is potential for significant water flow during storm events. Flowing water destabilizes natural substrate within structures. Options to minimize this include (1) designing an effective passage structure that does not require specific substrate, (2) increasing the size of substrate material (use of rocks large enough that they won't be dislodged by flood

waters) and infilling with finer material, or (3) implementing an inspection and maintenance program to replace lost substrate after storm events.

9.4. Terrestrial Passage Structures

Terrestrial passage structures include culverts and tunnels, wildlife underpasses and overpasses, and arboreal bridges and glider poles (see Figure 9.4 for examples). Several factors are important to the successful performance of terrestrial passage structures for smaller vertebrates; these involve the location and design of passage structures and directional fencing and guide walls. In general terms, the design features of any passage structure should mimic the conditions of the surrounding environment as much as possible because these are the characteristics favored by the species likely to use the structures.

Many small vertebrates travel only short distances, and it is therefore unlikely they can be diverted any significant distance in order to reach a passage structure. As a result, location is one of the most important factors influencing passage success for amphibians, reptiles, and small mammals. Given the short movement distances reported for many small vertebrates, it is important to consider whether multiple, small structures might provide better results than a single, larger structure (see also Case Study 10.3.4).

The most important rule for locating passage structures is to understand the biology and motivations for movement (e.g., searching for food, mates, or shelter; juvenile dispersal; seasonal migration) of the target species, taxa, or community. This information can be used to better understand potential habitat use and preferences, animal movement and migration paths, areas of concentration or congregation, and likely locations for occasional or frequent road crossings. Animal movement may be related to landscape features (e.g., use of habitat edges or ecotones, topographic relief, proximity to water) and this can play an important role in identifying potential road crossing locations (Chapter 8).

9.4.1. Aquatic-to-Upland Passage

Many amphibians and certain reptiles (e.g., turtles) move between terrestrial habitats and aquatic or wetland habitats as part of their life cycle. A road that runs

Figure 9.4. Examples of terrestrial passage structures: (a) salamander tunnel in Princeton, Massachusetts, United States; (b) large wildlife tunnel in Banff National Park, Alberta, Canada; (c) wildlife bridge (underpass) on Alligator Alley in Florida, United States; (d) wildlife overpass in Europe. *Credits: Michael D. Howard (a); Scott D. Jackson (b–d).*

along the boundary between upland and wetland habitats may act as a barrier and a source of mortality for a number of species. Although identifying wetlands or aquatic features in proximity to roads is a fairly routine task, determining the specific locations where crossing structures would be most effective is more difficult. This is aided at the project level by the analysis of existing data or predictions of specific movement paths of animals and particular habitat and topographic features (Chapter 8). In the absence of data on animal movements, field visits can reveal specific microhabitats and fine-scale topography that might serve to focus and guide animal movements to and from water features. Where roads run parallel to stream courses or wetlands, multiple structures interconnected by directional fences or guide walls may be needed to cover a number of crossing points and accommodate the full range of species likely to move between aquatic and upland areas.

9.4.2. Upland-to-Upland Passage

Some amphibians, most reptiles, and small mammals do not seek out wetland or aquatic environments but confine their movements to essentially terrestrial habitats. Without water features serving as discrete sources or destinations, crossing locations may be more difficult to identify. However, movements by certain terrestrial species are frequently concentrated where significant changes in topography or plant communities occur. For example, animal movements are often channeled along habitat edges (ecotones), valley bottoms, as well as draws, ridges, and slope formations in areas with significant variation in topography (Forman 1995).

Where topographic relief is negligible and habitat is homogeneous or continuous (i.e., distinct edges or ecotones are not present), animal movements may be evenly dispersed with no evident points of concentration. In this case, multiple structures interconnected by directional fences or guide walls to increase

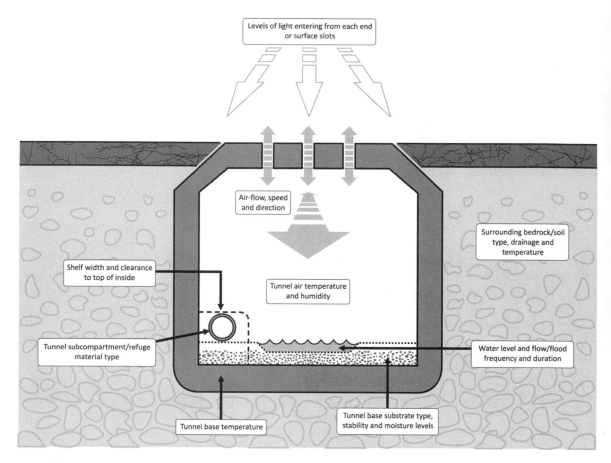

Figure 9.5. Generalized structure diagram of various feature considerations for accommodating small animals. *Credit: © Victor Young Illustration, adapted from original by Thomas E. S. Langton.*

ecological infrastructure along extended stretches of the road may be needed. The intervals at which these structures should be placed are a function of the target species' home range size and dispersal capabilities. Depending on the size of the area, it may prove economically impractical to perforate all road segments. A solution in this case could involve strategically placing either a single large structure, or clusters of small structures at wider intervals, simply to maintain sufficient connectivity to accommodate metapopulation dynamics (e.g., supplementation, gene flow, recolonization; Chapter 2).

9.5. Structure Design

Many considerations are important when designing effective wildlife crossing structures. These include physical size and shape, light, substrate, moisture,

drainage, and whether the top is open or closed. In passages that have constant water flow, there may not be the opportunity to manage certain factors, such as substrate. However, the considerations below can provide guidance for both aquatic and terrestrial structures. The requirements of the target species or group of species determine the parameters for these factors. If the design includes accommodation for multiple species with differing requirements, larger structures that can accommodate these various design features may be needed (e.g., Cavallaro 2005). Alternatively, the inclusion of multiple structures that can incorporate the specific requirements for each target species separately may be best (Figure 9.5).

9.5.1. Size and Shape

Many sizes and shapes of structures made from concrete, metal, or other materials are available. The primary structure types that apply to small animal passage are culverts and bridges. Culverts may be round, oval, arched, or rectangular in shape. Bridges can be constructed to almost any dimensions. It is critical to determine the minimum size of the structure for it to be effective for the target species. Though recent research efforts have begun to define these requirements in specific cases for certain species (e.g., Iuell et al. 2003; Clevenger and Huijser 2011), data are not available for most species of small animals. To address this lack of information, most project applications have used a focal or target species in determining size requirements. These focal species are generally representative of those present in the study area or are physically large enough to encompass the size requirements of many smaller species.

Size affects the openness and amount of light in the structure as well as airflow and potentially temperature. In simple terms, openness (defined in Reed et al. 1979 as the cross sectional area of the structure's opening divided by its length) has been considered a factor in the use or avoidance of structures by various species (Reed et al. 1979; Reed 1981; Foster and Humphrey 1995; MTPWWM 1995; Brudin 2003). Data from these studies suggest that the higher the openness value, the greater the potential for use, though this may not be true for all species.

There is evidence that weasels and other smaller mammals prefer small passage structures (Hunt et al. 1987; Rodriguez et al. 1996; Clevenger and Waltho 1999; Bellis 2008). Use of pipes within passage structures may be one way to accommodate species that prefer small, confined spaces. In a study of small mammal use of wildlife passage structures, Foresman (2002) found that meadow voles (*Microtus pennsylvanicus*) were abundant in areas within a few meters of the structure entrances but were never recorded using the passage structures. After the structures were modified to include "vole tubes" (Figure 9.6), meadow voles were documented moving freely through the passages (Foresman 2003). These vole tubes also were heavily used by weasels (*Mustela erminea*).

Structure length (related to road width) is also an important consideration. Given the more limited mobility of small vertebrates, it is important that crossing structures be kept as short as possible. Amphibians are vulnerable to dry conditions and both amphibians and

Figure 9.6. Crossing structure modified with a ramp (a) to facilitate use by small mammals during periods of high water, developed by Critter-Crossing Technology, LLC; section of the ramp showing the "vole tube" (b). *Credit: Reproduced by permission of Kerry R. Foresman.*

reptiles can be negatively affected by cold temperatures during migration. For example, if structures that cross a multi-lane highway or a highway with a median are too long, migrating individuals might be at risk when weather conditions change quickly and there is little opportunity to seek appropriate shelter. Minimizing the width of the road at designated crossing points or creation of stopover habitat (e.g., within the median) are some strategies that can reduce the risks of traveling through long passage structures (Jackson 1996). Additionally, refugia and cover substrate can be incorporated into longer passage designs to facilitate intermittent stops.

Shape and composition of the structure may also affect use by certain species. Preferences for particular shapes by certain species may be related to openness;

rectangular, oval, or arch pipes may be preferable to round pipes for some species simply because they are wider or have a flatter surface on the bottom. Some animals prefer rectangular culverts while others are more successful using circular pipe culverts (Rosell et al. 1997). Differences in composition and construction may also affect use. For example, the corrugated surface of a metal pipe may impede use by some species of small size and limited mobility (C. Rosell, Minuartia Inc., personal communication).

9.5.2. Light

Lighting within a crossing structure may be a critical factor for use by certain species. The amount of light needed is dependent on the preference of target species for dark or light environments. However, even species whose movements are generally conducted under dark, overcast conditions may be sensitive to light availability. In Amherst, Massachusetts, monitoring of two salamander tunnels revealed that spotted salamanders exhibited "tunnel hesitation," a reluctance to enter or otherwise pass through, due to insufficient light in the structures (Jackson and Tyning 1989; Jackson 1996). Observers noted that simply shining a flashlight into the tunnel (from either direction) was sufficient to eliminate tunnel hesitation. Research conducted with turtles suggests that open-top structures that provide more light are substantially more effective for turtle passage than larger structures with closed tops (e.g., Case Study 9.5.3). At Lake Brompton in Quebec, Canada, early tests of 200 mm polymer concrete tunnels revealed that a larger tunnel size that allows for greater light penetration was beneficial for use by at least nine species of amphibians (Bergeron 2012).

Light penetration into a typical crossing structure is regulated, in part, by the size of the openings at each end; the larger the openings, the farther the penetration and the greater the availability of light inside. Some structures have open tops or grates that can provide significant light penetration for species that prefer an open environment. Specially designed structures for amphibians may include a grate or opening that runs the length of the structure (to allow in light and moisture; Figure 9.7), while more standard crossing structures may only include a grate or skylight in the median of divided highways. Emerging technologies can transmit natural sunlight, and even moonlight, within

Figure 9.7. This slotted 500-mm surface tunnel design marketed by ACO Polymer Products was introduced in the 1980s to allow for more moisture and light to enter. Leaf litter along the bottom surface provides cover and protection from desiccation for a variety of small animals. *Credit: Thomas E. S. Langton.*

the structure through specialized light collectors and fiber optic cables.

On Route 7 in Connecticut, wildlife passage structures were constructed for two reptile species (eastern box turtles [*Terrapene carolina*] and eastern hog-nosed snakes [*Heterodon platirhinos*]) of conservation concern, as well as other wildlife using the area. In an effort to ensure adequate light for migrating animals, artificial lighting (in form of fluorescent lights) was installed along the roof of one of the structures (Figure 9.8). The lights were tied into the electrical grid and equipped with photoreceptors to turn them on and off to mimic natural lighting patterns. Post-construction monitoring documented use of the structure by a variety of wildlife, including box turtles, although not hog-nosed snakes. No controlled tests of the effects of the artificial lighting have been performed, and it is not known to what degree artificial lighting is affecting passage success (H. Gruner, Connecticut Science Center, personal communication).

Artificial light may also reduce the effectiveness of crossing structures. Scientists at Portland State University investigated the effects of artificial light on use of a bridge structure by amphibians, reptiles, birds, and small to medium-sized mammals. During periods when artificial lighting was being used, the lights were on day and night for one week. They found that connectivity

Figure 9.8. Woodchuck (*Marmota monax*) using an artificially lit passage structure on Route 7 in Connecticut, United States. *Credit: Dennis P. Quinn.*

for terrestrial wildlife was disrupted by the presence of artificial light and that the effects were strongest for crepuscular and nocturnal species (L. Bliss-Ketchum, Portland State University, personal communication). These results suggest that when artificial lights are used with crossing structures, it is important that they produce light levels that approximate ambient conditions and are programmed to simulate natural light-dark cycles.

Some species may seek out culverts because of the shade or shelter they provide. Mohave desert tortoises (*Gopherus agassizii*; Boarman et al. 1998) spend time in culverts, as do porcupines (*Erethizon dorsatum*; Griesemer et al. 1998). While open tops or larger sizes might make these structures less effective as refuges, they might be more effective if intended as passage structures.

9.5.3. Case Study: Effectiveness of Road Passage Structures for Freshwater Turtles

Paul R. Sievert, David J. Paulson, and Derek T. Yorks
Turtle populations are extremely vulnerable to road mortality as a result of their life history, which includes low annual recruitment, high adult survival, and delayed sexual maturity. Females are more vulnerable to vehicle strikes due to repeated nesting migrations and use of roadway shoulders for egg laying (Practical Example 1), and this may cause more rapid population declines. An increase in adult mortality of 2–3% due to vehicle collisions is enough to cause most turtle popu-

lations to decline, and a 10% increase in adult mortality of snapping turtles (*Chelydra serpentina*) has been shown to halve the number of adults in less than 20 years (Congdon et al. 1994).

Underpasses, in conjunction with barriers to guide animals into them, have the potential to greatly reduce road mortality of freshwater turtles; however, experimental studies evaluating the effectiveness of such structures are lacking. Use of passage structures by turtles may depend on characteristics of the tunnels, such as structure size, shape, length, amount of light within the structure (Section 9.5.2), and tunnel position relative to the surrounding landscape (i.e., whether tunnels are at grade or below grade). Barrier fencing used to block access to roads or directional fencing to guide turtles toward underpasses may vary in its effectiveness depending on whether turtles are able to see through the fencing material. To date, most evaluations of road passage systems (tunnel and fencing combinations) have focused on the effectiveness of single structures and thus are limited in their ability to provide general guidance and recommendations for designing passage systems for turtles.

The Massachusetts Department of Transportation, UMass, and the US Geological Survey collaborated in an experimental study to examine the effectiveness of passage structures for freshwater turtles in Massachusetts. Specifically, we (1) evaluated variation in tunnel height, width, length, and overhead lighting on movement behavior of eastern painted turtles (*Chrysemys picta picta*); (2) measured the effect of fence opacity on behavior of painted turtles; and (3) tested selected passage and fencing designs on the movements of eastern painted turtles, spotted turtles (*Clemmys guttata*), and Blanding's turtles (*Emydoidea blandingii*).

Over a 3-year period (2009–2011) we used outdoor laboratories to evaluate the behavior of approximately 740 eastern painted turtles, 50 spotted turtles, and 50 Blanding's turtles with respect to tunnels that varied in size and lighting, and fences that varied in opacity (Figure 9.9). Turtles were captured in nearby wetlands using live traps, brought to the laboratory for a single day of testing, and then returned to their point of capture. A factorial experimental design was used to examine the effects of tunnel length (12.1 m or 24.2 m), opening size (height × width: 0.6 m × 0.6 m, 1.2 m × 1.2 m, or 1.2 m × 2.4 m), and overhead lighting (0%, 75%,

Figure 9.9. Research array at University of Massachusetts Amherst, United States, used to test various tunnel designs for turtles. *Credit: Derek T. Yorks.*

or 100% of ambient) on movement of turtles through the tunnels. We constructed tunnels from plywood, manipulated light with shade cloth, and monitored turtle behavior using digital cameras and closed-circuit video cameras. For each turtle, we recorded whether it successfully passed through the tunnel, along with its rate of movement through the passage structure. To test turtles' response to fencing, we recorded their rate of travel as they moved along fences that were either opaque or translucent.

We found that freshwater turtles were most likely to pass through tunnels having large cross-sectional openings and those that transmitted 100% of overhead light. Tunnels with small cross-sectional openings and open tops performed as well as tunnels with large cross-sectional openings and closed tops. We installed artificial lights in some closed-top tunnels and found that turtles passed through them at rates similar to open-topped tunnels. While conducting a pilot study prior to this research, we found that turtles were more likely to pass through at-grade tunnels compared to embedded tunnels. Regarding fencing, turtles moved rapidly along opaque fences but slowed considerably along translucent ones. When turtles were able to see through fences, they would frequently stop and attempt to pass through the fence and this delayed their movement.

Our research yielded three principal findings with regard to tunnel design: (1) at-grade tunnels perform better than those that are embedded (below grade);

(2) larger tunnels outperformed smaller tunnels; and (3) light was a critical design feature that was more important than size for effective turtle passage. To maximize rates of turtle road passage, tunnels should be designed to provide overhead light and relatively large cross-sectional openings. Relatively small structures may be acceptable if significant amounts of overhead light are provided, either from natural or artificial sources. Our research suggests that if open tops or artificial light cannot be provided then passage structures will need to be significantly larger to achieve comparable results. Further, we found that translucent fencing may slow the rate of turtle movement (e.g., because they could see through the fence and attempted to cross through it), while opaque fencing may speed the rate of movement.

9.5.4. Substrate

The presence or absence of a particular substrate (e.g., soil, litter, or other base material) can be a significant factor in the use or avoidance of a crossing structure by some species. While most bridges are simply constructed over the natural surface, most culverts are structural enclosures that include a solid floor. This floor is usually made of the same material as the walls and ceiling of the structure (e.g., concrete, metal, or composite materials). When these types of culverts are installed, it is preferable to use an oversize culvert that can be placed below grade and filled with soil substrate to match the adjacent ground level. Installation of three-sided culverts and arches can be used as an alternative to standard culverts, eliminating the need to use fill to provide a floor of natural substrate.

A natural substrate composed of materials similar to that of the adjacent habitat is generally preferable. To facilitate movement through a structure by fossorial species (e.g., burrowing animals such as moles), the substrate depth and composition should be appropriate to allow movement through the structure by these animals without risking exposure at the surface. For many species, cover objects including rocks, rows of tree stumps, and debris piles help to reduce risk of predation and exposure to the environment. Some animals may not use structures that lack cover objects because of a behavioral reluctance to expose themselves to predators (Chapter 2).

Some species may have specialized substrate requirements that will need to be considered in the design of a passage structure. The mountain pygmy possum (*Burramys parvus*) of Australia lives in rock talus slopes and rarely ventures out in the open. Crossing structures for this species include a deep layer of appropriately sized rocks to allow individuals to move through the crossing using the interstitial spaces within the substrate (Mansergh and Scotts 1989). In North America, pikas (*Ochotona princeps*) and some species of lizards and snakes may have similar requirements.

In special circumstances, especially in the initial stages following construction of the crossing structure, attractants may be useful for encouraging use by certain species. For example, using scent lures or baits may attract and induce use by snakes that typically follow pheromone scent trails (Shine et al. 2004). Brehm (1989) suggests that washing substrate with water from breeding ponds may provide olfactory cues that could improve usage by migrating amphibians; however, this has not been tested for efficacy.

9.5.5. Moisture

The presence and retention of moisture is a critical factor for most amphibians and some reptiles. Desiccation is a significant factor in mortality of amphibians and some aquatic snake species during migration; therefore, both cover from environmental exposure and presence of moisture are important considerations in structure design. Providing adequate moisture is important in locations wherever amphibian use is expected, even if they are not the target species.

The most effective way to ensure adequate moisture is to use open-top structures. By providing moisture, open-top structures also maintain cues (wetness) used by amphibians (or other species) that typically migrate during rainy weather. An alternative is to direct stormwater into the structure with a mechanism to distribute it throughout the structure. This can be beneficial if the flow rate is not too high. If the flow rate is too high, erosion of desired substrate within the structure could occur. Some have proposed allowing stormwater to sprinkle down from a perforated conduit on the structure's ceiling, but it is not known whether any such system has ever been built.

An important concern about the use of stormwater runoff to provide moisture within the structure is the presence of toxic substances from vehicles and the road. This could be harmful to species such as amphibians, which are sensitive to roadside contaminants (Chapter 4). For these species, stormwater runoff needs to be treated so that only clean water flows into the structure.

When open-top structures are not desirable or possible, open-bottom structures (bridges or three-sided culverts) can be designed to allow for the presence of moisture from natural substrate and groundwater. The key in this situation is to construct the structure at an elevation relative to the water table that maintains moist soil conditions for those times of the year when amphibian passage is most needed.

In contrast to amphibians that require wetter environments, many species are adapted to dry conditions. In xeric habitats, light sandy soils allow rapid percolation of rainfall and drainage, and certain species such as the Florida sand skink (*Neoseps reynoldsi*) require this loose soil for mobility (Telford 1959). Heavy, compacted soils and water retention would inhibit their movement. For species adapted to dry, sandy soils the ground should be contoured to lead runoff away from the structure. It is important that these passages not serve any stormwater management function.

9.5.6. Water Management

While moisture retention may be an important factor with certain species (particularly amphibians), water management and flooding are critical issues for all species that may use terrestrial passages. Although some species traveling overland may swim through flooded passage structures (e.g., crocodilians, turtles, and snakes), many (e.g., most small mammals) are deterred by flooding or standing water (Janssen et al. 1997; Rosell et al. 1997; Santolini et al. 1997). Flowing water can be especially problematic. High velocities can prevent use by many species and result in erosion of substrate. The elevation and slope of a passage structure are key considerations for reducing these potential problems.

Unless the structure is specifically designed to use runoff to maintain appropriate moisture, wildlife passages should be managed to exclude stormwater. Where some amount of water is likely to enter the structure (runoff from the approaches or rain entering through open tops), the structure should be designed in such a way as to prevent pooling. Structures may be sloped

Figure 9.10. An example of incorrect design of an approach area; the roadside swale drains into an adjacent creek, resulting in significant backflow into a terrestrial passage. *Credit: Daniel J. Smith.*

gently from one end to the other or contoured such that the center is higher than the ends (e.g., tracking the topography of crowned roads) so that water can gently flow out. In some cases, accommodations will need to be made (e.g., dry wells) at one or both ends to prevent flooding of the entrances (Jackson and Tyning 1989).

The landscape surrounding the structure location plays an important role in water management as well. Steep slopes leading to the structure from adjacent approach areas would contribute to increased flooding and flow rates and should be avoided or contoured to reduce this effect (Figure 9.10). If a terrestrial passage is warranted, it is not recommended to place it where flooding and periodic increased flows occur; such sites are part of natural draws or at low elevations with poorly drained soils. These locations may be suitable for certain species to use at times when the structure is dry or simply wet, but not when water is flowing (e.g., Patrick et al. 2010). Also, retaining natural substrate within these structures might be difficult.

In certain cases, placement of multiple structures at the same location but at slightly different elevations (low and wet, high and dry) can provide opportunities for movement by species that prefer either wet or dry conditions. An example of this may be at a stream crossing where, in lieu of an extended bridge that could be quite expensive, a wet passage can be installed at the stream for aquatic and semi-aquatic species and a

smaller dry passage can be placed on each bank of the stream for terrestrial species.

9.5.7. Open versus Closed Tops

Open-top structures provide many advantages for small vertebrates and are especially beneficial to amphibians. Discoveries of dead, dried-out salamanders within crossing structures built for other wildlife (L. Rogers, Town of Concord Wildlife Passages Task Force, personal communication) confirm that even when conditions elsewhere are appropriate for amphibian movement, dry conditions within a passage structure can not only impede movement but may also contribute to mortality. Open-top structures allow rain to fall through the top and dampen the substrate on the bottom of the tunnel, maintaining the appropriate moist conditions for amphibians and cues that may be important for the continued movement of frogs and salamanders.

A number of designs have been used to achieve an open top. They include the use of storm drain grates over an open box culvert as well as pre-cast polymer concrete structures with slotted tops marketed by ACO Polymer Products (Figure 9.11). According to Brehm (1989), some of the advantages of ACO tunnels include their (1) very smooth internal surfaces, making them easy to clean; (2) low cost in comparison to conventional large tunnels; and (3) rate of wear, which is the same as the road surface. This last characteristic is particularly important for addressing concerns about safety and maintenance. If the structure did not wear down as fast as the road surface it would eventually protrude above the hardtop and present a threat to motorists and an obstacle to snow plows.

Open-top structures also provide more light and air flow within the structure. In England, a small difference in temperature between a tunnel and the outside air appeared to be responsible for delays in use of the structure by common toads (*Bufo bufo*; Langton 1989). Light is an important determinant of passage success for spotted salamanders (Jackson and Tyning 1989) and turtles (Yorks et al. 2012; Case Study 9.5.3). Furthermore, research on turtles indicates that it takes a very large structure to approach the passage success of relatively small (0.6 m × 0.6 m) tunnels with open tops (Yorks et al. 2012; Case Study 9.5.3).

Biologists have also raised concerns about open-top

Figure 9.11. Open-top structures for amphibians: (a) an open-top box culvert with storm drain grates in Princeton, Massachusetts, United States; (b) ACO Polymer Products tunnel with slotted top used at Henry Street in Amherst, Massachusetts. *Credits: Scott D. Jackson (a) and Tom Tyning (b).*

structures. These structures are more vulnerable to flooding and sediment accumulation, especially in areas where road sanding and salting occur in the winter. There are concerns about sound and vibration that might deter use of the structures by some species (Chapters 2 and 4). Open tops may facilitate the introduction and concentration of salt and other road-based contaminants, potentially affecting the health of animals that use the structure (Chapter 4). In arid environments, open-top structures with greater exposure to the air and sunlight can result in more rapid drying and the loss of moist soil conditions.

Highway departments generally are very reluctant to allow the use of open-top structures due to concerns related to safety and maintenance. These concerns include how grates might affect bicyclists, whether open-top structures will represent a dangerous discontinuity

in the road surface, the potential for grates to become dislodged and thereby threaten motorist safety, complications when the roads need to be resurfaced, and the potential for snow plows to get caught on tunnel edges.

There is no reason to believe that these are unsolvable problems, especially given the tremendous road engineering achievements that have taken place over the years. For example, one author (Jackson) has observed that the Henry Street salamander tunnels in Massachusetts, which used an early slotted-top design produced by ACO Polymer Products, were installed in 1987 and have functioned well without incident for over 25 years (Figure 9.12). Further, in Princeton, Massachusetts, an open-top structure using a three-sided box culvert with storm drain grates as the top has been in place for over 20 years and remains in good condition.

Figure 9.12. Salamander tunnel at Henry Street in Amherst, Massachusetts, United States, in 2013, 25 years after it was initially installed. *Credit: Scott D. Jackson.*

In this book, we emphasize that the design of passage structures, and implementation and monitoring of associated projects, should focus on as many species as possible. An open top may not be necessary to provide successful passage for some target species—turtles, for example. However, frogs and salamanders may also cross the road in the same places as turtles. A tunnel and barrier system designed for turtles but without an open top to meet the needs of amphibians may succeed in mitigating the effects of the road for one species or group while creating a significant impediment to movement for others that can no longer cross over the road because of barrier fencing or walls.

While open-top structures typically have not been installed, these passage designs can accommodate multiple species, and should be considered beyond the usual single-species conservation approach (Case Study 9.5.8). Wildlife professionals should pursue these options when engaging with transportation professionals. Research is urgently needed to better understand the threats and benefits of open-top systems and to develop designs that address the needs of small animals as well as the safety and maintenance concerns of transportation professionals.

9.5.8. Case Study: Structures for Turtles and other Herpetofauna on the Long Point Causeway, Ontario, Canada

Kari E. Gunson

Situated in a world biosphere reserve, the Long Point Causeway is deemed the fourth-deadliest highway for turtles in the world (http://www.lakejacksonturtles.org/top5.htm). The causeway was built in the 1920s to allow motorized access to the Long Point beaches, and consequently disconnected a low-lying marsh from a bay on Lake Erie. Several turtles and snakes designated species at risk in Canada inhabit the causeway, and it has been documented to be a zone of significant mortality for small animals, primarily amphibians and reptiles (Ashley and Robinson 1996). The causeway also poses a severe hydrologic barrier between the marsh and the bay because only one of the original four water connections remains.

In 2006, concerned experts and citizens along with local, provincial, and federal government agencies formed the Long Point Causeway Improvement Project Steering Committee, with a mandate to develop a habitat restoration plan (http://longpointcauseway.com/). Two years later, a feasibility study was initiated by the committee to improve water and wildlife passage at the causeway. After expert review and public consultation, it was recommended that at least 11 culverts or passages (with barrier walls) be built under the road, at a total estimated minimum cost of C$8.1 million. The local council endorsed the plan and the committee raised sufficient funds from national conservation and government funding programs and corporate sponsors to initiate the effort. The funding covered installation of 4 km of barrier fencing, continuation of road mortality studies, and promotion of key environmental issues through public awareness activities. In 2011, the committee continued implementation of the plan after persuading Norfolk County, which manages the road, to undertake a municipal environmental assessment (EA) to install three of the required culverts (two terrestrial and one aquatic).

During consultation in the EA process, a proposal was drafted to include different culvert designs that would accommodate both hydrological flow and amphibian and reptile passage. A larger hydrologic structure (precast box culvert, with a height and width of about 2 m) was recommended for aquatic passage, in addition to two smaller structures for semi-aquatic and terrestrial passage (precast box culverts, with a height and width of about 1 m). In addition, it was suggested that one of the smaller culverts include openings in the top slab to create ambient conditions (e.g., light, moisture, and temperature) within the culvert, given that increased lighting can facilitate use by turtles (Yorks et al. 2012; Case Study 9.5.3). The inclusion of three different structure designs in the mitigation plan is important from an experimental perspective because little is known about the types of culverts that turtles will use to cross under a road. As such, results from this project can inform future design considerations for similar species and types of conditions. Additional funding was received in 2013, which will allow for more structures to be built per the recommendations of the plan.

9.6. Case Study: Steering Snakes, the Effects of Road Type, Canopy Closure, and Culvert Type

Bruce A. Kingsbury, Bryan C. Eads, and Lindsey Hayter
Evidence is growing that different species of wildlife vary in their propensity to cross roads. Of course, not all roads are created equal, most obviously with respect to width and traffic volume. We were interested in how snakes would respond to roads as a potential barrier, and specifically, how factors such as the road surface or surrounding habitat influence the behavior of animals encountering a road. Our work was motivated by determining how to mitigate the fragmenting effect of roads on populations of copper-bellied watersnakes (*Nerodia erythrogaster*), listed as threatened by US Fish and Wildlife Service in the northern part of their range. Recovery of copper-bellied watersnake populations will require providing populations with landscapes of substantial size and connectivity to ensure population viability. Because these landscapes must extend for miles to provide sufficient habitat, roads provide a major challenge to achieving that goal; even where substantial habitat is available, it is fragmented by these pervasive barriers.

There are two perspectives to mitigating the impacts of roads. Most obviously, our primary objective is to avoid road mortality and injury. To do so, we want to reduce the time animals spend on the road, which in part can be achieved by preventing them from crossing the road surface. On the other hand, we also want to encourage interbreeding among potentially isolated populations. As a result, we aim to encourage crossing where the net benefit of breeding between populations outweighs any losses due to negative encounters with vehicles.

We investigated the response to roads by a suite of snakes—copper-bellied and midland watersnakes (*Nerodia sipedon pleuralis*), eastern gartersnakes (*Thamnophis sirtalis sirtalis*), and eastern ribbonsnakes (*T. sauritus sauritus*)—to explore the effects of road surface (paved vs. gravel) and canopy closure (tree canopy present or not) on road crossing attempts. Anecdotal observations suggested that snakes might be more likely to cross coarser road surfaces and that canopy cover might diminish their sense of vulnerability to predators.

Our approach, adapted from Andrews and Gibbons (2005), was to place snakes under a bucket positioned half on the road and half off the road. Trials were conducted by an observer hidden behind a camouflaged hunting blind who remotely raised the bucket about 0.5 m from the ground, releasing the snake. Releases occurred from both sides of the road and in multiple locations along roads to reduce any potential biases that might occur based on directionality, influences of roadside habitat, or scent trails from previous snakes.

All four species showed the greatest avoidance of paved roads without canopy cover. Copper-bellied watersnakes did not show a preference between the other three road treatments. Midland watersnakes, gartersnakes, and ribbonsnakes had an intermediate inclination to cross gravel roads without canopy cover and were least likely to cross paved roads without canopy cover. Midland watersnakes crossed roads of any type.

Ultimately, the issue becomes how to design culverts under the road to allow for safe passage. While we expect snakes to use large underpasses, the likelihood of using smaller passageways is poorly understood. We examined the use by three of our study species of existing culverts and artificial culverts, the latter of which we could manipulate. For the existing culvert experiments, snakes were released from a bucket placed within one meter of the culvert entrance. Trials were conducted from both sides of the road or culvert to reduce any potential bias based on nearby habitat characteristics. We tested four culverts in this experiment: culvert 1 was 10 m long and had a soil substrate and a diameter of approximately 0.33 m, culvert 2 was 10 m long and had a water substrate and also had a diameter of approximately 0.33 m, culvert 3 was 8 m long and had a soil substrate with a diameter of approximately 1 m, and culvert 4 was 11 m long and had a water substrate and a culvert diameter of approximately 1 m.

Copper-bellied watersnakes were only tested in culverts 1, 2, and 4. They showed a preference for passing through the larger culverts (40%) compared to the smaller culvert (8%). Midland watersnakes readily crossed through all 4 existing culverts, with the lowest passage frequency being 81%. Ribbonsnakes crossed through both larger diameter culvert types at much higher frequencies (≥76%) than the 2 smaller diameter culvert types (≤34%).

We also employed artificial culverts, in addition to those already in place, so that we could manipulate conditions more easily. They were constructed of plywood and pond liner and consisted of a 4.9 m covered runway with adjustable culvert widths and substrates. This length simulates that of a culvert under a single-lane road, though other studies have already shown that watersnakes will use longer culverts (Dodd et al. 2004). Snakes were released into an enclosure at one end of the runway and had the option of passing into and through the runway to "escape." The first experiment was to determine whether or not culvert width (0.33, 0.66, 1.0, 1.3 m) influenced snake crossings through culverts. All culverts in the first experiment had a soil substrate. We then performed a second experiment wherein we conducted trials with a shallow water substrate versus a soil substrate (for 0.33 m and 1.3 m wide culverts).

Culvert width and substrate had little effect on whether or not snakes crossed through them. Ribbonsnakes were the only species that differed in passage frequency, using the 0.33 m culvert less frequently (68%), compared to >90% for all other species and culvert widths. Additionally, substrate had little effect on whether snakes crossed through artificial culverts. Only ribbonsnakes differed in passage frequency, crossing less frequently through the 0.33 wet (70%) and dry (48%) culverts than the 2 larger culverts (93% and 100%).

The results of these experiments support the conclusion that we can influence if, when, and where some species of snakes will cross roads. In areas where we want to encourage road crossings, such as in areas of low traffic and between adjacent requisite habitat components, vegetative cover should be allowed to grow as close to the road edges as allowable, including canopy cover over the roads. On the other hand, where we do not want snakes to cross, we can follow common practices of woody management and mowing to expose and expand the width of the road to make it even less attractive.

Then, in areas where the danger of crossing roads warrants it, use of culverts may be a viable option for mitigation because our results show that snakes will use them. Further, though snakes will use relatively small culverts, we recommend that they be as large as practicable to facilitate greater use by other taxa. Lastly, vegetation and canopy adjacent to the road can be managed to attract snakes to safer crossing points where culverts and other underpasses are located.

9.7. Structure Enhancement

Almost as important as the structures themselves, the characteristics of approach areas and the use of directional fencing and walls are important considerations for the design of wildlife passage systems. Although species that prefer to move through water will not have much trouble finding stream and other aquatic crossings, terrestrial or semi-aquatic animals will typically need directional fencing or guide walls to prevent them from crossing over the road surface and to help them find passage structures.

9.7.1. Approach Areas

The design of approach areas is important for helping animals find the passage structure. On most roadsides, there is a "clear zone" that is designed to provide an obstruction-free area to lower the collision risk should a vehicle run off the road. This also allows for driver recovery of the vehicle, clear sight lines, and improved motorist safety. Because approach areas are often located in these clear zones, this constrains, to some extent, design options for enhancing wildlife movement. Despite the constraints of working in a clear zone, there are some characteristics of approach areas that can be designed to improve passage success.

9.7.1.1. Terrain

The gradient leading to a passage structure should generally mimic that of the adjacent terrain using natural (but gradual) contours to draw animals toward the structure. If these natural contours are not present, then some earthwork may be beneficial to enhance the movement of animals toward the structure. Gradual slopes are preferred; steep slopes should be avoided because of drainage, flooding, and erosion problems.

Swales, ditches, or canals adjacent to the road, especially those that have steep slopes or are paved concrete, may act as impediments to movement by many species. These drainage structures should be modified at locations leading to passage structures. Use of drainage pipes covered by soil near the entrances of the crossing structure can convey stormwater and provide a level and unimpeded approach to the passage structure.

9.7.1.2. Vegetation and Cover Objects

The use of vegetation, rocks, and logs on approaches may enhance use of crossing structures by a variety of small and mid-sized mammals (Hunt et al. 1987; Rodriguez et al. 1996; Rosell et al. 1997; Santolini et al. 1997; Clevenger and Waltho 1999). In conjunction with the slope of the terrain, vegetation can be used to assist animals in finding the passage structure. Native plant species should be used whenever possible. Because of the safety issues mentioned above, trees would likely be prohibited in the clear zone of the right-of-way. Plantings should be done in a random and natural fashion, not in rows or blocks. The target species for the passage need to be considered when determining choices for vegetation. Certain species prefer thicker, dense groundcover and shrub layers, while others prefer more open ground. Shrub and groundcover plantings can be arranged at approaches to structures intended for use by both types of species (e.g., providing multiple clear pathways among strips of groundcover and shrub layers). Canopy cover at passage openings (Case Study 9.6), as well as accommodations for arboreal animals (Case Study 9.8), are relatively new considerations that may also be important for particular species (see also recent studies on passage structure use by bats, e.g., Bach and Limpens 2004; Boonman 2011; Abbott et al. 2012; Berthinussen and Altringham 2012). Standard construction practices for culverts and bridges often require the use of riprap at entrances and around pilings to prevent scour and erosion and maintain sta-

bility of the structure. Riprap can be a significant impediment to many species. The design of the structure should maintain clear pathways into the structure for animals, while recognizing that the use of a stabilizing material may be needed to maintain the integrity of the structure. The amount of riprap and size of the rocks used should be carefully considered, along with the use of alternative materials that serve the same purpose. Another option might be to fill the spaces within the riprap with smaller material to provide a more even substrate for small animal movement. In the approach area, earthen berms and shrub plantings can be used along the fence line leading to the passage to reduce the potential negative effects of vehicle headlights and traffic noise.

9.7.1.3. Specialized Features

Certain structures have been specially designed to assist in funneling animals into passage structures. Wing walls are commonly used with culverts to prevent erosion of the banks adjacent to streams and abutting the road. These angled walls also act as a funnel to direct animals toward the entrance to passage structures. Another device is a septum fence, a Y-shaped structure at the tunnel entrance to prevent animals from bypassing the structure (Figure 9.13a). Brehm (1989) reported that amphibians have difficulty finding tunnel entrances when directional fencing is not angled but instead runs parallel to the road. When it is not possible to angle the fences, he recommended use of a "swallowtail barrier" (i.e., septum) at the tunnel entrance (Figure 9.13b). ACO Polymer Products manufactures a custom tunnel entrance for amphibian tunnels that includes a septum (http://www.acowildlife.us/).

9.7.2. Directional Fencing and Guide Walls

The two primary types of barriers used to keep animals off the road surface and direct them to the passage structures are directional fencing and guide walls (Pepper et al. 2006). Fences are generally free-standing structures supported by posts that protrude from the ground and often (although not always) present a barrier to movement in both directions. Guide walls are structures built into an embankment, or are built as part of road-supporting retaining walls, that present a vertical face on one side but are typically flush with the ground surface on the other (Figure 9.14). Directional fences or guide walls are an essential element to the performance of terrestrial animal passages. Many types and sizes of both fencing and walls have been used for different applications.

Compared to guide walls, directional fencing is a considerably cheaper application to construct, but is vulnerable to ongoing damage and, depending on the material, its usable life span is much shorter (Table 10.2). Construction costs that are saved up front must be made up for in higher maintenance. Walls,

where they can be used, are almost always preferable to fences. From a maintenance perspective, guide walls require less effort and expense and are more durable. They resist damage from cars, snow plows, and falling trees. Most guide walls are solid and impenetrable, whereas directional fences may be semipermeable (e.g., juvenile animals may be able to pass through a fence), depending on the fencing type and mesh size (Smith and Noss 2011). No barrier is perfect, and sometimes animals manage to get over or around the fence or wall (trespass) and get access to the road. Some walls provide a barrier in only one direction when they are installed flush with the ground surface on the other side; animals that do get access to the road are not trapped along the roadside between a fence and the road. When directional fences are used, it is important to consider using one-way gates (see Figure 9.2) or elevated breaches (i.e., jump-outs) to prevent animals from becoming trapped on the road.

9.7.2.1. Materials

Fences have been constructed from a variety of materials including wire or plastic small-mesh fencing, silt fencing, fabric, aluminum flashing, rigid plastic, wood,

a

b

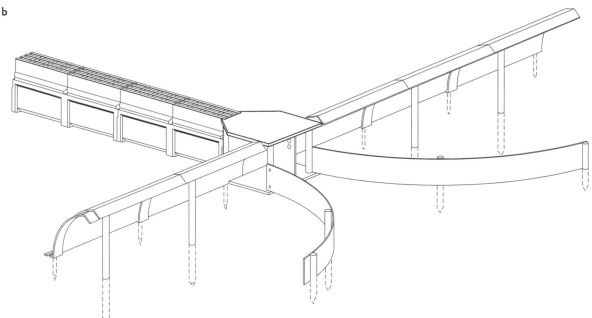

Figure 9.13. Two barrier and passage systems for amphibians in Europe: (a) guide wall with septum and (b) curved guide fence with swallowtail septum. *Credits: Thomas E. S. Langton (a) and © Victor Young Illustration, adapted from original by ACO USA (b).*

Figure 9.14. Barrier wall and passage structure, both manufactured by ACO Polymer Products, Inc. *Credit: Thomas E. S. Langton.*

Figure 9.15. Highway fencing modified with additional material at the bottom to serve as guide fencing for small vertebrates: (a) small-mesh fencing and (b) rigid plastic. *Credits: Scott D. Jackson.*

and metal (Evink 2002; Pepper et al. 2006; see also Table 10.2). Sometimes directional fencing for small vertebrates is created by affixing small-mesh fencing (but see Box 11.1 for risks associated with various mesh materials) or other material to the bottom portion of typical roadside wire or chain-link fences (Figure 9.15). Other specialized materials that have been used as barriers or guide fencing include single- or multiple-stacked guardrail and railway rails (Evink 2002; Pepper et al. 2006). It is important that all joints are properly sealed and that the fence is attached to the passage structure in a way that does not allow animals to circumvent the barrier. Posts used to support fencing warrant special attention because they can create opportunities for animals to climb over the barrier.

Guide walls can be constructed of concrete, granite, wood, metal, brick, and potentially other materials. Texture is an important consideration in the choice of wall material. Rough and textured materials are easier for small vertebrates to climb than smooth-faced walls.

9.7.2.2. Geometry

In the approaches to passage structures, it is important to use a directional angle (perhaps 45°) to bring the fence or wall in from the outer edge of the road right-of-way to the abutment of the passage structure. Angled fences and walls work better than those that run parallel to the road and require animals to turn at an abrupt angle (e.g., 90°) to enter the structure (Brehm 1989; Ryser and Grossenbacher 1989; Allaback and Laabs 2003; Pagnucco et al. 2012). Angled fencing effectively deflects animals toward the structure rather than blocking movement and forcing animals to move parallel to the road. Angled fencing is only practical to

Figure 9.16. Angled guide fencing associated with sala-mander tunnels in Amherst, Massachusetts, United States. *Credit: Scott D. Jackson.*

use at or near the approaches or when the application calls for a short segment of fence or wall—otherwise the fences would extend too far out into adjacent habitat. Multiple, closely spaced passage structures connected by short sections of angled fencing or walls represent an optimal arrangement (Figure 9.16).

9.7.2.3. Opaqueness

For certain species (particularly turtles), opaque fencing or walls are preferred over those that are transparent (e.g., mesh; Case Study 9.5.3). Desert tortoises were found to repeatedly (to the point of exhaustion) try to penetrate a transparent fence to get to the other side (Ruby et al. 1994). One advantage of mesh fencing is that in steep, rocky, or frozen conditions, runoff can flow through the fence rather than being directed (as with walls or solid fencing) into the passage structure (see also Section 9.7.2.6). Mesh fencing with a covering to create a visual barrier might be one solution for these types of settings.

9.7.2.4. Height

The height of the fence or wall is dependent on target species and the potential for a buildup of vegetation or debris against the barrier. As much as 1 m above the ground may be needed to prevent various species of snakes and some turtles from climbing over (Boarman et al. 1997; Griffin and Pletscher 2006; Woltz et al. 2008; Smith and Noss 2011) or certain frogs from jumping over (Woltz et al. 2008; Smith and Noss 2011). Maintenance is a significant consideration when deter-

mining the height of the fence or wall. Growth of vegetation and accumulation of leaf litter can allow various animals to climb over the barrier, thus compromising its effectiveness (Dodd et al. 2004).

9.7.2.5. Base

A significant problem with standard fencing is gaps at the base that allow animals to gain access to the road surface. Directional fencing, especially for small animals, must be buried; recommendations range from 15 cm to 30 cm (Pepper et al. 2006; Smith and Noss 2011). Periodic maintenance inspections are needed to detect and repair the fence where washouts have occurred and in spots where digging by large animals has created gaps. A flat platform 20 cm wide at the base of a guide wall (on the habitat side) may prevent vegetation from growing in immediate contact with the wall and may reduce likelihood of trespass by climbing species. This design feature is required in Europe by the Federal Ministry of Transport, Building and Housing, Road Engineering and Road Traffic (FMTBHRERT 2000).

9.7.2.6. Texture and Mesh Size

Smooth-textured walls or fencing material are best for deterring trespass. For fencing, the mesh size is important in determining what can penetrate the fence and what cannot. Of generally available mesh, a size of 1.3 cm or larger allows many species of vertebrates to penetrate (Clevenger and Huijser 2011; but see Table 10.2 and Box 11.1 for species entrapment risks associated with certain mesh). Conversely, a 0.65 cm mesh prevents most species from passing through (Smith and Noss 2011). However, certain fencing materials with smaller mesh sizes may be more fragile (Table 10.2). One problem with any solid fencing or small mesh is poor drainage, causing retention and buildup of debris from stormwater runoff. The smaller the mesh size, the more problems are likely to occur with water flow; the larger the mesh size, the more problems with species trespass or entrapment (Box 11.1).

9.7.2.7. Top

On some fences, an angled extension on the top edges has been used to prevent trespass. Several variations of this have been used, including bending over the top of a flexible fence (Figure 9.17), providing an angled roof over the top of the fence, and adding an extended lip at

Figure 9.17. Plastic mesh fencing used at salamander tunnels in Amherst, Massachusetts, United States, with top bent to prevent animals from climbing over. Mesh fencing also prevented runoff from the hillside flooding the tunnels. *Credit: Scott D. Jackson.*

the top of concrete retaining walls (Pepper et al. 2006; Clevenger and Huijser 2011; Figure 9.18). Some extensions angle downward on the animal's side of the fence while others extend straight out parallel to the ground. The lengths of the extensions also vary. It has been suggested that modifications at the top (i.e., an extension angled downward) can allow for reduced overall height of the fence. In the absence of standards for these measures, designs should be based on the climbing or jumping abilities of the target species.

An alternative method for deterring trespass by climbing is to use a barrier that curves upward away from the road surface. ACO Polymer Products produces a curved fencing material that allows soil to be piled up on the back side to create a one-way barrier. Curved walls have also been used (Figure 9.19).

Figure 9.18. Examples of guide walls with lips at the top to prevent animals from climbing over. *Credit: Thomas E. S. Langton.*

a

b

Figure 9.19. Curved barriers to prevent animals from climbing over curved guide fence (a) and curved guide wall (b). *Credits: Thomas E. S. Langton.*

9.8. Case Study: Living the High Life— Restoring Connectivity across Roads for Arboreal Species

Rodney van der Ree

Arboreal species of wildlife spend all (or most) of their time in trees. They usually prefer to move across the landscape by moving from tree to tree, either by climbing (or small jumps) via an interconnecting canopy or by gliding. The length of the glide depends on the weight and size of the animal, its glide capability (which is determined by its shape and aeronautical design), and its launch height. The extent to which different arboreal species are willing to leave the trees to travel across the ground varies significantly—some species are strictly arboreal and rarely leave the canopy, while others regularly come to the ground. However, when arboreal species do come to the ground, many appear clumsy and are at increased risk of predation and mortality due to vehicle strikes.

Options for mitigating the barrier effects of roads and other linear infrastructure for arboreal species include the maintenance of an interconnecting tree canopy, ensuring the size of the canopy gap is within the gliding range of the target species, and installation of artificial structures such as canopy bridges or glider poles (essentially timber poles that act as artificial trees to shorten the glide distance [Taylor and Goldingay 2011]; Figure 9.20). Traditional underpasses and overpasses for wildlife can be modified for arboreal species by providing structures to make them more suitable. Trees can be planted or glider poles can be installed on wildlife overpasses. For underpasses, timber rails can be provided that pass through the structure and then are connected to trees at each end. In some situations semi-arboreal species that are willing to travel across the ground can be accommodated by providing suitable habitat structure (e.g., branches and other woody debris) within wildlife crossing structures designed for other species.

One of the major challenges for arboreal species is funneling them toward crossing structures. They spend their entire lives in trees, and most of the fences that are built to keep terrestrial wildlife off roads are often ineffective for arboreal wildlife because of their climbing ability, a situation that is often exacerbated by the presence of overhanging trees. One approach is to use

Figure 9.20. Examples of crossing structures for arboreal animals in Australia: goanna (*Varanus sp.*) on a rope bridge (a) and squirrel glider (*Petaurus norfolcensis*) on a glider pole (b). *Credit: Rodney van der Ree.*

strategic planting of trees within road corridors to funnel arboreal animals toward the crossing structures. Alternative fence designs adapted to the behavior of the target arboreal species may also be effective, but more research is required.

The design of roads for gliding species requires detailed knowledge of the glide capability and typical glide behavior of the target species. While glider poles can restore movement, they may actually increase mortality by placing animals in the path of oncoming vehicles if the glide ratios have been incorrectly calculated. It is essential to know the average height (and range of heights) of launching positions within trees, as well as the rate at which animals fall during the glide. These launch details and glide angles can be used to calculate whether a target species can successfully make it across the planned road. A highway in southeastern Australia was to be widened beyond the glide capability of squirrel gliders (*Petaurus norfolcensis*), and a decision was made to install glider poles and rope bridges (van der Ree et al. 2010; Soanes et al. 2013; van der Ree et al. 2015). To determine the height and spacing of the glider poles, we estimated projected glides based on a conservative glide ratio and a requirement that gliders be able to pass at least 2 m above the maximum height of the tallest trucks using the road.

There is evidence from around the world that some of the impact of roads on arboreal species can be mitigated. Canopy bridges have been installed for dormice (*Gliridae* spp.) in the UK, samango monkeys (*Cercopithecus albogularis*) in Uganda, possums and gliders in Australia, and howler monkeys (*Alouatta* spp.) in Brazil. Glider poles have been installed for squirrel gliders, sugar gliders (*Petaurus breviceps*), yellow-bellied gliders (*Petaurus australis*) and greater gliders (*Petauroides volans*) in eastern Australia, and northern flying squirrels (*Glaucomys sabrinus*) in North America. It is still unclear to what extent arboreal movement can be fully restored and mortality reduced. Before designing the appropriate mitigation structure, it is important to clearly identify the impact (or potential impact) of the road on arboreal species. It is highly recommended that these mitigation measures be thoroughly tested before their widespread deployment. Additional information on arboreal crossings is covered in van der Ree and colleagues (2015).

9.9. Conclusions

There are many choices to be made when designing wildlife passage systems for small vertebrates. In making these decisions one must consider the characteristics of both target and nontarget species, as well as the particular terrain and landscape in which the system will be located. Often compromises must be made, and it is rare that the resources allocated for a project are sufficient to fully achieve the conservation objectives. The goal of mitigation design is often to achieve the greatest benefit possible given a set of financial and logistical constraints. Combining the expertise of both biologists and engineers is an effective strategy for creating a design that is both practical to implement and effective for achieving the conservation objectives.

9.10. Key Points

- Signs are an inexpensive way to encourage motorist awareness and speed reduction on roads. However, their long-term effectiveness at reducing animal road mortality is not known. To increase the effectiveness of wildlife warning signs consider the use of novel designs, selective placement, inclusion of information about distances, and seasonality of deployment.

- Reduced speed zones, bucket brigades, and temporary road closures may be effective in the short term or under certain circumstances, but there are questions about their overall effectiveness for long-term maintenance of small vertebrate populations.

- Barrier fencing when used without crossing structures can reduce road mortality but may disrupt landscape connectivity and contribute to the fragmentation of populations and habitat. Mesh size for certain barrier fence materials should also be considered to prevent trespass or entanglement of some species.

- Habitat modifications can be used to reduce the number of animals that trespass onto the road surface or to maintain access to vital habitats when other mitigation techniques are impractical.

- Even roads with low traffic volumes and risks of wildlife mortality can have significant effects on small vertebrates if curbing blocks their movement or funnels them into catch basins. Possible solutions include use of country drainage (no curbing) or use of gently sloping curbs (sometimes referred to as "mountable curbs" or "Cape Cod berms") instead of more traditional asphalt or granite curbing.

- Stream simulation is an approach to designing stream passages that simulate conditions found in the natural stream; these include water depth, water velocity, and substrate and channel characteristics. It is generally considered the most effective technique for facilitating passage for aquatic vertebrates and semi-aquatic wildlife, such as stream salamanders, snakes, turtles, and small mammals.

- Important design considerations for aquatic crossing structures include water velocity, water depth, substrate, bank edge areas, light, and openness.

- For terrestrial crossing structures it is important to consider the following: location, size, shape, light, substrate, moisture, water management, approaches, vegetation and cover objects, and specialized features (wing walls, swallowtail barrier, or septum).

- An important consideration for terrestrial crossing structures is whether to use an open- or closed-top design. Open-top structures maintain moist conditions for amphibians during rain events and provide more light and air flow within the structure. Highway departments generally are very reluctant to allow the use of open-top structures. There are also concerns about sound and vibration that might deter use of the structures by some species. In addition, open tops may facilitate the introduction and concentration of salt and other road-based contaminants.

- In general, wildlife passage structures are much more effective when used in conjunction with appropriately designed directional fencing or guide walls. Opaque fencing materials may increase effectiveness of directional fencing for some species.

- Many types and sizes of mitigation measures, including aquatic, terrestrial, and arboreal structures, can be customized for multiple applications, species, and habitats.

LITERATURE CITED

Abbott, I. M., S. Harrison, and F. Butler. 2012. Clutter-adaptation of bat species predicts their use of under motorway passageways of contrasting sizes—a natural experiment. Journal of Zoology 287:124–132.

Allaback, M. L., and D. M. Laabs. 2003. Effectiveness of road tunnels for the Santa Cruz long-toed salamander. 2002–2003 Transactions of the Western Section of the Wildlife Society 38/39:5–8.

Andrews, K. M., and J. W. Gibbons. 2005. How do highways influence snake movement? Behavioral responses to roads and vehicles. Copeia 2005:772–782.

Andrews, K. M., J. W. Gibbons, and D. M. Jochimsen. 2008. Ecological effects of roads on amphibians and reptiles: a literature review. Pages 121–143 in Urban herpetology. J. C. Mitchell, R. E. Jung Brown, and B. Bartholomew, editors. Herpetological Conservation Vol. 3. Society for the Study of Amphibians and Reptiles, Salt Lake City, Utah, USA.

Andrews, K. M., T. A. Langen, and R.P.J.H. Struijk. 2015. Reptiles: overlooked but often at risk from roads. Pages 271–280 in A handbook of road ecology. R. van der Ree,

C. Grilo, and D. J. Smith, editors. John Wiley and Sons, Hoboken, New Jersey, USA.

Aresco, M. J. 2005. Mitigation measures to reduce highway mortality of turtles and other herpetofauna at a North Florida lake. Journal of Wildlife Management 69:549–560.

Ashley, E. P., and J. T. Robinson. 1996. Road mortality of amphibians, reptiles and other wildlife on the Long Point Causeway, Lake Erie, Ontario. Canadian Field Naturalist 110:403–412.

Bach, L., and H.J.G.A. Limpens. 2004. Tunnels as a possibility to connect bat habitats. Mammalia 689:411–420.

Bellis, M. A. 2008. Evaluating the effectiveness of wildlife crossing structures in southern Vermont. Master's thesis, University of Massachusetts, Amherst, Massachusetts, USA.

Bergeron, D. 2012. Les enfants à la rescousse des grenouilles tuées sur les routes. Le Naturaliste Canadien 136:72–75. [In French.]

Berthinussen, A., and J. Altringham. 2012. Do bat gantries and underpasses help bats cross roads safely? PLoS ONE 7:e38775.

Boarman, W. I., M. L. Beigel, G. C. Goodlett, and M. Sazaki. 1998. A passive integrated transponder system for tracking animal movements. Wildlife Society Bulletin 26:886–891.

Boarman, W. I., M. Sazaki, and W. B. Jennings. 1997. The effects of roads, barrier fences, and culverts on desert tortoise populations in California, USA. Pages 54–58 in Proceedings of the conservation, restoration and management of tortoises and turtles—an international conference. J. van Abbema, editor. Turtle and Tortoise Society, New York, New York, USA.

Boonman, M. 2011. Factors determining the use of culverts underneath highways and railway tracks by bats in lowland areas. Lutra 54:3–16.

Brehm, K. 1989. The acceptance of 0.2-metre tunnels by amphibians during their migration to the breeding site. Pages 29–42 in Amphibians and roads: proceedings of the toad tunnel conference, Rendsburg, Federal Republic of Germany. T.E.S. Langton, editor. ACO Polymer Products, Shefford, UK.

Brudin, C. O. 2003. Wildlife use of existing culverts and bridges in North Central Pennsylvania. Pages 344–352 in Proceedings of the 2003 international conference on ecology and transportation. C. L. Irwin, P. Garrett, and K. P. McDermott, editors. Center for Transportation and the Environment, North Carolina State University, Raleigh, North Carolina, USA.

Cavallaro, L., K. Sanden, J. Schellhase, and M. Tanaka. 2005. Designing road crossings for safe wildlife passage: Ventura County guidelines. Unpublished report in partial satisfaction for a Master's thesis. University of California, Santa Barbara, USA. http://www.bren.ucsb.edu/research/documents/corridors_final.pdf/. Accessed 17 September 2014.

Clevenger, A. P., B. L. Cypher, A. Ford, M. Huijser, B. F. Leeson, B. Walder, and C. Walters. 2008. Wildlife-vehicle collision reduction study: Report to Congress. FHWA-HRT-08-034. US Department of Transportation, Washington, DC, USA.

Clevenger, A. P., and M. P. Huijser. 2011. Handbook for design and evaluation of wildlife crossing structures in North America. Publication No. FHWA-CFL/TD-11-003. US Department of Transportation, Federal Highway Administration, Washington, DC, USA.

Clevenger, A. P., and N. Waltho. 1999. Dry drainage culvert use and design considerations for small and medium sized mammal movement across a major transportation corridor. Pages 263–277 in Proceedings of the third international conference on wildlife ecology and transportation. G. L. Evink, P. Garrett, and D. Zeigler, editors. Center for Transportation and the Environment, North Carolina State University, Raleigh, North Carolina, USA.

Congdon, J. D., A. E. Dunham, and R. C. van Loben Sels. 1994. Demographics of common snapping turtles (Chelydra serpentina): implications for conservation and management of long-lived organisms. American Zoologist 34:397–408.

Cristoffer, C. 1991. Road mortalities of northern Florida vertebrates. Florida Scientist 54:65–68.

Dodd, C. K., Jr., W. J. Barichivich, and L. L. Smith. 2004. Effectiveness of a barrier wall and culverts in reducing wildlife mortality on a heavily traveled highway in Florida. Biological Conservation 118:619–631.

Evink, G. L. 2002. Interaction between roadways and wildlife ecology: A synthesis of highway practice. National Cooperative Highway Research Program, Synthesis 305. Transportation Research Board. National Academies Press, Washington, DC, USA.

FMTBHRERT (Federal Ministry of Transport, Building and Housing, Road Engineering and Road Traffic). 2000. Guidelines for amphibian protection on roads. Amphibian conservation in Alps-Adriatic Region. FMTBHRERT, Vienna, Austria. [In German.]

Foresman, K. R. 2002. Small mammal use of modified culverts on the Lolo South Project of Western Montana. Pages 581–582 in Proceedings of the 2001 international conference on ecology and transportation. Center for Transportation and the Environment, North Carolina State University, Raleigh, North Carolina, USA.

Foresman, K. R. 2003. Small mammal use of modified culverts on the Lolo South Project of Western Montana—an update. Pages 342–343 in Proceedings of the 2003 international conference on ecology and transportation. C. L. Irwin, P. Garrett, and K. P. McDermott, editors. Center for Transportation and the Environment, North Carolina State University, Raleigh, North Carolina, USA.

Forman, R.T.T. 1995. Land mosaics: The ecology of landscapes and regions. Cambridge University Press, Cambridge, UK.

Foster, M. L., and S. R. Humphrey. 1995. Use of highway underpasses by Florida panthers and other wildlife. Wildlife Society Bulletin 23:95–100.

Griesemer, S. J., T. K. Fuller, and R. M. DeGraff. 1998. Habitat use by porcupines (Erethizon dorsatum) in central Massachusetts: effects of topography and forest composition. American Midland Naturalist 140:271–279.

Griffin, K. A., and D. H. Pletscher. 2006. Potential effects of highway mortality and habitat fragmentation on a population of painted turtles in Montana. Report No. FHWA/MT-06-010-8169. Montana Department of Transportation and Federal Highway Administration, Washington, DC, USA.

Gunson, K. E., G. Mountrakis, and L. J. Quackenbush. 2011. Spatial wildlife-vehicle collision models: a review of current work and its application to transportation mitigation projects. Journal of Environmental Management 92:1074–1082.

Huijser, M. P., A. V. Kociolek, P. McGowen, A. Hardy, A. P. Clevenger, and R. Ament. 2007. Wildlife-vehicle collision and crossing mitigation measures: a toolbox for the Montana Department of Transportation. Final Report submitted. Western Transportation Institute, Bozeman, Montana, USA.

Huijser, M. P., P. McGowen, A. P. Clevenger, and R. Ament. 2008. Wildlife-vehicle collision reduction study: best practices manual. Report to Congress. US Department of Transportation, Federal Highway Administration, Washington, DC, USA.

Hunt, A. H., J. Dickens, and R. J. Whelan. 1987. Movement of mammals through tunnels under railway lines. Australian Zoologist 24:89–93.

Iuell, B., G. J. Bekker, R. Cuperus, J. Dufek, G. Fry, C. Hicks, V. Hlavac, et al. 2003. COST 341: Habitat fragmentation due to transportation infrastructure; Wildlife and traffic: a European handbook for identifying conflicts and designing solutions. European Co-operation in the Field of Scientific and Technical Research, Brussels, Belgium.

Jackson, S. D. 1996. Underpass systems for amphibians. Pages 255–260 in Trends in addressing transportation related wildlife mortality: proceedings of the transportation related wildlife mortality seminar. G. L. Evink, P. Garrett, D. Zeigler, and J. Berry, editors. FL-ER-58-96. State of Florida Department of Transportation, Tallahassee, Florida, USA.

Jackson, S. D., and T. F. Tyning. 1989. Effectiveness of drift fences and tunnels for moving spotted salamanders Ambystoma maculatum under roads. Pages 93–99 in Amphibians and roads: proceedings of the toad tunnel conference, Rendsburg, Federal Republic of Germany. T.E.S. Langton, editor. ACO Polymer Products, Shefford, UK.

Janssen, A.A.A.W., H.J.R. Lenders, and R.S.E.W. Leuven. 1997. Technical state and maintenance of underpasses for badgers in the Netherlands. Pages 362–366 in Habitat fragmentation and infrastructure, proceedings of the international conference on habitat fragmentation, infrastructure and the role of ecological engineering. K. Canters, editor. Ministry of Transport, Public Works and Water Management, Delft, The Netherlands.

Jones, M. 2000. Road upgrade, road mortality and remedial measures: impacts on a population of eastern quolls and Tasmanian devils. Wildlife Research 27:289–296.

Kornilev, Y. V., S. J. Price, and M. E. Dorcas. 2006. Between a rock and a hard place: responses of Eastern Box Turtles (*Terrapene carolina*) when trapped between railroad tracks. Herpetological Review 37:145–148.

Langton, A.E.S. 2002. Measures to protect amphibians and reptiles from road traffic. Pages 223–248 in Wildlife and roads: The ecological impacts. B. Sherwood, D. Cutler, and J. Burton, editors. Imperial College Press, Covent Garden, London, UK.

Langton, T.E.S. 1989. Tunnels and temperature: results from a study of a drift fence and tunnel system at Henley-on-Thames, Buckinghamshire, England. Pages 145–152 in Amphibians and roads: proceedings of the toad tunnel conference, Rendsburg, Federal Republic of Germany. T.E.S. Langton, editor. ACO Polymer Products, Shefford, UK.

Mansergh, I. M., and D. J. Scotts. 1989. Habitat continuity and social organization of the mountain pygmy-possum restored by tunnel. Journal of Wildlife Management 53:701–707.

MTPWWM (Ministry of Transport, Public Works and Water Management). 1995. Wildlife crossings for roads and waterways. Ministry of Transport, Public Works and Water Management, Road and Hydraulic Engineering Division, Delft, The Netherlands.

Pagnucco, K. S., C. A. Paszkowski, and G. J. Scrimgeour. 2012. Characterizing movement patterns and spatio-temporal use of under-road tunnels by long-toed salamanders in Waterton Lakes National Park, Canada. Copeia 2012: 331–340.

Patrick, D. A., C. M. Schalk, J. P. Gibbs, and H. W. Woltz. 2010. Effective culvert placement and design to facilitate passage of amphibians across roads. Journal of Herpetology 44:618–626.

Pepper, H. W., M. Holland, and R. Trout. 2006. Wildlife fencing design guide. CIRIA, Classic House, London, UK.

Podloucky, R. 1989. Protection of amphibians on roads—examples and experiences from Lower Saxony. Pages 15–28 in Amphibians and roads: proceedings of the toad tunnel conference. T.E.S. Langton, editor. ACO Polymer Products, Shefford, UK.

Reed, D. F. 1981. Mule deer behavior at a highway underpass exit. Journal of Wildlife Management 45:542–543.

Reed, D. F., T. N. Woodard, and T. D. Beck. 1979. Regional deer-vehicle accident research. Report No. FHWA-RD-79-11. Federal Highway Administration, Washington, DC, USA.

Rodriguez, A., G. Crema, and M. Delibes. 1996. Use of non-wildlife passages across a high speed railway by terrestrial vertebrates. Journal of Applied Ecology 33:1527–1540.

Rosell, C., J. Parpal, R. Campeny, S. Jove, A. Pasquina, and J. M. Velasco. 1997. Mitigation of barrier effect of linear infrastructures on wildlife. Pages 367–372 in Habitat frag-

mentation and infrastructure, proceedings of the international conference on habitat fragmentation, infrastructure and the role of ecological engineering. K. Canters, editor. Ministry of Transport, Public Works and Water Management, Road and Hydraulic Engineering Division, Delft, The Netherlands.

Ruby, D. E., J. R. Spotila, S. K. Martin, and S. J. Kemp. 1994. Behavioral responses to barriers by desert tortoises: implications for wildlife management. Herpetological Monographs 8:144–160.

Ryser, J., and K. Grossenbacher. 1989. A survey of amphibian preservation at roads in Switzerland. Pages 7–13 in Amphibians and roads: proceedings of the toad tunnel conference, Rendsburg, Federal Republic of Germany. T.E.S. Langton, editor. ACO Polymer Products, Shefford, UK.

Santolini, R., G. Sauli, S. Malcevschi, and F. Perco. 1997. The relationship between infrastructure and wildlife: problems, possible solutions and finished works in Italy. Pages 202–212 in Habitat fragmentation and infrastructure, proceedings of the international conference on habitat fragmentation, infrastructure and the role of ecological engineering. K. Canters, editor. Ministry of Transport, Public Works and Water Management, Road and Hydraulic Engineering Division, Delft, The Netherlands.

SETRA (Service d'Études Techniques des Routes et Autoroutes). 2005. Facilities and measures for small fauna—technical guide. Ministère de l'Ecologie du développement et de l'aménagement durables, Chambéry, France. [In French.]

Shine, R., M. Lemaster, M. Wall, T. Langkilde, and R. Mason. 2004. Why did the snake cross the road? Effects of roads on movement and location of mates by garter snakes (*Thamnophis sirtalis parietalis*). Ecology and Society 9:9. http://www.ecologyandsociety.org/vol9/iss1/art9/. Accessed 29 August 2014.

Smith, D. J., and R. F. Noss. 2011. A reconnaissance study of actual and potential wildlife crossing structures in central Florida, Final Report. University of Central Florida–Florida Department of Transportation, Contract No. BDB-10. Florida Department of Transportation, District Five, Deland, Florida, USA.

Soanes, K., M. Carmody Lobo, P. A. Vesk, M. A. McCarthy, J. L. Moore, and R. van der Ree. 2013. Movement re-established but not restored: inferring the effectiveness of road-crossing mitigation for a gliding mammal by monitoring use. Biological Conservation 159:434–441.

Taylor, B. D., and R. L. Goldingay. 2011. Restoring connectivity in landscapes fragmented by major roads: a case study using wooden poles as "stepping stones" for gliding mammals. Restoration Ecology 20:671–678.

Telford, S.R.T., Jr. 1959. A study of the sand skink, *Neoseps reynoldsi*. Copeia 1959:110–119.

USFSSSWG (US Forest Service Stream Simulation Working Group). 2008. Stream simulation: an ecological approach to providing passage for aquatic organisms at road-stream crossings. US Forest Service Technology and Development Program, San Dimas, California, USA.

van der Ree, R., S. Cesarini, P. Sunnucks, J. L. Moore, and A. Taylor. 2010. Large gaps in canopy reduce road crossing by a gliding mammal. Ecology and Society 15:35. http://www.ecologyandsociety.org/issues/article.php/3759. Accessed 29 August 2014.

van der Ree, R., C. Grilo, and D. J. Smith, editors. In press. Handbook of road ecology. John Wiley and Sons, Hoboken, New Jersey, USA.

Woltz, H. W., J. P. Gibbs, and P. K. Ducey. 2008. Road crossing structures for amphibians and reptiles: informing design through behavioral analysis. Biological Conservation 141:2745–2750.

Yorks, D. T., P. R. Sievert, and D. J. Paulson. 2012. Experimental tests of tunnel and barrier options for reducing road mortalities of freshwater turtles. Page 1034 *in* Proceedings of the 2011 international conference on ecology and transportation. P. J. Wagner, D. Nelson, and E. Murray, editors. Center for Transportation and the Environment, North Carolina State University, Raleigh, North Carolina, USA.

10

Daniel J. Smith,
Julia Kintsch,
Patricia Cramer,
Sandra L. Jacobson,
and Stephen Tonjes

Modifying Structures on Existing Roads to Enhance Wildlife Passage

10.1. Introduction to Retrofitting: Why and When to Modify Existing Structures

Our understanding of the impacts of roads on all wildlife (both large and small species) and on wildlife habitat has increased significantly over the last 25 years (Andrews 1990; Spellerberg 1998; Forman et al. 2003; Andrews et al. 2008). During this period, many wildlife crossing structures have been built on newly constructed roads and on existing roads when improvement projects have been warranted (Bissonette and Cramer 2008). Unfortunately, most of the existing road networks were constructed before this knowledge base had accumulated, and consequently, few measures were previously taken to mitigate for negative effects to wildlife. Forman (2000) estimated that at least 20% of the land area in the United States is directly affected ecologically by roads.

Over the last decade, opportunities to remedy impacts on small animals across existing road networks have emerged with increasing frequency. One means of mitigation is to "retrofit," or modify existing structures originally designed for other purposes, such as cross drainage (i.e., allowing for movement of water from one side of the road to the other). Retrofit or retrofitting is defined as follows (Merriam-Webster 2013): "To furnish with or install new or modified materials or features not available or considered necessary at the time of construction; to adapt to a new purpose or need." Some transportation departments consider installing new structures in an existing road a form of retrofit-

ting; however, in this chapter the term is limited to modifying existing structures.

10.1.1. Why Retrofit?

New wildlife crossing structures that are specifically designed for the species of interest and combined with fencing or barrier walls offer the greatest opportunity for mitigating the negative impacts of roads. Whether placed before or after the original construction of the road, their design and placement may involve a lengthy process and significant financial resources that may not be immediately available. However, where roads are already in place, retrofitting existing transportation infrastructure offers opportunities to reduce wildlife mortality and habitat fragmentation. In many cases, improvements to existing structures (those not initially intended for wildlife use) not only accommodate the movement needs of wildlife, but are publicly less controversial. This is because they typically cost much less than constructing new structures and can be completed in shorter time frames with less disruption to normal traffic flow.

Retrofitting an existing structure (e.g., culvert, bridge, or traffic interchange) to accommodate small animal passage is usually much cheaper than building a dedicated structure, but the amount of money that can be saved depends on (1) the type of retrofit needed; and (2) whether the modified existing structure is to be replaced in the near future, or simply extended as part of the road project. For example, a simple bolt-on shelf, such as those produced by Critter-Crossing Technology, LLC (Missoula, Montana, United States; Figure 10.1),

Figure 10.1. Installation of a bolt-on metal shelf, produced by Critter-Crossing Technology, LLC, for small animal passage within a drainage culvert, as shown by drawing (based on an actual image) of a short-tailed weasel (*Mustela erminea*) crossing the shelf (inset). *Credits: Reproduced by permission of Kerry R. Foresman; © Victor Young Illustration (inset, adapted from original photo by Kerry R. Foresman).*

can be installed in almost any existing structure with little modification, provided the structure's hydrologic function is not compromised. Larger shelves can only be cast into new structures before they are installed, so the existing structures must already be planned for replacement in order to realize any cost savings; this makes retrofitting a less cost-effective option. Retrofitting is not feasible in cases where a wildlife-appropriate design requires a bridge to be longer or higher or a culvert to be wider or higher; the only cost-effective approach to enhancing wildlife passage would be replacement of the existing structure.

10.1.2. When to Retrofit

A process to determine when retrofitting existing structures is a viable measure for improving habitat connectivity should include the following evaluations, made in this order: biological, structural, and cost-benefit. First,

a Rapid Assessment (Section 10.6.1) can be performed to determine whether a biological need is present at any existing structure location. Second, an evaluation of the ability of the existing structure to accommodate passage by the target species should be conducted. Finally, a cost-benefit analysis is necessary to determine if it is more cost effective to retrofit the existing structure or replace it with a new structure.

Retrofitting opportunities fall into three categories:

1. When barrier or mortality effects on small animals are chronic and so severe that it is inadvisable to wait for a full highway project to construct a wildlife-specific structure, retrofitting may be possible to minimize, at least temporarily, the negative effects in a much shorter time period.

2. When highway projects are in the planning stage, there are often opportunities to include lower-cost

retrofits of existing structures as part of the entire package.

3. When funding is not allocated for any existing road projects, but a structure is suitable for retrofitting to accommodate small animals, this can sometimes be accomplished as a stand-alone project; such a retrofit can be done at a lower cost than would be possible if the structure were replaced with a large animal wildlife passage.

10.2. Policies Supporting Retrofits

Few, if any, policies exist in the United States or Canada that are directly related to retrofitting existing structures. Two examples of programs that encourage retrofitting existing structures are in the states of Massachusetts and Washington. The Massachusetts Department of Transportation (MassDOT) has produced two documents that provide guidelines for mitigation: "Design of Bridges and Culverts for Wildlife Passage at Freshwater Streams" and "GreenDOT Implementation Plan." Although neither refers to retrofits by name, they address the promotion of fish (and other aquatic organism) passage via maintenance and replacement of existing structures.

In Washington, an executive order signed by the state Secretary of Transportation in 2011 is the only policy measure that suggests support for work that has primarily ecological benefits. This order makes reference to priorities for maintaining and enhancing habitat connectivity. One important emphasis is restoration of fish passage in multiple jurisdictions, including retrofit projects throughout the state. In certain cases, measures that promote ecological infrastructure have been included. Although not specific to retrofits for terrestrial species, the policies are supportive and allow creative coupling of project elements to benefit both aquatic and terrestrial species.

10.3. Species Considerations

Small animals do not have the body size constraints of large animals. However, a number of factors are important in creating functional passages for small animals. Such factors include the biology of the species (Kintsch and Cramer 2011; Chapter 2) as well as the characteristics of the external environment (Section 10.4).

Differences in mobility are important for many small animals (Case Study 10.3.4). Other biological factors include the following, in order of general importance:

- Migration between seasonal habitats and breeding areas
- Access to suitable habitat and resources
- Establishment of new territories
- Avoidance of human pressures
- Predator avoidance and defense

10.3.1. Focal Species and Ecological Groups

Given that wildlife-crossing design considerations are highly dependent upon species-specific needs, the first step in designing or retrofitting crossing structures is to identify important focal species, or ecological groups (e.g., functional groups, Cavallaro et al. 2005; movement guilds, Case Study 10.3.4). Species whose movement paths are intersected by roads are highly susceptible to habitat fragmentation. These effects are further magnified when they involve species with federal or state threatened, endangered, sensitive, or other special status. There are existing models for species prioritization; for example, Lambeck (1997) developed a useful approach for species prioritization that involves categorizing focal species as area-, dispersal-, resource-, or process-limited.

10.3.2. Indicator, Umbrella, and Flagship Species

Indicator, umbrella, and flagship species are sometimes referred to collectively as "surrogate species" (Caro and O'Doherty 1999). Indicator species are important for evaluating levels of environmental degradation, population trends, and threats to biodiversity. Umbrella species have been used to determine minimum areal extent and habitat composition for conservation areas as well as for site selection of reserve networks (Roberge and Angelstam 2004). Flagship species have similar purposes to umbrella species, but are also important in raising public awareness for conservation (Caro and O'Doherty 1999).

As an example, umbrella species are those whose resource requirements are spatially broad; thus, they provide an umbrella of protection for many other species (Wilcox 1984; Noss and Cooperider 1994). One major benefit of using umbrella species includes reducing the number of species that need to be monitored to gauge success, which translates into reductions in time and cost for evaluation and monitoring (Roberge and Angelstam 2004). A systematic process is recommended for setting goals and objectives and determining selection criteria for evaluating suitability of a candidate species or group as an effective umbrella species (Noss et al. 1996; Caro and O'Doherty 1999; Roberge and Angelstam 2004).

10.3.3. Addressing Specializations

There are many examples of mitigating effects for species with special requirements. It is important to identify key species of conservation interest that have specializations requiring uncommon mitigation planning measures. Two such examples in Florida include the federally threatened, endemic sand skink (*Plestiodon reynoldsi*), and the rare and endemic Florida scrub lizard (*Sceloporus woodi*). These species are both xeric oriented species that prefer open (not vegetated), sandy areas exposed to full sunlight (Ashton and Ashton 1991; Branch et al. 2003). They are both limited in range and mobility so habitat fragmentation by roads is a serious concern. These requirements pose challenges to mitigation. A dark, cool, and moist culvert would likely not be suitable (P. Moler, Florida Fish and Wildlife Conservation Commission [retired], personal communication), and only existing tall bridges (generally not present in these habitat areas) would likely allow enough sunlight penetration to provide the warm soil conditions preferred by sand skinks. Management would have to include providing open, continuous pathways under the structure and on the approach areas. On low traffic volume roads, lighting grates could be imbedded in the pavement to allow light penetration inside existing culverts (Brehm 1989). The ideal mitigation measure for these species would be an overpass, ultimately an expensive solution.

Examples of other specialized needs would be certain small mammals requiring dense cover specific to their habitats. Thus far, designing underpasses or small culverts for small animal use has not included plant growth, which means these rodents will be less likely to use them. Culverts that allow light penetration and have soil as substrate and vegetation for cover, high open bridges, or overpasses vegetated with native plants all facilitate natural ecosystem dynamics, enhancing small animal passage, invertebrate movement, and seed dispersal. Consideration of these specializations can be applied in road mitigation planning, including the determination of retrofit priorities.

10.3.4. Case Study: Small Animal Movement Guilds

Julia Kintsch, Patricia Cramer, and Sandra L. Jacobson
Lack of knowledge about species' ecology frequently stymies adequate consideration of species' movement needs. In this case, a lack of sufficient data often results in inaction or inadequate designs. Recognizing this problem, Kintsch and Cramer (2011) developed a "Species Movement Guild" classification. These guilds categorize wildlife based on their body size, mode of locomotion, and preferred crossing structure characteristics. These preferences are largely based on predator avoidance behaviors and the need for continuous habitat conditions through a crossing structure. The classification system facilitates an understanding of the influential features that render a structure functional or nonfunctional for different types of wildlife. Further, a classification approach allows transportation biologists to evaluate the physical and environmental conditions and potential constraints on movement from the perspective of groups of species.

The species movement guilds comprehensively address how wildlife use different types of structures to cross under and over roads, and allow any wildlife species to be categorized into one of eight classes—the three guilds that include small animals are presented below.

Low Mobility Fauna. This guild is based on smaller size, slower modes of movement, smaller home ranges, and the need for ambient conditions. Some frogs and salamanders are included in this guild because they need ambient conditions of light, moisture, and natural substrates (Jackson 1996). Crossings proven effective for amphibians are as short as possible and lined with soil (Jackson 1996; Aresco 2005a). Grates on top of these passages allow for ample light, rain, and air, thus mimicking local conditions (Brehm 1989), but the passages should also be situated to prevent flooding (Puky and Vogel 2004; Chapter 9). While distances between crossing structures should vary with site-specific habitat features and mitigation and planning objectives, recent studies suggest approximately 60 m or less is best for amphibian and reptile use (e.g., Ryser and Grossenbacher 1989; Jackson 2003). Specific types of fencing are necessary to guide these animals to these passages (e.g., Dodd et al. 2004; Aresco 2005a; Pepper et al. 2006).

Moderate Mobility Fauna. This guild includes small- to medium-sized mammals, snakes, and some salamanders and turtles that are fairly adaptable to a variety of structure types. Foresman (2004) found these animals could move under roads on metal shelves placed inside smaller culverts. McDonald and St. Clair (2004) found that small mammals had much higher success moving through shorter rather than longer structures. Also, a lack of natural vegetative cover at the approaches to crossing structures limited use among small mammals. Overall, these animals crossed more readily in culverts that were shorter in length (approximately 90 m or less), had vegetation close to the openings, were free of water year-round, or included a shelf structure above the water level to facilitate movement.

High Mobility Fauna. This guild is comprised of medium-sized mammals and large reptiles that are naturally accustomed to enclosed spaces and are somewhat tolerant of more enclosed underpasses. Examples of this group include bobcat (*Lynx rufus*), gray fox (*Urocyon cinereoargenteus*), Canada lynx (*Lynx canadensis*), North American river otter (*Lontra canadensis*), and American alligators (*Alligator mississippiensis*). Bobcats will use existing crossing structures (Ng et al. 2004), although in some cases they are less prone to using them and prefer to cross roads at grade, even when fences are present (Cain et al. 2003). While Cain et al. (2003) found bobcats using more open structures in southern Texas, Smith (2003b) recorded passage use by fox, coyote (*Canis latrans*), and bobcat through several medium-sized box culverts (2.4 to 3.1 m wide, 1.5 to 1.8 m high) under two-lane roads in Florida. River otters, American alligators, and cottonmouths (*Agkistrodon piscivorus*) used box culverts (1.8 m wide × 1.8 m high) to cross under a four-lane highway in Florida (Barichovich and Dodd 2002).

10.4. Habitat Considerations

In addition to considering species biology and vagility, it is most important to determine the habitats on which they rely, and therefore the highest priority locations for retrofitting. Additional considerations should include environmental factors such as the range in temperature and moisture associated with particular habitats important for target species groups. In lieu of performing field-based assessments of species presence, habitat or ecosystem characteristics can be used to predict presence (Chapter 8). Species or habitat information can then be used to determine both priority locations for retrofitting and the design of the structures and approach areas; these determinations should mimic the conditions of the adjacent habitat and improve functionality for the target species. Further discussion of how to prioritize locations for retrofitting can be found in Chapter 8.

10.5. Structural Retrofit Considerations

Ideally, retrofitting should only occur as a function of adaptive management following the proactive design and construction of appropriate features intended to address ecological infrastructure. In other words, retrofitting is done to further enhance existing features that minimize road impacts. The process of identifying retrofitting opportunities to enhance wildlife passage should include (1) selecting the target species or group and confirming their presence, (2) having a basic understanding of the structure and site characteristics that may enhance usage by the particular target species or group at any given location (see also Chapter 9), (3) developing an inventory and conducting evaluations of existing structures, and (4) determining feasibility of applicable solutions and devising a retrofitting strategy. Generally, there are two ways to retrofit existing structures: external or internal enhancements to structures (Sections 10.5.1 and 10.5.2).

10.5.1. External Enhancements to Structures

Several enhancements can be made external to the structure to improve suitability and function of that structure (Table 10.1; see also Friedman 1997; Smith 2003b; Pepper et al. 2006; Smith 2011; Smith and Noss 2011). In general, these enhancements should act as elements of the natural habitat that would have been present before the road was in place. When incorporating any of the features described in Table 10.1 into a project, maintenance needs should also be included in

Table 10.1. Guidance for, and benefits of, external enhancements to structures to improve suitability and function

Enhancement	Guidance	Benefits
Soil	Replace roadbed aggregate with topsoil typical of the adjacent habitat. Use topsoil from the existing location if possible; otherwise use similar soil additives to improve soil quality.	Fosters growth of vegetation similar to adjacent habitat areas. Increases suitability for species that prefer vegetative cover or move below the soil surface. Using local soil sources minimizes introduction of invasive species.
Contouring	Lessen the gradient of severe slopes. Reshape and smooth overly rugged terrain adjacent to the passage entrance.	Creates a natural funnel toward the structure. Enhances animals' ability to move toward and find the entrance.
Drainage	Contour approach areas to direct stormwater away from the road and the structure (special structural features to reduce flow rates and direction may be necessary along with native plantings). Replace adjacent roadway drainage swales with soil-covered culverts and native plantings to provide continuous dry access (see Figure 10.2).	Eliminates accumulation of stormwater and prevents creation of depression pools or flooding within the structure, and reduces erosion (which may reduce maintenance costs). Enhances use by species hesitant to cross areas inundated by water.
Natural cover materials	Add natural materials in approach areas typical of the adjacent habitat—for example, stumps, logs or brush-piles, bark, leaf litter, or rocks. Plant native vegetation, including appropriate groundcover, shrub, and tree species (see Figure 10.3).	Encourages use by species that prefer cover and those that are susceptible to desiccation from light or temperature extremes. Reduces exposure to predators.
Screening	Use hedgerows or densely planted shrubs, or artificial materials such as walls or opaque fencing to separate or buffer the structure and approach areas from the road.	Reduces visual impact of the road and improves suitability for species sensitive to noise and lights from highway traffic.
Fencing	Add directional fencing to guide small animals to structure entrances (Section 9.7.2; see also Table 10.2). Important factors include the following: 1. height of the fence (based on jumping and climbing abilities of the target species or group); 2. characteristics of the material used to block access (e.g., level of transparency, mesh size); 3. presence of a "lip" or angled-in top edge (to prevent animals from climbing over); 4. depth that the fence is buried (to prevent the target species from tunneling under); and 5. material longevity (affects maintenance and replacement costs).	Increases use of the structure and prevents access to the road by small animals. May reduce visual impact of the road.

the budget to adaptively manage the needs of the target species (or those species benefitting the most from the structure). Specifically, many of these features may require replacement, supplementation, or continued maintenance on an annual, or seasonal, basis. More detail about maintenance considerations is provided in Chapter 11, and adaptive management options with respect to post-construction monitoring are provided in Chapter 12.

Fence materials are discussed in Table 10.2 (see also Sections 9.2.5 and 9.7.2); of these, all can trap snow or debris and eventually impede flow of stormwater if placed in inappropriate locations or not maintained.

Further, all require periodic maintenance to remove vegetation, thus preventing animals from climbing over the fence. These materials are easily climbable for treefrogs and most lizards (with the exception of aluminum flashing for lizards). To maximize effectiveness, fencing should be at least 90 cm above ground and 30 cm below ground.

The most cost-effective of these materials is 0.25 in (0.65 cm) metal mesh (Smith and Noss 2011). Specifically, hot-dipped, galvanized metal mesh of at least 23-gauge and with at least 29% weight in zinc is among the best for longevity, durability, and low maintenance; it allows stormwater penetration and prevents most

Figure 10.2. An example of creation of a land bridge over a swale and re-contouring of approach area to improve access, facilitate drainage, and reduce pooling within the passage (inset). *Credits: Daniel J. Smith.*

animals from trespassing and climbing. Its primary limitation is that it does allow for the visual impact of the road. Some species (e.g., frogs, salamanders, turtles, snakes) may be negatively affected by this characteristic. Aluminum flashing is the most effective biologically, but is cost prohibitive and potentially may cause significant stormwater drainage issues or flooding on the road.

10.5.2. Internal Enhancements to Structures

There are several enhancements associated with modifying a structure that can improve suitability and function (Table 10.3; see also Krikowski 1989; Langton 1989; Jackson 1996; Friedman 1997; Iuell et al. 2003; Smith 2003b; Foresman 2004; Aresco 2005a; Righetti et al. 2008; Smith 2011; Smith and Noss 2011; Trocmé and Righetti 2012; Yorks et al. 2012). All characteristics should be designed according to a target species' or group's physiological and ecological needs and preferences. Further, because each potential structure available for modification will vary with respect to surrounding ecological infrastructure, customized structure designs and maintenance regimes should be incorporated into the construction contract (Chapter 11).

10.6. Site-Based Assessments

Field assessments of potential modifications to existing infrastructure can be conducted well in advance of transportation planning schedules. This proactive evaluation allows planners to address ecological infrastructure considerations in retrofit project designs and budgets. In addition, such assessments may reveal small

Figure 10.3. Planting of native trees, shrubs, and ground cover can enhance suitability of approach areas to crossing structures. *Credit: Daniel J. Smith.*

modifications (Tables 10.1 and 10.3) that can be made outside of major construction project planning as part of ongoing operations and maintenance (Chapter 11).

Typically, these assessments are conducted as needed on a project-by-project basis, are rarely standardized, comprehensive, or transparent, and often only focus on large species. Recent years have seen a shift toward evaluating existing infrastructure in a more systematic manner. In Colorado, a three-year study of Interstate 70 developed a basic framework for evaluating wildlife passage at existing bridges and culverts (Kintsch and Cramer 2011). MassDOT (2010) published a report that presents considerations for evaluating existing structures at freshwater streams and offers best practices for accommodating wildlife passage at these road-stream crossings. However, while these contributions are valuable, a comprehensive

methodology is still needed for evaluating how various structure and landscape characteristics promote or inhibit passage with specific consideration of the varying needs of different species.

10.6.1. Rapid Assessments

If cost and time are significant limiting factors, a Rapid Assessment process can be a practical and effective approach (Chapters 6 and 8; see http://www .rapidassessment.net for additional information). This is a collaborative process that may involve a diverse group of public and private stakeholders or experts from a single consulting firm; they typically review existing data and information on the project site. They then use this information as well as local knowledge to form professional opinions on species and habitats affected, identify ecological hotspots, and develop

Table 10.2. Pros and cons of different directional and barrier fencing materials for small animals (see also Sections 9.2.5 and 9.7.2), including engineering and maintenance considerations for various temporary or permanent retrofit options

Material	Pros	Cons
Aluminum flashing	Does not allow trespass Small animals cannot climb Opaque, negligible visual impact of road to animals Longevity—20+ yrs High durability, little long-term maintenance	Prevents normal stormwater flow, trapping runoff and potentially causing flooding on the roadway Increased visual impact to residents and drivers Most expensive material
Shade cloth	Prevents most trespass Few small animals can climb Minimal visual impact of road to animals Longevity—10 to 15 years Least expensive material	Severely impedes normal stormwater flow, trapping runoff and potentially causing flooding on the road Increased visual impact to residents and drivers Durability is poor, subject to damage from maintenance equipment, may require frequent repairs for tears or holes Requires rigid backing for support, such as wire fencing which increases initial costs
Superscreen™	Prevents most trespass Few small animals can climb Minimal visual impact of road to small animals Longevity—10 to 20 years Moderate material cost	Severely impedes normal stormwater flow, trapping runoff and potentially causing flooding on the roadway Increased visual impact to residents and drivers Durability is poor, subject to damage from maintenance equipment, may require frequent repairs for tears or holes Requires rigid backing for support, such as wire fencing, which increases initial costs
0.65 cm (0.25 in) plastic mesh	Very little trespass Few small animals can climb Allows stormwater flow through Some visual impact to residents and drivers Less expensive than most other fencing materials presented in this table (exception is shade cloth) Moderate material cost	Increased visual impact of road to small animals Longevity—only 5 to 7 years due to weakening from sunlight exposure Moderate durability, may be damaged by maintenance equipment and may require repairs to perforations Requires rigid backing for support such as wire fencing, which increases initial costs
0.65 cm (0.25 in) metal mesh*	Very little trespass Few small animals can climb Pliable, allows manipulation to create an extended lip or angled edge to deter climbing and jumping Allows stormwater flow-through Some visual impact to residents or drivers Longevity—+20 years High durability, little long-term maintenance required Moderate material cost	Increased visual impact of road to small animals
1.25 cm (0.5 in) metal mesh*	Pliable, allows manipulation to create an extended lip or angled edge to deter climbing and jumping Allows stormwater flow through Reduced visual impact to residents or drivers Longevity—+20 years High durability, little long-term maintenance required Moderate material cost	Allows frequent trespass for many species Easy for many small animals to climb, including some species of turtles Some entanglement issues for small snakes that may partially fit through the openings (see also Box 11.1) Increased visual impact of road to small animals

* Metal mesh materials specifically refer to hot-dipped, galvanized metal mesh (sometimes called "hardware cloth") that is at least 23 gauge and at least 29% weight in zinc.

Table 10.3. Guidance for, and benefits of, internal enhancements to structures

Enhancement	Guidance	Benefits
Shelves (Culverts)	For culverts associated with streams, wetlands, lakes, or other water features that may be seasonally or permanently inundated, add shelving (e.g., hanging metal, Figure 10.1) above the high water line. Ensure vertical clearance above the shelf is sufficient for use by the target species. For arboreal species, an elevated bolt-on shelf within the culvert may be useful as a walkway.	Encourages passage by terrestrial species in structures designed for conveyance of water.
Ledges (Bridges)	Arrange (or rearrange) bank and structure stabilization or scour prevention materials (e.g., large rocks or rubble riprap, gabions [wire baskets filled with rocks]) to create a clear, level pathway or ledge. Earthen ledges may require retaining walls made of concrete, soil-cement, or geotextile material to protect against erosion or destabilization. Ledge substrate should consist of aggregate bedding materials of variable size mixed with soil or pavement blocks filled with soil or bedding materials (i.e., materials resistant to erosion).	Creates a dry pathway above the high water line under structures that span water features (e.g., rivers or creeks) and encourages passage by terrestrial species. Proper arrangement (or rearrangement) of materials will prevent the inadvertent barrier to species that prefer smooth surfaces (Figure 10.4).
Soil substrate	Use native soils as the substrate within the structure. Consider the soil type in the surrounding habitat along with moisture needs of target species when determining soil type to place inside the structure. Maintain soil depth based on needs of target species, particularly fossorial (burrowing) species.	Increases small animal passage by providing a more natural and continuous surface throughout the passage. Use of appropriate soil types and soil depths can ensure proper moisture retention, which in turn will encourage species passage based on moisture or dryness needs. Use of native soils can also facilitate growth of native plants, which provides cover and forage for many small animals.
Rocks, debris, and tubes	Include rocks or woody debris along the walls of the structure. Use metal or PVC tubes (along the base of the wall) covered by rocks or woody debris to provide a concealed movement pathway for small mammals.	Rocks and woody debris encourage passage by those species that require cover or protection from predators. Tubes provide additional cover as well as a means of escape from predators for many herpetofauna and small mammals.
Lighting	Provide natural lighting (via grates or fiber-optic solar technology, see also Section 9.5.2) within the structure, particularly in longer passages (e.g., multi-lane divided highways). Place culvert lighting grates, when situated in medians of divided highways, on the upslope of median swales or raised slightly above grade if placed at the lowest elevation (to reduce stormwater flow into the structure).	Species that are deterred by darkness, or diurnal species, may be more likely to pass through structures with some lighting. Use of natural light via grates requires less maintenance than use of man-made lighting technology. Grates placed above grade can help minimize stormwater flow into the structure, thereby reducing erosion or washout of substrate within the structure and reducing animal exposure to runoff pollutants.
Removal of obstructions	Selectively remove or thin vegetation near structure entrances, or within structures, where it may have become too dense and may be obscuring the opening or preventing animal passage. Remove other obstructions, such as fence material or storm-washed debris, at the entrance of, or within, the structure.	Increases likelihood of passage for small animals, particularly those that are deterred by low visibility conditions created by dense vegetation. Increases visibility of and access to the structure entrance.

Figure 10.4. Examples of improper (a) and proper (b) placement of riprap to allow terrestrial wildlife movement adjacent to waterways. *Credit: Daniel J. Smith.*

<table>
<tr><td>

Box 10.1. Rapid Assessment to Retrofit for Fish Passage

In the northern Rockies, aquatic biologists developed a protocol for assessing fish passage through existing culverts (Ruediger and Lloyd 2003; Box 6.1). The process was designed to (1) determine if the existing culvert was a barrier to fish passage, and (2) prioritize the need for retrofitting (i.e., replacement with a more appropriately designed culvert) among the hundreds to thousands of other culverts needing replacement. In this case, a low-cost Rapid Assessment was possible because ample information had been collected on fish-culvert priorities for this area over the past 40 years. The cost was significantly less because few new field assessments were necessary due to this existing knowledge base. Each new field assessment cost about $3,000 per culvert.

</td></tr>
</table>

with high stream densities (e.g., Pacific Northwest; Box 10.2), or where wetland systems are extensive (e.g., southeastern United States) and high densities of herpetofauna and numerous water conveyance structures exist.

recommendations for retrofit needs (see Table 3.1 in Cavallaro et al. 2005 for an example). Additional details are provided in Box 10.1. Participants should strive to keep the process simple and include only what is necessary to solve the problem in a timely and cost-efficient manner (Chapter 6).

10.6.2. System-Wide Facility Evaluations

Transportation agencies can use system-wide facility evaluations to conduct more effective mitigation planning. By proactively evaluating existing structures over an entire or partial road network, the agency has the information necessary to easily incorporate retrofits or new mitigation measures into proposed road improvement projects. These assessments are another example of the type of information that is especially important to introduce early in the transportation planning process (Chapter 6). System-wide facility evaluations may be particularly relevant in regions

10.6.3. Case Study: Florida DOT Statewide Inventory and Prioritization Project

Daniel J. Smith

A systematic approach used in Florida combined modeling and field site assessments (Smith 1999, 2003a, 2006) to identify and prioritize areas where the existing road network compromised ecological infrastructure (see Chapter 8 and references therein). Specifically, chronic roadkill sites, focal species hotspots, greenway linkages, presence of listed species, and strategic habitat conservation areas (a designation identified by the Florida Fish and Wildlife Conservation Commission to indicate the likely occurrence of suitable habitat for certain species that may be required to maintain viable populations) strongly influenced modeling results. Over 15,000 road segments were prioritized; 81% were located in conservation areas or designated greenways. Ninety-five scheduled road construction projects coincided with modeled high-priority road segments and

Box 10.2. Washington State Passage
Assessment System

In the State of Washington, an assessment methodology, called the Passage Assessment System (PAS), was developed to assist in the evaluation of existing bridges and culverts and identification of potential retrofit measures that would enhance the structures' ability to function as wildlife passages (Kintsch and Cramer 2011). The PAS guides users through a series of standardized evaluation questions; these include questions about a number of variables that influence the likelihood of successful passage. Methods for assessing how different crossing characteristics may affect various species are also provided. The PAS may be adapted to any location and is intended for use by ecologists and biologists familiar with local wildlife needs and transportation infrastructure and planning. It provides an objective tool to determine if structures are suitable for retrofits to improve their functionality as wildlife passages or, if no such retrofits are appropriate, to identify structure replacement options for improved ecological infrastructure.

Table 10.4. Thresholds for size of existing drainage culverts based on 90% vs. 60% of detections of herpetofauna and small mammals (Smith 2003*b*). Culverts in the study ranged in width from 61–610 cm, in length from 8–61 m, and in height from 30–365 cm

Thresholds: 90% (upper), 60% (lower)	Herpetofauna (n = 580)	Small mammals (n = 270)
openness index value (w × h / l), m	0.28	0.42
	0.23	0.08
width, m	3.4	3.1
	2.7	1.5
length, m	11.6	11.6
	14.5	18.3
height, m	1.5	1.5
	1.4	1.2

Note: Herpetofauna and small mammal groups included certain species that frequented culverts as part of their home range rather than solely as movement corridors.

thus presented opportunities for addressing ecological infrastructure.

Field site assessments were conducted to (1) verify accuracy of modeling results, (2) characterize ecological conditions of road segments identified in the prioritization model, and (3) evaluate wildlife-passage functions of existing structures and provide recommendations for mitigation and restoration. The model correctly predicted areas where ecological infrastructure was compromised 87% of the time. Based on the field assessments, recommended mitigation measures ranged from installation of passage structures to minor measures. These minor measures included signage, speed restriction, fencing, landscaping, and re-contouring of approach areas to encourage use of existing drainage structures.

Finally, monitoring was conducted to determine the capacity of existing highway drainage structures to function as wildlife passages (Smith 2003*b*). Fifty-five

different species and taxonomic groups (from 36,870 individual records) were identified using 67 bridges and 223 culverts on 37 different roads of varying traffic levels and road widths; these were represented in 7 general landscape types. Analyses suggested that culvert width and height were important, and use by herpetofauna and small mammals generally increased as the dimensions increased. However, culvert length, distance between adjacent structures, and road verge width (i.e., distance to the habitat) had the opposite effect, so that use by these animals decreased as these dimensions increased (Smith 2003*b*). The results also suggested that use of structures decreased as frequency of human presence or domestic predator presence surrounding the mitigation site increased. Thresholds for faunal use (Tables 10.4 and 10.5) were suggested by the results, and these were then used to identify opportunities and make recommendations for retrofitting existing structures, such as adding fencing, restoring local plant communities and landscape features, screening of traffic noise and lights, and restricting access by humans and domestic predators.

10.7. Applying Retrofitting Solutions

Once a structure is determined to be a candidate for retrofitting to enhance wildlife passage, the next step is

Table 10.5. Thresholds for other parameters associated with existing drainage culverts based on 90% versus 60% of detections of herpetofauna and small mammals (Smith 2003*b*)

Thresholds: 90% (upper), 60% (lower)	Herpetofauna (n = 1,428)	Small mammals (n = 888)
distance to habitat (road verge width), m	3.7	3.7
	7.8	7.9
distance between adjacent structures, m	260	250
	875	585
human presence, no./yr	3	1
	46	46
domestic predators, no./yr	1	1
	1	20

to determine how to do this. Given variable objectives, and the unique characteristics of every structure and site, there is no simple answer. An array of possible retrofits can be implemented at existing culvert locations to promote passage by different species. Kintsch and Cramer (2011) initiated a database of retrofit options, largely focusing on options for larger species that also included a number of ideas for enhancing structures for small mammals and herpetofauna. This living database is a useful resource for practitioners seeking to learn from similar efforts in other locations. Other syntheses and compilations of solutions that address small mammals and herpetofauna include Iuell et al. (2003), SETRA (2005), van der Ree et al. (2007, 2015), and Proceedings of the International Conference of Ecology and Transportation (http://www.icoet.net/), and this book.

These resources provide a number of enhancement options for commonly encountered situations. For example, at a location with a culvert that conveys water but could also be used by smaller animals, shelves can be installed inside the culvert to facilitate wildlife movement above the flow of water (Figure 10.1; Table 10.3). Simply adding guide walls may suffice in guiding fauna toward an existing culvert (Table 10.1). Other enhancements focus on maintenance efforts, such as clearing debris and installing a sediment trap, modifying the drainage system, or restoring a perched culvert (Chapter 11). Still others may involve enhancing native vegetation in the approaches or installing

woody debris throughout the culvert to provide cover for small species (Tables 10.1 and 10.3). For example, Connolly-Newman and authors (2013) found that including dead tree branches in underpasses increased use by small mammals by almost 43%. As no two situations are identical, practitioners are encouraged to consider the range of possible enhancements and how they may be implemented and adapted for a given site.

Following any retrofit application, targeted monitoring of such enhancements is important. Monitoring can help determine whether adjustments are needed as part of an adaptive management process (Chapter 12). This, in turn, will create a positive feedback loop for designing improved enhancements at future locations (Section 12.7).

10.7.1. Temporary versus Permanent Measures

In some cases it may be appropriate to initiate temporary measures. As implementation of permanent measures can take considerable time, temporary measures can be taken to partially alleviate the impact and generate the necessary support from the public and transportation agencies (Box 10.3). This may include erecting a temporary barrier to prevent roadkill, imposing temporary road closures during migration periods, and using volunteers to move animals from one side of the road to the other (e.g., bucket brigades, Chapter 9).

Where some of these measures may be suitable on a longer-term basis on unpaved roads or those with very low traffic, they still require continued maintenance and constant participation by volunteers. On paved roads with moderate to high traffic, these are not ideal solutions to the problem and even moderate traffic levels can limit their effectiveness. More permanent, low-maintenance solutions are more appropriate for the long term, even where traffic levels are moderate.

10.8. Applying Retrofits to Single Sites, Single Roads, and Road Networks

10.8.1. Case Study: Single Site—Turtle Use of Retrofitted Culverts in an Urban Setting in Ontario, Canada

Brennan Caverhill

Seven of eight native turtle species in Ontario are at risk of extinction, primarily due to habitat loss and

Box 10.3. Lake Jackson

Aresco (2005a) studied the impact of traffic on turtles attempting to cross US 27, a four-lane highway that bisects Lake Jackson, Florida, United States. This mortality was so extensive (e.g., 343 turtles along 700 m of road in 40 days) that turtles could not successfully cross all four lanes of the highway. This decreased probability of crossing success was validated through a model developed by Hels and Buchwald (2001) that indicated less than 2% success based on crossing speed of the animal and traffic volume (Aresco 2005a). A temporary drift fence system was installed to lead turtles to an existing cross-drainage culvert while Aresco and a supportive public lobbied the state for a permanent structure. The drift fence was 0.61 m (2 ft) high and made from woven vinyl erosion-control material typically used in construction projects. The fence was buried to prevent animals from burrowing under.

In 3 years, over 8,800 turtles were recorded either attempting to cross the road or along the temporary drift fences. Prior to installation of the fence, turtle mortality was approximately 12/km/day. Following installation of the fence, turtle mortality was approximately 0.1/km/day.

The use of frequently (daily or more) monitored, temporary drift fences greatly reduced roadkills and enhanced habitat connectivity through the use of a cross-drainage culvert. This study presented the highest attempted road-crossing rate (1,263/km/year) published to date for turtles and was one of the first to document male-biased sex ratios adjacent to the road as a consequence of hundreds of reproductive adult females being lost from the population due to road mortality (Aresco 2005b).

Despite the incredible level of mortality observed, it took over 10 years for advocates to gain the political and financial support for the project. Because of lesser concern for human safety and economic damage, in many cases, road mitigation projects with small vertebrates do not take place without political pressure and public support. Lake Jackson represents one of the greatest successes in applying retrofits to reptiles. Additionally, it demonstrates the tremendous effort required to educate public officials and taxpayers on the value of incorporating ecological infrastructure for small animals in transportation projects.

---

road mortality (Ireland and Karch 2010). Situated in the Greater Golden Horseshoe area with more than 5 million people, provincial Highway 24 carries more than 400 vehicles/hour, as determined in 2011 Traffic Inventory Counts by the Ontario Ministry of Transportation (MTO); the highway also bisects several wetlands that comprise important turtle habitat. In April 2008, a concerned citizen reported to the Toronto Zoo's Ontario Turtle Tally program eight dead-on-road Blanding's turtles (*Emydoidea blandingii*) and two snapping turtles (*Chelydra serpentina*), both at-risk species, at a single location on this highway. After field visits by the Toronto Zoo and MTO, a 25 m long, 1.8 m diameter corrugated steel drainage culvert was found at the swamp-creek crossing where turtles were found dead on the road. In August 2008, MTO installed temporary silt fencing extending approximately 75 m in both directions from the culvert along both sides of the

road. Following this installation, no turtle mortalities occurred at the site in 2009, so a permanent chain link fence was installed in March 2010 with the goal of continuing to prevent turtle-vehicle collisions by guiding animals to the culvert.

From April 2010 to October 2011, researchers from the Toronto Zoo's Adopt-A-Pond Wetland Conservation Programme used visual surveys, radio tracking, and live trapping to study resident turtle populations and their response to the retrofitted culvert. Data from the 19 radio-tagged Blanding's turtles showed 70 road crossings where each individual crossed 1–17 times. Researchers observed these turtles traveling through the retrofitted culvert at least 15 times. Based on the Gibbs and Shriver (2002) model, the probability of turtles being killed by crossing on the road was 93%. Because no roadkills were documented, it was assumed that all other crossings occurred through the aquatic culvert.

Further observations of the turtles indicated that the fence successfully prevented their access to the road.

Although researchers also observed (directly as well as via images from a remote camera installed near the mouth of the culvert) other wildlife including fish, amphibians, birds, and mammals using the retrofitted culvert, the chain link guide fence was not successful in preventing all animals from accessing the road. For example, daily road surveys in 2010 along the approximately 200 m stretch of road spanning the culvert resulted in observations of 76 individuals from 11 vertebrate species (mostly frogs, birds, and mammals) found dead in the study area. Monitoring is ongoing for this project. Funding was generously provided by the Toronto Zoo, Environment Canada's Habitat Stewardship Fund, the Ontario Ministry of Natural Resource's Species at Risk Stewardship Fund, MTO, and the Ontario Road Ecology Group.

10.8.2. Case Study: Single Road—The Effect of Retrofits on Habitat Connectivity at Paynes Prairie, Florida, United States

Daniel J. Smith

High numbers of roadkill herpetofauna were recorded over several decades on US 441 (a 4-lane divided highway) that bisects Paynes Prairie State Preserve (a 5,000 ha freshwater marsh) in north-central Florida, United States. After much pressure from citizens, park staff, and local elected officials to institute mitigation measures, the Florida Department of Transportation (FDOT) developed plans in 1996 to construct a system of culverts and guide walls to improve habitat connectivity and reduce wildlife mortality (Dodd et al. 2004). This project was the first mitigation project in the United States driven by large numbers of herpetofauna being killed, many of which were snakes. Prior to construction of the passages and fencing, the site had the highest documented level of persistent snake mortality. It is an example of using existing structures and complementing them with new features to improve overall performance.

The system of culverts and guide walls (Figure 10.5) is a designed retrofit that includes (1) the use of four existing concrete box culverts (two are 1.8 × 1.8 m, located at higher elevations toward the perimeter of the basin and two are 2.4 × 2.4 m, located at lower elevations toward the center of the basin); (2) the addition

Figure 10.5. The Paynes Prairie wall and culvert passage system, Florida, United States. *Credit: Daniel J. Smith.*

of four concrete round pipe culverts (0.9 m wide each, placed between each of the existing 2.4 × 2.4 m box culverts); and (3) a concrete guide wall extending through the length of the basin (approximately 2.5 km) that is 1.1 m high with a 15.2 cm lip at the top that extends out away from the road to help prevent small animal from climbing over (Dodd et al. 2004). The culverts are 44 m in length and spaced 200–300 m apart (with 2 separated by 400–500 m) across the basin.

Prior to construction, 2,411 roadkills were recorded over the course of 1 year (12 months); following construction, only 158 animals were counted, again over 12 months (excluding hylid treefrogs, whose roadkill counts increased post-construction given their ability to climb over the guide wall; Smith and Dodd 2003; Dodd et al. 2004). Within the prairie basin, mortality was reduced by 93.5% (65% when hylids were included). Almost 66.6% of the post-construction roadkills occurred at gaps where the barrier wall was absent to allow access to private properties. Of the 1,891 vertebrate roadkills recorded in the post-construction survey (excluding hylids), 73% occurred in the habitat-transitional (ecotones) areas and adjacent uplands beyond the terminus of the guide wall.

Pre- and post-construction monitoring were conducted to evaluate change in use of the culverts by wildlife (Dodd et al. 2004). Only 28 vertebrate species were found in the 4 culverts prior to construction, while 42 vertebrates (excluding fish) were recorded in the 8 culverts following construction. Use of culverts increased from 5 amphibian species to 13 following construction.

Overall, the retrofits demonstrated significant improvement over pre-construction conditions, dramatically reducing road mortality and increasing wildlife use of culverts.

However, several problems emerged, some of which were able to be addressed, while others were not. The fencing created a tourist attraction for people to observe wildlife, which resulted in a safety hazard due to the feeding of alligators. Fencing was added to separate the public from alligators in the adjacent wetlands. Road mortality continued and even increased at gaps at access roads and along adjacent private land. Siltation and pooling of water occurred in the culverts and could be an issue for terrestrial animals. Additionally, growth of vegetation on the wall facilitated movement over the wall by treefrogs and other climbing species. The issue with trespass by arboreal species is a pervasive one with current road mitigation designs. Solutions to this problem may include use of other materials that prevent climbing or use of other crossing structures designed for arboreal species such as treefrogs (Chapter 9). Most of the climbing issues at Paynes Prairie were addressed with scheduled maintenance operations (e.g., more frequent mowing or vegetation removal along or near the guide wall) and certain design modifications (e.g., a 30.5 cm wide strip of aluminum flashing along the top edge of the wall at an upward 45 degree angle toward the habitat side). While improved design features are needed to increase the breadth of effectiveness, no single or suite of mitigation features works for all wildlife species in a given location. Paynes Prairie is particularly notable for (1) being the first retrofitting project in the United States that targeted a small animal wildlife community and involved such extensive construction, (2) its success in reducing mortality and increasing connectivity in that location, and (3) serving as a model for other retrofit projects.

10.8.3. Case Study: Road Network— Adaptive Management Associated with Retrofits on Two Major Road Corridors in Central Florida, United States

Stephen Tonjes and Daniel J. Smith

In 2004, data collection began for a study of 13 wildlife crossings in central Florida, United States; this was commissioned by the FDOT. Two major roads in rural areas, consisting mostly of natural and range lands,

were upgraded from two to four lanes. Rather than constructing large dedicated crossings, the wildlife agencies concurred with retrofitting several of the existing water conveyance structures with wildlife ledges during scheduled highway widening.

Following construction, researchers quickly discovered significant functional problems with the design and construction of many of these structures. Numerous species (particularly herpetofauna) continued to be killed on the road, and passage use by terrestrial species was significantly less than expected. As a result, FDOT established a Wildlife Crossing Design Team (WCDT) to evaluate and devise corrections that addressed the deficiencies. These improvements required additional funding through a separate design project. Rather than waiting for formal plan review, the WCDT and the project manager met several times, including on the site, during both design and permitting. An unfavorable permitting review initiated another round of adaptive management meetings. Because of the flexibility of the WCDT, the permitting setback was quickly resolved with a much less expensive and more functional design.

Five structures were programmed for modifications to improve their function, which included two culverts (one with shelves, one without) and three bridges with ledges. Modifications to the culverts included replacement of temporary wooden ramps with permanent concrete ramps (Figure 10.6), rehabilitation of the median lighting grate, and addition of an earthen bridge over a drainage swale. Two bridges required removal and rearrangement of riprap (Figure 10.7) and construction of ledges made of soil and bedding stone aggregate. Efforts at the other bridge included addressing erosion problems, filling gaps within submerged wire-basket gabions to prevent ensnaring aquatic turtles, and the addition of PVC tubes to provide covered pathways through lengthy open areas for small mammals. All five structures included additions to existing directional and barrier fencing for herpetofauna, including the installation of 1.2 m high, 0.635 cm mesh, galvanized hardware cloth buried to 0.3 m depth.

10.8.4. System-Wide Implementation: Examples from Arizona and Washington

Two examples of using retrofits to improve connectivity statewide are initiatives by Arizona and Washington transportation agencies. Arizona DOT implemented a

Figure 10.6. Following a road-widening project, this culvert was retrofitted to include ledges to facilitate small animal passage, initially with temporary wooden ramps (a). Due to structural and functional problems, modifications were made to replace the wooden ramps with permanent concrete ramps (b). *Credit: Daniel J. Smith.*

Figure 10.7. Boulders of this extent and size (a) are typically used across the United States, yet they are difficult for many wildlife species to cross. Retrofits to this bridge included removal of large rocks to create continuous paths of different sized bedding stone aggregates mixed with soil (b). *Credits: Daniel J. Smith (a) and Stephen Tonjes (b).*

program that would prioritize existing structures for retrofits rather than pursuing new construction of more expensive wildlife crossing structures (Nordhaugen et al. 2006; Eilerts and Nordhaugen 2010). Priorities were set by cross-referencing the Arizona Wildlife Linkages Assessment (Nordhaugen et al. 2006) with a statewide inventory and evaluation of existing structures that had the potential to be modified for wildlife passage. For structures in areas not scheduled for major upgrades, low-cost retrofits were planned, including addition of, or changes to, directional fencing, removal of riprap, and construction of ramps or pathways. For

example, some of these measures have been used to increase Mohave desert tortoise (*Gopherus agassizii*) use of seasonal drainage culverts.

In Washington, Washington State DOT (WSDOT), as part of a multi-organizational team, began the development of a statewide habitat connectivity retrofit program (McAllister 2010). This project occurred as a result of the WSDOT Secretary's Executive Order, "Protections and Connections for High Quality Natural Habitats." This large-scale, multi-year project identified valuable wildlife movement corridors between large blocks of high-quality habitat using least-cost distance and circuit-theory models (Section 8.4) for 16 focal species. The results of these models will be used to develop priorities for future retrofit projects to improve habitat connectivity. At a finer scale, the program will assess existing structures to determine retrofit needs and opportunities where habitat connectivity can be enhanced. As part of this effort, WSDOT commissioned the development of a retrofit inventory and analysis tool, called the Passage Assessment System (Box 10.2).

10.9. Key Points

- After roads are built, retrofitting the existing transportation infrastructure is the only option for restoring ecological infrastructure (i.e., reducing wildlife mortality, maintaining habitat connectivity). Given the vast existing road network, the majority of wildlife mitigation measures are included retroactively. This may have the illusion of being a reduced cost option over installing new structures; however, when compared to proactively incorporating the necessary enhancement features in the initial planning phase, retrofits often end up being more costly.

- Ideally, proactive design and construction of appropriate features that address ecological infrastructure occur when the road is originally built; retrofitting then occurs only as a function of adaptive management.

- Transportation agencies and the public should support policies and legislation that facilitate the inclusion of retrofits directly in construction project budgets (vs. maintenance budgets).

- Whether through the identification of focal species, umbrella species, or species' specializations, it is important to consider species' biological needs and patterns when considering types and locations of retrofits.

- Grouping animals based on movement guilds can benefit a greater number of species within a wildlife community and can be more effective in targeting appropriate retrofit designs.

- Following considerations of species' biology, retrofit locations should be selected in conjunction with the habitats on which these species rely.

- After assessing which species are moving through

surrounding habitats, design of structures, both internally and externally, should reflect the biological conditions that are being targeted for restoration and maintenance.

- Several retrofit options are available and range from lower-cost minor enhancements (e.g., signage, contouring approach areas) to higher-cost major improvements (e.g., directional and barrier fencing, structure modifications).

- As transportation and ecological infrastructure vary among locations, retrofit designs should be site based. Field assessments are necessary to identify retrofit options at specific sites.

- Rapid Assessments are an option when cost and time are limited.

- Any planned retrofit project should include measures for monitoring and maintenance, which in turn will allow for adaptive management to design and test retrofits for improved ecological function.

- Temporary measures can be taken to alleviate impacts while generating financial resources and public support for more permanent measures.

LITERATURE CITED

Andrews, A. 1990. Fragmentation of habitat by roads and utility corridors: a review. Australian Zoologist 26:130–141.

Andrews, K. M., J. W. Gibbons, and D. M. Jochimsen. 2008. Ecological effects of roads on amphibians and reptiles: a literature review. Pages 121–143 in Urban herpetology. J. C. Mitchell, R. E. Jung Brown, and B. Bartholomew, editors. Herpetological Conservation Vol. 3. Society for the Study of Amphibians and Reptiles, Salt Lake City, Utah, USA.

Aresco, M. J. 2005a. Mitigation measures to reduce highway mortality of turtles and other herpetofauna at a north Florida lake. Journal of Wildlife Management 69:549–560.

Aresco, M. J. 2005b. The effect of sex-specific terrestrial movements and roads on the sex ratio of freshwater turtles. Biological Conservation 123:37–44.

Ashton, R. E., and P. S. Ashton. 1991. Handbook of reptiles and amphibians of Florida, Part two: Lizards, turtles and crocodilians. Windward Publishing. Miami, Florida, USA.

Barichovich, W. J., and C. K. Dodd Jr. 2002. The effectiveness of wildlife barriers and underpasses on U.S. Highway 441 across Paynes Prairie State Preserve, Alachua County, Florida. Phase II post-construction Final Report, contract no. BB-854. Florida Department of Transportation, Tallahassee, Florida, USA.

Bissonette, J. A., and P. C. Cramer. 2008. Evaluation of the use and effectiveness of wildlife crossings. NCHRP Report No. 615. National Cooperative Highway Research Program, Transportation Research Board. National Academies Press,

Washington, DC, USA. www.trb.org/Main/Public/Blurbs/160108.aspx. Accessed 2 September 2014.

Branch, L. C., A. M. Clark, P. E. Moler, and B. W. Bowen. 2003. Fragmented landscapes, habitat specificity, and conservation genetics of three lizards in Florida scrub. Conservation Genetics 4:199–212.

Brehm, K. 1989. The acceptance of 0.2-metre tunnels by amphibians during their migration to the breeding site. Pages 29–41 in Amphibians and roads. Proceedings of the toad tunnel conference, Rendsburg, Federal Republic of Germany. T.E.S. Langton, editor. ACO Polymer Products Ltd., Shefford, UK.

Cain, A. T., V. R. Tuovila, D. G. Hewitt, and M. E. Tewes. 2003. Effects of a highway and mitigation projects on bobcats in southern Texas. Biological Conservation 114:89–197.

Caro, T. M., and G. O'Doherty. 1999. On the use of surrogate species in conservation biology. Conservation Biology 13:805–814.

Cavallaro, L., K. Sanden, J. Schellhase, and M. Tanaka. 2005. Designing road crossings for safe wildlife passage: Ventura County guidelines. Unpublished report in partial satisfaction for a Master's thesis. University of California, Santa Barbara, USA. http://www.bren.ucsb.edu/research/documents/corridors_final.pdf/. Accessed 17 September 2014.

Connolly-Newman, H. R., M. P. Huijser, L. Broberg, C. R. Nelson, and W. Camel-Means. 2013. Effect of cover on small mammal movements through wildlife underpasses along US Highway 93 North, Montana, USA. In Proceedings of the 2013 international conference on ecology and transportation. http://www.icoet.net/icoet_2013/documents/papers/ICOET2013_Paper401C_ConnollyNewman_et_al.pdf. Accessed 16 September 2014.

Dodd C. K., Jr., W. J. Barichivich, and L. L. Smith. 2004. Effectiveness of a barrier wall and culverts in reducing wildlife mortality on a heavily traveled highway in Florida. Biological Conservation 118:619–631.

Eilerts, B. D., and S. E. Nordhaugen. 2010. Looking to the future with retrofit options from lessons learned. Pages 25–29 in Proceedings of the 2009 international conference on ecology and transportation. P. J. Wagner, D. Nelson, and E. Murray, editors. Center for Transportation and Environment, North Carolina State University, Raleigh, North Carolina, USA.

Foresman, K. R. 2004. The effects of highways on fragmentation of small mammal populations and modifications of crossing structures to mitigate such impacts. Final Report to Montana Department of Transportation, Helena, Montana, USA. http://www.mdt.mt.gov/other/research/external/docs/research_proj/animal_use/phaseII/final_report.pdf. Accessed 2 September 2014.

Forman, R.T.T. 2000. Estimate of the area affected ecologically by the road system in the United States. Conservation Biology 14:31–35.

Forman, R.T.T., D. Sperling, J. A. Bissonette, A. P. Clevenger, C. D. Cutshall, V. H. Dale, L. Fahrig, et al. 2003. Road ecology. Science and solutions. Island Press, Washington, DC, USA.

Friedman, D. S. 1997. Nature and infrastructure: the National Ecological Network and wildlife crossing structures in the Netherlands. Report No. 138. DLO Winand Staring Centre, Wageningen, The Netherlands.

Gibbs, J. P., and W. G. Shriver. 2002. Estimating the effects of road mortality on turtle populations. Conservation Biology 16:1647–1652.

Hels, T., and E. Buchwald. 2001. The effect of road kills on amphibian populations. Biological Conservation 99:331–340.

Ireland, D. H., and M. Karch. 2010. A guide to road ecology in Ontario. Ontario Road Ecology Group, Toronto Zoo, Scarborough, Ontario, Canada.

Iuell, B., G. J. Bekker, R. Cuperus, J. Dufek, G. Fry, C. Hicks, V. Hlavac, et al. 2003. COST 341: Habitat fragmentation due to transportation infrastructure; Wildlife and traffic: a European handbook for identifying conflicts and designing solutions. European Co-operation in the Field of Scientific and Technical Research, Brussels, Belgium.

Jackson, S. D. 1996. Underpass systems for amphibians. Pages 224–227 in Trends in addressing transportation related wildlife mortality, proceedings of the transportation-related wildlife mortality seminar. G. L. Evink, P. Garrett, D. Seigler, and J. Berry, editors. FL-ER-58-96. Florida Department of Transportation, Tallahassee, Florida, USA.

Jackson, S. D. 2003. Proposed design and considerations for use of amphibian and reptile tunnels in New England. University of Massachusetts, Amherst, Massachusetts, USA.

Kintsch, J., and P. C. Cramer. 2011. Permeability of existing structures for wildlife: developing a Passage Assessment System. Final Report to Washington Department of Transportation, Report WA-RD 777.1. Washington Department of Transportation, Olympia, Washington, USA. http://www.wsdot.wa.gov/research/reports/fullreports/777.1.pdf. Accessed 2 September 2014.

Krikowski, L. 1989. The "light and dark zones": two examples of tunnel and fence systems. Pages 89–92 in Amphibians and roads. Proceedings of the toad tunnel conference, Rendsburg, Federal Republic of Germany. T.E.S. Langton, editor. ACO Polymer, Ltd., Shefford, UK.

Lambeck, R. J. 1997. Focal species: a multi-species umbrella for nature conservation. Conservation Biology 11:849–857.

Langton, T.E.S. 1989. Amphibians and roads. Proceedings of the toad tunnel conference, Rendsburg, Federal Republic of Germany. ACO Polymer Products, Shefford, UK.

MassDOT (Massachusetts Department of Transportation). 2010. Design of bridges and culverts for wildlife passage at freshwater streams. Massachusetts Department of Transportation, Highway Division, Boston, Massachusetts, USA. http://www.transwildalliance.org/resources/2011113172748.pdf. Accessed 2 September 2014.

McAllister, K. R. 2010. Washington's habitat connectivity highway retrofit initiative. Pages 363–365 in Proceedings of the

2009 international conference on ecology and transportation. P. J. Wagner, D. Nelson, and E. Murray, editors. Center for Transportation and Environment, North Carolina State University, Raleigh, North Carolina, USA.

McDonald, W., and C. C. St. Clair. 2004. Elements that promote highway crossing structure use by small mammals in Banff National Park. Journal of Applied Ecology 41:82–93.

Merriam-Webster. 2013. Definition of retrofit. http://www.merriam-webster.com/dictionary/retrofit. Accessed 2 September 2014.

Ng, S. J., J. W. Dole, R. M. Sauvajot, S.P.D. Riley, and T. J. Valone. 2004. Use of highway undercrossings by wildlife in southern California. Biological Conservation 115:499–507.

Nordhaugen, S. E., E. Erlandsen, P. Beier, B. D. Eilerts, R. Schweinsburg, T. Brennan, T. Cordery, et al. 2006. Arizona's wildlife linkages assessment. Arizona Wildlife Linkages Workgroup. Arizona Department of Transportation, Phoenix, Arizona, USA. http://www.azdot.gov/business/environmental-services-and-planning/programs/wildlife-linkages. Accessed 2 September 2014.

Noss, R. F., and A.Y. Cooperrider. 1994. Saving nature's legacy: Protecting and restoring biodiversity. Island Press, Washington, DC, USA.

Noss, R. F., H. B. Quigley, M. G. Hornocker, T. Merrill, and P. C. Paquet. 1996. Conservation biology and carnivore conservation in the Rocky Mountains. Conservation Biology 10:949–963.

Pepper, H. W., M. Holland, and R. Trout. 2006. Wildlife fencing design guide. CIRIA, Classic House, London, UK.

Puky, M., and Z. Vogel. 2004. Amphibian mitigation measures on Hungarian roads: design, efficiency, problems and possible improvement, need for coordinated European environmental education strategy. Pages 1–13 in Proceedings of the 2003 IENE conference on habitat fragmentation due to transportation infrastructure and Cost 341 action. Infra Eco Network Europe, Brussels, Belgium.

Righetti, A., J. Müller, A. Wegelin, A. Martin, P. Drollinger, V. Mason, S. Zumbach, and A. Meyer. 2008. Adapting existing culverts for the use by terrestrial and aquatic fauna. Civil Engineering Department of the Canton of Aargau, Swiss Association of Road and Transportation Experts (VSS), Bern, Switzerland.

Roberge, J. M., and P. Angelstam. 2004. Usefulness of the umbrella species concept as a conservation tool. Conservation Biology 18:76–85.

Ruediger, B., and J. Lloyd. 2003. A rapid assessment process for determining potential wildlife, fish and plant linkages for highways. Pages 217–225 in Proceedings of the 2003 international conference on ecology and transportation, C. L. Irwin, P. Garrett, and K. P. McDermott, editors. Center for Transportation and the Environment, North Carolina State University, Raleigh, North Carolina, USA.

Ryser, J., and K. Grossenbacher. 1989. A survey of amphibian preservation at roads in Switzerland. Pages 7–13 in Amphibians and roads. Proceedings of the toad tunnel conference, Rendsburg, Federal Republic of Germany. T.E.S. Langton, editor. ACO Polymer Products Ltd., Shefford, UK.

SETRA (Service d'Etudes techniques des routes et autoroutes). 2005. Facilities and measures for small fauna, technical guide. Ministere de I'Ecologie du Developpment et de I'Amenagement durables, Chambéry, France. [In French.]

Smith, D. J. 1999. Identification and prioritization of ecological interface zones on state highways in Florida. Pages 209–229 in Proceedings of the third international conference on wildlife, ecology and transportation. G. L. Evink, P. Garrett, and D. Zeigler, editors. Florida Department of Transportation. Tallahassee, Florida, USA.

Smith, D. J. 2003a. The ecological effects of roads: theory, analysis, management, and planning considerations. PhD dissertation, University of Florida, Gainesville, Florida, USA.

Smith, D. J. 2003b. Monitoring wildlife use and determining standards for culvert design. Final Report, Contract No. BC354-34. Florida Department of Transportation, Tallahassee, Florida, USA. http://www.dot.state.fl.us/research-center/Completed_Proj/Summary_EMO/FDOT_BC354_34_rpt.pdf. Accessed 2 September 2014.

Smith, D. J. 2006. Incorporating results from the prioritized "ecological hotspots" model into the Efficient Transportation Decision Making (ETDM) process in Florida. Pages 127–137 in Proceedings of the 2005 international conference on ecology and transportation. C. L. Irwin, P. Garrett, and K. P. McDermott, editors. Center for Transportation and the Environment, North Carolina State University, Raleigh, North Carolina, USA.

Smith, D. J. 2011. Cost effective wildlife crossing structures which minimize the highway barrier effects on wildlife and improve highway safety along US 64, Tyrrell County, NC. Final Report, No. FHWA/NC/2009-26. North Carolina Department of Transportation, Raleigh, North Carolina, USA.

Smith, D. J., and R. F. Noss. 2011. A reconnaissance study of actual and potential wildlife crossing structures in central Florida. Final Report, Contract No. BDB-10. University of Central Florida–Florida Department of Transportation, Orlando, Florida, USA.

Smith, L. L., and C. K. Dodd Jr. 2003. Wildlife mortality on U.S. Highway 441 across Paynes Prairie, Alachua County, Florida. Florida Scientist 66:128–140.

Spellerberg, I. F. 1998. Ecological effects of roads and traffic: a literature review. Global Ecology and Biogeography Letters 7:317–333.

Trocmé, M., and A. Righetti. 2012. Standards for fauna friendly culverts. Pages 557–560 in Proceedings of the 2011 international conference on ecology and transportation. P. J. Wagner, D. Nelson, and E. Murray, editors. Center for Transportation and the Environment, North Carolina State University, Raleigh, North Carolina, USA.

van der Ree, R., C. Grilo, and D. J. Smith, editors. 2015. Handbook of road ecology. John Wiley and Sons, Hoboken, New Jersey, USA.

van der Ree, R., E. van der Grift, C. Mata, and F. Suarez. 2007.

Overcoming the barrier effect of roads: how effective are mitigation strategies? An international review of the use and effectiveness of underpasses and overpasses designed to increase the permeability of roads for wildlife. Pages 423–431 *in* Proceedings of the 2007 international conference on ecology and transportation. C. L. Irwin, D. Nelson, and K. P. McDermott, editors. Center for Transportation and the Environment, North Carolina State University, Raleigh, North Carolina, USA.

Wilcox, B. A. 1984. In situ conservation of genetic resources: determinants of minimum area requirements. Pages

639–647 *in* National parks, conservation, and development: The role of protected areas in sustaining society. J. A. McNeely, and K. R. Miller, editors. Smithsonian Institution Press, Washington, DC, USA.

Yorks, D. T., P. R. Sievert, and D. J. Paulson. 2012. Experimental tests of tunnel and barrier options for reducing road mortalities of freshwater turtles. Page 1034 *in* Proceedings of the 2011 international conference on ecology and transportation. P. J. Wagner, D. Nelson, and E. Murray, editors. Center for Transportation and the Environment, North Carolina State University, Raleigh, North Carolina, USA.

11 — Construction and Maintenance

SANDRA L. JACOBSON
AND STEPHEN TONJES

11.1. Introduction

The planning, preliminary engineering, and detailed design phases of a road project, which are described in Chapter 6, often require many years. Once the final plans are completed and the project is approved for construction, stakeholders such as agencies, organizations, and citizens who have been involved in project planning and design sometimes feel their job is done and the road builders can take it from there. However, numerous challenges associated with the construction phase may result in unintended interpretations of, or changes to, the project plans that decrease the effectiveness of mitigation measures for road effects on wildlife. Because of these challenges, and because construction schedules are tight and delays are costly, involvement of wildlife experts throughout the construction phase can help identify and minimize or even prevent these unintended consequences.

The effectiveness of even the most well-designed mitigation measures and excellent design decisions, including those for retrofits, can be unintentionally undone in a single pass of a bulldozer during the construction phase. Common challenges during construction include cost overruns, unexpected engineering challenges, archeological finds that require changes in structure locations, the inability to locate specified materials, and errors or omissions in plans. Several of these challenges may require immediate consultation with wildlife experts and other stakeholders involved in the design phase, in order to avoid misinterpretations. Regular contact among transportation professionals, construction professionals, and other stakeholders, as well as proper documentation of decisions made in the design phase, can greatly reduce the negative impacts of these challenges.

Notably, the longest phase in the life of a road begins once construction is complete. Roads need to be maintained and many of the most important mitigation methods used for small animals depend on adequate maintenance for continued effectiveness. The need for maintenance continues for the life of the structure, which can often approach a century, so it is imperative to formulate a good strategy that includes funding for maintenance, preferably while the project is being developed. Considerations of durability of structure components can save money in the long run, especially because maintenance budgets related to wildlife are typically insufficient.

11.2. Construction

Once the final set of construction plans and specifications is completed in the design phase, the project is advertised for construction, and companies submit bids for the contract. Building a project takes a different set of skills from those needed to design a project; hence, usually a different company, or a different division of a company, does the building. Unfortunately, there may be little communication between the two entities. The winning construction firms, and sometimes even the bidding firms, often are allowed to propose changes they think would save money after the plans have been finalized—this process is called value engineering. A similar process is also often employed in earlier design phases as well (Section 6.7.2). These changes could benefit greatly from scrutiny by wildlife experts if they pertain to or potentially affect mitigation measures.

It is important to clearly document specific mitigation measures in the early stages of project development due to regulatory requirements for some projects. Projects subject to the National Environmental Policy Act (NEPA) require specific environmental reports where these features should be explicitly described (Section 6.3.1.2). For projects not subject to NEPA (Section 6.5.2), it may be necessary to work with the project manager to compile a written record of environmental decisions; this can facilitate the construction manager's compliance with the environmental components of the design. Each state and federal agency has different procedures and organizational structures, so it is important to determine a single point of contact in each involved agency early in the construction process to ensure planning agreements are carried out during construction. The Federal Highway Administration's (FWHA) Eco-Logical process can provide a framework for stakeholder team identification and continuity (FHWA 2006).

Adding mitigation measures or changing design details in these measures after construction plans have been finalized is much more difficult, and much less cost-effective and efficient, than making changes to the iterative draft plans that are circulated during the design phase (Section 6.7). Time is especially critical once equipment, materials, and workers are deployed, as delays can cost thousands of dollars per day. In addition to the cost of delays, the cost to construct the changes also can be high because there is no time to write another contract and advertise for competitive bidding; therefore, there is no opportunity for another company to offer a lower price. For these reasons, the early incorporation of mitigation measures and the involvement of wildlife experts throughout the design process are critical for both logistical and economic efficiency of the ultimate construction of the project (Chapter 6).

11.2.1. Challenges during Construction

11.2.1.1. Biological Challenges

During the construction phase, there are threats to small animals with respect to physical risks, move-ment disruption, and habitat damage. Physical threats include direct mortality from construction or entrapment of snakes and other animals in erosion control netting (Box 11.1). Disruption to animal movement patterns can occur as a result of temporary fencing that creates barriers to key habitats such as wetlands. Habitat damage frequently includes negative impacts on the quality of water sources (e.g., siltation of small streams and wetlands) and the quantity of water (e.g., using local water sources to spray around construction sites for dust abatement), thus affecting potential breeding or watering sources for small animals. Further, construction debris and litter (the removal process or failure to remove items) can result in all of these effects. Chapters 2–4 address the susceptibility of small animals to these disturbances based on their physiology and behaviors. Unfortunately, even with full compliance with existing regulations, impacts to small animals may not be precluded.

11.2.1.2. Process Challenges

The translation from the design phase plans to implementation in the construction phase is vulnerable to the same communication difficulties encountered within the design phase (Section 6.7). Because the design process is so complex, plans almost always need adjustments during the construction phase once actual conditions are considered in the field. Changes to a project component, even those that are completely unrelated to mitigation measures, may result in unintentional impacts to small animals or reductions of the effectiveness of mitigation measures or retrofits. Engineers and construction managers typically do not understand the rationale for some features and may make design changes without realizing that these could affect functionality. To increase effectiveness, plans should include a brief but descriptive statement of the objective of a given mitigation measure or retrofit (e.g., "native soil for wildlife substrate"). This may reduce the potential for inadvertent changes that compromise the intent of the mitigation measure; however, there is no substitute for monitoring the project and engaging a wildlife expert in the review of any proposed changes, as is ideal in the design phase.

Almost all engineering aspects of a project (e.g., pavement composition and thickness, survey tolerances,

Box 11.1. Construction Agreement Reduces Risk of Mesh Entanglement

Snakes and other small animals are vulnerable to entanglement, and subsequent mortality, in the mesh of erosion control netting and other temporary products (e.g., certain fence materials; Mitchell et al. 2006; Table 10.2; Figure 11.1). The widespread use of sod that is harvested in large rolls backed with plastic netting has emerged as a threat to small animals. Large sod rolls are easier and less expensive to apply in large quantities, and therefore are more popular than the standard pallets of individual sod squares. Specifically, rolled sod, netting and all, can be unrolled from a vehicle in a continuous swath while sod squares must be placed by hand. It is impossible to prevent some of the netting from being exposed, either at carelessly matched seams and uneven edges during construction, or later, when errant vehicles and mowers gouge the sod, utility work or other excavation is performed, or the grass dies. Biodegradable netting is available but can take years to degrade under the sod, as degradation requires several months of direct exposure to sunlight and oxygen.

Certain materials can be prohibited as part of the terms of a construction agreement for projects where small animals, particularly rare or sensitive species, are likely to be present. As construction agreements are legally binding, prohibited products or actions should be stated explicitly in the contract. Any prohibitions within a construction agreement must apply both to construction as well as to ongoing shoulder maintenance projects. The Washington State Department of Transportation (DOT) prohibits the use of erosion control matting with nylon netting when working in areas with warm, dry climates where snakes are prevalent (Washington State DOT 2008). Vermont DOT specifies biodegradable, temporary erosion control materials. Other states without this standard construction provision can achieve a similar result if project designers specify biodegradable erosion control netting and prohibit rolled sod in the construction agreement for the project. Recent studies indicate certain rolled erosion control products made of unwoven, organic fibers or open-weave textiles with loose, freely moving weave may have a lower risk of entrapment than with fused plastic mesh products (Kapfer and Paloski 2011, and references therein).

Figure 11.1. Snakes frequently get entangled in various types of netting and are unable to extract themselves; thus, they either starve, overheat, or are easily preyed upon. *Credit: CRESO.*

concrete strength, pile depth) can be measured. These are continually monitored by engineering technicians during construction so that adjustments can be made immediately to avoid faulty work. Mitigation measures generally do not receive the same level of oversight by wildlife experts during construction, and there are no analogous wildlife resources to provide guidance or numerical references for wildlife or habitat connectivity that engineers could consult. As a result, problems during the construction of mitigation measures may remain undetected until after construction is complete and the road is opened to traffic (see also Section 6.8).

11.2.2. Solutions during Construction

Open and frequent communication among agencies ensures that the hard work accomplished during the planning and design phases results in a product that meets the original intentions during the construction and maintenance phases. The following practices have proven to produce more efficient and effective outcomes:

1. Identify a single contact person at each of the stakeholder agencies and the contractor, and keep detailed notes for continuity as personnel changes. A contact person who follows a project through design, construction, and maintenance phases can maintain continuity but is rare due to the long duration of these phases.

2. Maintain access to all agreements (electronic or hard copy) for all stakeholders. These include both the design commitments discussed in Section 11.2 and subsequent agreements made during construction. This communication is important for disclosure and open access to information for all interested parties and for the identification of problems before construction is completed.

3. Ensure all agreements to be implemented in the field are signed on construction plans or change orders. Because projects can take many years to complete, agreements that are not officially documented or that do not result in signed construction plans (i.e., handshake agreements) are nonbinding and much more easily forgotten.

4. Schedule periodic (e.g., monthly or quarterly) meetings among the agency contacts to update all stakeholders on the status of construction. Construction contractors generally hold weekly progress meetings with the transportation agency project supervisors, and these meetings could include updates for or from wildlife experts. Additionally, visits to the project site by the stakeholders can be invaluable in finding and solving problems before they escalate beyond what can be prevented or mitigated.

5. Request a change order if something goes wrong during project construction. Also request that the change order be processed as soon as the problem is noticed, because change orders are expensive and carry political costs.

6. Use standardized contract specifications where applicable (e.g., Section 6.7). Standardized contract items are specific and instructive, and thus easier for construction managers to interpret and implement. However, for the reasons discussed in Section 6.7, mitigation measures generally require more adaptive management and site-specific adjustments than typical road construction; thus, custom contract specifications are more likely to be needed. In these instances, contract specifications from other contracts can be used as a starting template to be adapted for the given project.

Translating road ecology research into design and construction guidance is a huge challenge. While a number of manuals and guidelines have been published, including this book, these are not always readily available to the transportation agency offices involved with individual projects. Further, it is daunting to convert conceptual drawings and examples from these publications into project-specific procedures and pay items (established list of construction procedures that can be contracted outside of departments of transportation [DOTs]). There are states that have developed true specifications (e.g., passage bench [a shelf used with bridges to allow terrestrial passage at peak flow], Minnesota Department of Transportation 2011; erosion netting, Box 11.1). Such standards should be applied where available, yet these applications remain uncommon despite availability. Accordingly, involving wildlife experts who can facilitate the application of research findings to design and construction in this stage and all stages of the transportation planning process can be mutually beneficial (Chapter 6).

11.3. Maintenance

Transportation projects are big and exciting, with hundreds of people involved in planning, funding, designing, and constructing a large, tangible infrastructure. After the ribbon is cut, everybody leaves with a feeling of accomplishment, thinking the job is done. Maintaining these projects after construction, on the other hand, is not exciting. Much like housework, nobody notices unless it is not done, and nobody wants to be responsible for the ongoing tasks. While everybody feels good about a job well done, reporters rarely show up to trumpet the completion of a repaving project or a culvert replacement. If a project is built correctly, the maintenance phase lasts many decades longer than *all* the phases combined in developing the project; yet, budgets are tighter for maintenance departments than for other sections within most transportation agencies. Although more transportation money is spent on maintenance than on new construction, maintenance budgets rarely keep pace with increasing demands over the life of the infrastructure element. With less public exposure and support, maintenance is the first place targeted for cost cutting, as reflected by the increasing size and deteriorating condition of transportation infrastructure in the United States.

11.3.1. Include Maintenance in the Design Phase

One strategy for addressing maintenance challenges is to engage maintenance personnel in the design phase of a project, including the design of a retrofit. Maintenance personnel may offer recommendations for designs that are less burdensome to maintain. Consider inviting maintenance department personnel to participate in the development and review of project concepts and design plans, preferably as members of a wildlife crossing design task team as described in Section 6.7. Early consultation with maintenance personnel as procedures are being established is helpful in minimizing resistance to new and unfamiliar responsibilities and protocols. Establishing maintenance protocols prior to completion of the construction phase can greatly increase the likelihood of properly maintaining the mitigation measures associated with the infrastructure (Box 11.2).

Few state DOTs currently have strategies for ensuring the continued functionality of wildlife-related measures, such as crossing structures and fencing. Thus, wildlife experts can make a significant contribution in the design phase by establishing detailed maintenance protocols to retain effectiveness for wildlife. Protocols should apply to mitigation measures that are either new or part of a retrofit, and may cover critical design features both external and internal to a structure, such as substrates, lighting, interior cover, fencing and walls, landscaping, and water flow (Section 2.3, but see also Chapters 9 and 10, and Tables 10.1 and 10.2). If maintenance protocols for mitigation measures have not already been established during the design phase (Section 11.3.1), then agency biologists can work with the maintenance department during the construction phase to establish clear and reasonable procedures. The most effective protocols are those incorporated into existing road maintenance procedures and criteria.

Noting different time frames for maintenance of various features will help maintenance crews to schedule maintenance more effectively. Maintenance phase time frames may be substantially different for structures intended for small animals than for large animals. Compared to the generally robust fencing for larger animals, fencing for small animals is much more fragile and prone to damage by water flow, snow and snow removal, vegetation growth, or humans, so maintenance intervals may be quite short. Vegetation management may need to be monitored for several years to determine the optimal management strategy (e.g., mowing interval), and the time frame will vary according to the regional climate.

Local maintenance departments are often subject to periodic inspections from a central maintenance authority. If mowing is the norm for the road, then the local unit will receive negative ratings for not mowing in the approaches to the wildlife crossing structure unless these areas have been specifically identified in the rating procedure. An existing maintenance rating procedure could be modified not only to withhold mowing penalties, but to reward compliance with specialized protocols developed for identified mitigation measures.

Box 11.2. Considerations for Developing a Maintenance Protocol for Small Animal Mitigation Structures

Structure:
Identify the structure type (e.g., State Route 41 underpass, small animal fence and ledge). Each feature with a different maintenance schedule will need a separate maintenance protocol.

Map:
Include precise location and image of structure in good functional condition (e.g., Mile Post 51.4).

Objectives:
- Identify target species, taxa, or habitat.
- Identify existing site features that contribute to functionality.
- Include other important notes (e.g., small animal fence is designed to keep snakes, turtles, and small rodents from accessing the road, with a bent top "lip" to keep climbing species from overtopping it, small mesh to keep small animals from pushing through holes, and a buried bottom edge to prevent gaps at ground level from erosion or burrowing animals).

Key Maintenance Requirements:
- Desired Condition
Describe the desired condition for the mitigation measure. For example: Fence is firmly

attached to underpass abutment with no gaps, mesh is intact, fully buried at base, and top lip is intact and facing away from highway. Gaps are defined as spaces at attachments or tears in mesh twice as large as original mesh size.
- Schedule
Include a separate item for each aspect of the structure, especially if the maintenance schedule is different. For example:
1. Inspect fence at least once annually.
2. Repair to original specifications within two weeks of noted failure.
3. If repair costs exceed $5,000, refer to project Memorandum of Understanding (MOU) for funding options.
- Responsibility
Include any agreements that establish (a) responsible parties for maintenance or monitoring, (b) pertinent dates, and (c) special funding sources. For example: According to the 2013–2018 State DOT/State Resource Agency MOU for Project X, the maintenance requirements listed above will be accomplished and funded by the State DOT.

Likewise, a bridge inspection procedure could be modified such that it institutionalizes maintenance for continued effectiveness for wildlife use.

11.3.2. Challenges to Maintenance Operations

Increased attention by wildlife experts to the maintenance phase of highway operations can better ensure that mitigation measures continue to function properly. When mitigation measures are included in road projects, additional funding is usually required to develop and implement programs for monitoring their effectiveness and subsequently modifying their maintenance (i.e., adaptive management, Sections 12.7.2.2

and 12.7.3). Unfortunately, for the reasons stated above, funding may be difficult to find for maintaining mitigation measures. It is therefore unrealistic to expect the design department to develop maintenance protocols as described above, or the maintenance department to budget the additional resources needed, without a strong environmental commitment on the part of the transportation agency overall. There are several examples of state transportation agencies that have implemented programs to maintain and monitor mitigation measures (Box 11.3; Section 12.7.3).

Natural resource and transportation agencies tend to disagree over whose responsibility it is to maintain and monitor the continued functioning of wildlife-related mitigation measures on highways. Transportation planners usually incorporate the philosophy that form benefits function (i.e., roads are built for the benefit of transporting people and goods). If these benefits to people result in harm to wildlife, the transportation agency is responsible for mitigating negative effects to the ecological infrastructure and subsequently maintaining the functionality of those measures.

Projects that include monitoring studies for mitigation measures ironically may develop maintenance issues after the monitoring program has been completed, because maintenance needs are ongoing and agencies rarely plan for these continuing needs. A formal agreement for continued maintenance can provide support and legal documentation regarding the relative responsibilities of transportation and natural resource agencies in this process. Some natural resource agencies have attempted to develop maintenance partnerships with volunteer organizations; however, volunteers may not be reliably available for the decades of maintenance needed for wildlife crossing structures. Ideally, long-term interagency maintenance agreements are adaptive by including periodic reviews during the first decade to incorporate lessons learned by both maintenance crews and environmental specialists, including the provision for opportunities to incorporate new procedures as standard operations.

11.3.2.1. Specific Maintenance Considerations for Mitigation Measures

Experience with mitigation structures for small animals is still unusual in most maintenance crews. Therefore, it is likely that most maintenance personnel will require some guidance about the objectives and specific needs of these mitigation measures. For example, as noted above, there are specific requirements for rights-of-way along wildlife crossing structures that are usually different from most roadsides or shoulders. The pathway between the entrance to a crossing structure and the surrounding habitat (approach) usually requires an optimum amount and type of vegetative cover that needs to be maintained more frequently in warmer and wetter climates where vegetation can grow rapidly and obstruct the entrance (see also Sections 9.7.1 and 10.5.1). Further, periodic removal of invasive plant species to favor native plant species can also increase crossing structure use, though care should be taken with herbicide use for these purposes to prevent harm to species that may use the structure (Chapter 4). The opposite problem of insufficient cover may occur due to frequent mowing or deposition of harsh substrates from the road surface. The structure approaches must be clearly distinguished from the rest of the shoulder to protect them from ordinary roadside maintenance. This demarcation can be facilitated by directional fencing or walls designed to funnel animals to the crossing entrance; these are much more effective than signage on the shoulder or instructions on paper.

Carefully designed features (e.g., landscaping, waterways, ramps; Chapter 9) that are essential to accommodate the target species may require specific maintenance. Ledges in culverts may need periodic debris removal and hydrological monitoring to ensure functionality. Boulders and logs placed in dry culverts as cover for small animals may need to be replenished.

Small animal fencing is more likely to be ignored during maintenance phases than large animal fencing, because small animals usually constitute less of a road safety hazard to motorists and are consequently less

Figure 11.2. Regularly removing climbing vegetation from the barrier wall on US 441 across Paynes Prairie, Florida, United States, is key to reducing mass mortality of amphibians and reptiles. *Credit: Courtesy of Florida Department of Transportation.*

of a concern to DOTs. Additionally, fencing for small animal mitigation measures often involves more fragile materials. Yet, proper maintenance of fencing is key to proper functionality of small animal structures (Prudon and Creemers 2004); most small animals rely on guide or barrier fences more than large species because they move comparatively shorter distances and orient to their landscapes at a finer spatial scale. Directional or barrier fencing for small animals can be made of diverse materials (e.g., solid concrete barrier, wire mesh; Chapter 9), and maintenance needs and time frames vary by the type of fencing (Chapter 10, Table 10.2). For example, climbing vegetation may compromise the functionality of a barrier fence, allowing amphibians and reptiles to surmount the barrier (Figure 11.2). However, in Germany, the Federal Ministry of Transport, Building and Housing, Road Construction Department, and Road Traffic requires that barrier walls for wildlife include a flat platform (20 cm wide) along the habitat side of the wall to prevent vegetation growing up the barrier; this further reduces ongoing maintenance costs to trim vegetation during its growing season (FMTBHRCDRT 2000). Another maintenance consideration is that small animals readily exploit gaps under fencing; frequent breaches by digging animals may require repairs or reburying the bottom edge of the fence.

11.3.2.2. Modification of Standard Maintenance Approaches

Several standard maintenance procedures can be counterproductive to small animals when used on wildlife crossing structures. There are existing manuals that describe recommended techniques and best practices to reduce impacts to wildlife as part of road maintenance activities (Venner et al. 2004; FHWA 2013). For example, routine maintenance or repairs slated for bridges that also serve as underpasses, particularly over aquatic habitat, could be scheduled outside of known aquatic species breeding seasons (e.g., fish or aquatic salamander nesting or egg incubation). These and other practices can be incorporated into written maintenance protocols that clearly describe the intended function of the mitigation measures; such explicit protocols can help reduce inadvertent maintenance errors.

Culverts are often used along with fencing to accommodate small animal movement, requiring ongoing maintenance according to the specific microhabitat requirements of the target animals. In some instances, wildlife crossing culverts intentionally designed and installed with a sandy substrate necessary for the target species (such as frogs or salamanders) have been cleaned to bare concrete—a standard maintenance treatment for drainage culverts, but one that defeats the purpose of a wildlife crossing culvert. Likewise, drainage and water management considerations for wildlife crossing culverts must also be considered as part of maintenance. For example, high flow rates in these culverts also can cause erosion or scouring of substrate, blockage, and pooling, which can reduce usage by certain terrestrial species (Jackson 1996; Puky and Vogel 2004). However, presence of standing water or even flowing water may be desired for certain aquatic or semi-aquatic species. Ongoing maintenance to sustain acceptable water flow rates based on the target species can help preserve functionality. Additionally, erosion at a culvert outlet may result in a perched culvert, blocking passage for both aquatic organisms when water is flowing and for small terrestrial animals during drier periods (Figure 11.3). In these situations, ongoing erosion control is an essential maintenance need for continued wildlife passage.

Deicing or dust abatement chemicals used over many years can accumulate in roadside soils, creating a potential toxic hazard as well as changes in vegetation

Figure 11.3. Perched culvert resulting from erosion at its outlet, preventing passage by both aquatic and terrestrial organisms and causing disruption of the natural stream flow. *Credit: Jane Winn, courtesy of Berkshire Environmental Action Team.*

Figure 11.4. Deicing cinders thrown by snow blowers collect to 1 m deep in this Oregon Cascades marsh. *Credit: Sandra L Jacobson, courtesy of USDA Forest Service.*

that increase the width (and thus the barrier effect) of the road effect zone (Section 3.2.4, and Chapter 4). In areas with heavy snow, deicing sand or cinders may accumulate over time up to a meter in depth and extend many meters beyond the road, blocking entry into culverts (Figure 11.4). Specialized culverts, such as those with slotted drains used for amphibian passage, may become filled with contaminated water from runoff where deicing agents are used on roads. Thus, periodic maintenance is needed to flush out salts or sands and refill with appropriate substrate. Delineated areas to

avoid application of deicing agents can minimize toxic runoff in key areas and reduce maintenance of affected culverts. Planting living snow fences near crossing structures or other sensitive areas can also reduce the need for deicing agents. Some types of culvert repairs (e.g., cured-in-place pipe rehabilitation [in which a liner saturated with epoxy resin is inserted into the pipe and expanded into place with air or water pressure]) may release materials with extremely toxic compounds (styrene); these are not typically monitored in waterways without protected fish species, though they may be harmful to many other taxa (Donaldson 2009). Improper use of herbicides along roadsides to control invasive species can also expose amphibians and other species to toxic chemicals (Chapter 4). To fully mitigate for the negative impacts of standard road maintenance, these factors must be accounted for in the monitoring and adaptive management plans referred to in Section 11.3.2 (see also Chapter 12).

11.4. Conclusion

It is important for wildlife experts to partner with transportation agencies during the construction and maintenance phases in the life of a road as well as during the planning phase. Construction errors can undo well-planned designs very quickly (Figure 11.5). Not all of these errors can be fixed, but there is at least a chance to do so when bulldozers and crews are still

Figure 11.5. A construction error destroyed the delineated, protected riparian area (darker center of image) with an excessive blast. *Credit: Sandra L. Jacobson, courtesy of USDA Forest Service.*

mobilized and construction accounts are still active. Once road construction is completed, it may be decades before heavy equipment arrives at the site again, but the maintenance crews will continue their regular chores. Until wildlife crossing structures and fencing become so established that they are no longer considered non-traditional "additions," well-documented procedures that transcend job transfers and retirements will be necessary to protect the functionality of these important investments for the ecological infrastructure and for small animal populations. Lastly, the installation of current wildlife crossing structure designs and materials, and the development of new applications that do not require as much maintenance, will help to relieve the current pressure of labor and material investment in the initial construction phase and ongoing maintenance phase.

11.5. Key Points

- Regular, scheduled meetings among all stakeholders that continue from the planning period through construction can forestall and minimize problems.
- Identify a single contact person at each stakeholder agency and keep detailed notes for continuity as personnel changes. A contact person who follows a project through design, construction, and maintenance phases can maintain continuity.
- Maintain access to all agreements (electronic or hard copy) for all stakeholders; these agreements include both the design commitments discussed in Section 11.2.2 and subsequent agreements made during construction. This is important for disclosure to all interested parties, but also facilitates the identification of problems during construction rather than after it is over.
- By participating actively in the construction and maintenance phases, wildlife experts can help ensure that critical design details of mitigation measures are not changed in ways that reduce or negate effectiveness.
- Transportation ecologists would benefit by sharing tested, standardized specifications widely. Construction specifications are needed that are based on transportation ecology research, are standard-

ized as much as practicable, but are also flexible enough to adapt to specific sites and target species.
- The maintenance phase of a project lasts longer than all other phases, but it may not receive the attention and funding needed to ensure that mitigation measures continue to function effectively. Mitigation measures often require different maintenance procedures from the standard components of the road.
- Invite maintenance department personnel to participate in the development and review of project concepts and design plans.
- Wildlife experts can help ensure that necessary maintenance is performed to maintain effectiveness of mitigation measures over time by assisting in developing maintenance protocols.
- Most maintenance personnel will require guidance on the objectives and intended functions of mitigation measures for small animals; this guidance can help reduce inadvertent maintenance errors.
- A formal maintenance agreement can provide support and legal documentation regarding the relative responsibilities of transportation and natural resource agencies.
- Explicitly written maintenance protocols that clearly describe the intended function of the mitigation measures can help reduce inadvertent maintenance errors.
- Durable and low maintenance mitigation measures are likely to retain their intended functions over the long term.

LITERATURE CITED

Donaldson, B. 2009. Environmental implications of cured-in-place pipe rehabilitation technology. Transportation Research Record: Journal of Transportation Research 2123:172–179.

FHWA (Federal Highway Administration). 2006. Eco-Logical: an ecosystem approach to developing infrastructure projects. FHWA-HEP-06-011. US Department of Transportation, Washington, DC, USA.

FHWA (Federal Highway Administration). 2013. "Keeping it Simple." http://www.fhwa.dot.gov/environment/wildlife_protection/index.cfm. Accessed 4 September 2014.

FMTBHRCDRT (Federal Ministry of Transport, Building and Housing, Road Construction Department, Road Traffic), eds. 2000. Instruction sheet for protecting amphibians on roads (MAMS). Forschungsgesellschaft für Straßen- und

Verkehrswesen (FSGV; Research Association for Roads and Transport [FSGV]), Cologne, Germany. [In German.]

Jackson, S. D. 1996. Underpass systems for amphibians. Pages 224–227 *in* Trends in addressing transportation related wildlife mortality, proceedings of the transportation related wildlife mortality seminar. G. L. Evink, P. Garrett, D. Zeigler, and J. Berry, editors. FL-ER-58-96. Florida Department of Transportation, Tallahassee, Florida, USA.

Kapfer, J. M., and R. A. Paloski. 2011. On the threat to snakes of mesh deployed for erosion control and wildlife exclusion. Herpetological Conservation and Biology 6:1–9.

Minnesota Department of Transportation. 2011. Best practices for meeting DNR general public water work permit GP 2004-0001. Division of Ecological and Water Resources, St. Paul, Minnesota, USA. http://www.dnr.state.mn.us /waters/watermgmt_section/pwpermits/gp_2004_0001 _manual.html. Accessed 4 September 2014.

Mitchell, J. C., J. D. Gibson, D. Yeatts, and C. Yeatts. 2006. Observations on snake entanglement and mortality in plastic and horticultural netting in Virginia. Catesbeiana 26:64–69.

Prudon, B., and R.C.M. Creemers. 2004. Safely to the other side. A critical approach to the construction and maintenance of amphibian tunnels. RAVON, Nijmegen, The Netherlands. [In Dutch.]

Puky, M., and Z. Vogel. 2004. Amphibian mitigation measures on Hungarian roads: design, efficiency, problems and possible improvement, need for coordinated European environmental education strategy. Pages 1–13 *in* Proceedings of the 2003 IENE conference on habitat fragmentation due to transportation infrastructure and Cost 341 action. Infra Eco Network Europe, Brussels, Belgium.

Venner, M., Venner Consulting, and Parsons Brinckerhoff. 2004. Environmental stewardship practices, procedures, and policies for highway construction and maintenance. Transportation Research Board, Washington, DC, USA.

Washington State DOT (WSDOT). 2008. I-90 Snoqualmie Pass East final environmental impact statement and section 4(f) evaluation. Federal Highway Administration Report No. FHWA-WA-EIS-05-01-F. Federal Highway Administration, Washington, DC, USA.

12

Daniel J. Smith, David M.
Marsh, Kari E. Gunson,
and Stephen Tonjes

Monitoring Road Effects and Mitigation Measures and Applying Adaptive Management

12.1. Introduction

Monitoring is essential for identifying the location and measuring the severity of effects of roads and traffic on wildlife (pre-construction). Further, it is critical for evaluating the performance of mitigation measures intended to remedy these effects (both during and following construction). Whether it is a short-term monitoring project of a single (new or existing) road or a long-term monitoring program on several roads, success is enhanced when a structured process is followed. This structured process should include (1) developing a set of goals and objectives, including whether to assess road mortality, habitat connectivity, or both; and (2) constructing an effective study design suitable for pre- and post-construction phases, and during construction where feasible (Section 11.2.1). This study design should (a) address the scope, timing, and duration of the monitoring project; (b) include achievable methods for data collection and statistical analysis; and (c) incorporate adaptive management considerations.

The goal of any mitigation measure, no matter the level of sophistication or cost, is to function at its optimal performance level. Monitoring can increase our knowledge of the type and extent of road and traffic effects on wildlife as well as the performance of various mitigation measures. Importantly, long-term performance of mitigation measures is most effective when adaptive management frameworks are used to identify deficiencies and opportunities for improvement. Adaptive management is a process of applying adjustments and changes necessary to ensure and improve future performance as research and monitoring provide new information (Holling 1978).

12.2. Setting Goals and Objectives for Monitoring

Setting goals (end points) and objectives (tasks necessary to achieve each goal) is essential to any successful monitoring effort. In developing these goals and objectives, it is important to pinpoint the conflicts between the road and both the species and adjacent habitat areas. The process should outline several questions that define the focus of the project (e.g., potential locations for mitigation and species affected; Box 12.1). To be effective in answering these questions, the monitoring objectives should be attainable and measurable. In summary, the goals and objectives resulting from this process should guide the expected outcomes of the study.

12.2.1. Legal Requirements and Interagency Agreements

In many cases, the goals and objectives of a monitoring effort may be dictated or limited by legal requirements, mutual agreements, or land use or transportation plans. Legal requirements vary according to each jurisdiction and whether it is designated a federal, state, or local road. There are two different phases in the transportation planning process during which there are opportunities available to require monitoring and adaptive management if the transportation agency has not already addressed these (Sections 6.3.1 and 6.7).

In a reiterative process of developing goals and
objectives, including adaptive management
considerations, potential steps and important
issues to consider include the following for
establishing road effects before construction (steps 1–4) and for monitoring mitigation measures during and after construction
(steps 5–7):

1. Is the goal to improve habitat connectivity or reduce wildlife-vehicle collisions,
   or both?
2. Is the road a barrier to breeding migration
   or dispersal pathways? Is the areal extent of
   the barrier effects localized or broad?
3. What are the impacts to rare and imperiled
   fauna and flora or other species of conservation interest? Of the species present in
   the study area, which are particularly susceptible to the negative effects of roads?
4. What type of mitigation measure might be
   appropriate to alleviate the impacts?
5. Are there structural or site-related issues
   that need correction to ensure or improve
   performance?
6. Has the mitigation measure reduced mortality and increased connectivity?
7. Where are animal crossing attempts still occurring along unmitigated road segments?

As part of the adaptive management process,
if issues are observed in steps 5–7, return to
step 1.

The first opportunity is during the preliminary engineering phase (Section 6.3.1). Under the National Environmental Protection Act (NEPA) and as required by
many other federal and state laws, projects that are either federally funded, or occur on federal lands, require
concurrences from federal and state natural resource
agencies that environmental impacts have been adequately minimized or mitigated. It is important to work
with these agencies to develop written agreements or
documentation (e.g., memoranda of understanding, or

related written communications) that ensure effective
designs and that require a monitoring and adaptive
management program (Chapter 6).

The second opportunity is during the detailed design phase (Section 6.7). Many projects require permits
from the Army Corps of Engineers and state natural
resource agencies for impacts to wetlands and navigable waters (e.g., stream and river crossings). These
requirements usually include avoiding or mitigating
impacts to wetland-dependent wildlife, and the permit applications may be reviewed by the same wildlife
resource agencies that review the NEPA documents as
described above.

12.2.2. The Importance of Collecting Data Pre- and Post-construction

An important aspect of monitoring should be to
evaluate the performance of the mitigation measure
in achieving the goal of the project (e.g., reducing
vehicle-related mortality, restoring habitat connectivity for wildlife populations, and ideally, improving
population viability). To quantitatively assess the performance of the mitigation measure, it is important to
collect pre-construction baseline data for comparison
to post-construction data (Section 12.4). In addition,
collecting data during construction, particularly in the
absence of pre-construction data, can further inform
post-construction analyses.

12.2.3. The Importance of Evaluating Spatial and Temporal Variation

It is important to characterize how road effects vary
among locations (i.e., spatially) and across time (i.e.,
temporally). For many species, habitat use and movement patterns are governed by environmental factors
(e.g., temperature, rainfall, water levels; Chapter 2)
that change by season and by year. As a result, the frequency at which animals encounter roads may vary
significantly over space and time.

12.2.4. What to Do When Funding Is Limited

Funding is a key factor in the types of monitoring methods that can be employed and the frequency with which
data are collected. Goals and objectives must reflect
the available funding and level of effort that can be expected in the monitoring plan. When funding and time

are limited, Rapid Assessments are typically performed (Ruediger and Lloyd 2003; Box 6.2; Section 10.6.1). Rapid Assessments generally include short periods of field data collection and employ pre-existing field data or expert opinion. If the amount of time to collect field data is limited, it should coincide with periods when effects are expected to be greatest (e.g., temperature or rainfall induced species movement). Relevant wildlife data or reports from previous or ongoing studies and existing remote-sensing and Geographic Information System (GIS) data can be used to identify road segments where significant effects can be expected (Chapter 8). This alternative can save money in relation to long-term, labor-intensive field data collection.

12.3. Steps to Develop a Monitoring Program

There are many important considerations in the design of a monitoring plan for determining road impacts and the effectiveness of mitigation measures. These are applicable whether it is a long-term monitoring program involving several roads or a single road, or whether it involves complex solutions involving multiple mitigation measures or simple solutions that require few mitigation measures.

1. *Identify and Coordinate with Stakeholders.* Stakeholders are important in obtaining many types of support for the project, including community (public or private), management, and financial support. Insurance companies may be an important stakeholder where vehicle damage and human injury are an issue. Stakeholders also can provide valuable information on the study location or affected fauna and flora (from previous or current studies) that can be used to help design the monitoring project. Broad inclusion of constituents, even those who may oppose the project, can help increase buy-in; the perception by potentially oppositional constituents of being overlooked can hinder project progress. Thus, striving for inclusivity can avoid conflicts that might otherwise prevent the project from moving forward.

2. *Determining the Scope of the Monitoring Project.* The scope of a monitoring project is defined by both the transportation infrastructure (i.e., the characteristics of the road, the traffic, the right-of-way [ROW], and the mitigation measure) and ecological infrastructure (i.e., the connections among necessary habitat components that are potentially affected by the road). One of the greatest challenges in determining the scope of the monitoring project is the number and complexity of the parameters related to both transportation and ecological infrastructure. For example, considerations for ecological infrastructure could include all biodiversity; certain focal or target species (representative of floral and faunal diversity of the study area); endangered, threatened, or rare species; or a single species of conservation interest (Chapter 10). Because it is difficult to monitor for all faunal diversity, it is typical to establish a target species, or a species list. For an ecosystem- or community-level approach, target species can be grouped according to similar characteristics with respect to likelihood of using an existing or proposed mitigation measure (see also Section 10.3).

3. *Identify the Most Effective Methods for Monitoring the Target Species.* There are many monitoring methods that have been employed in road impact and mitigation evaluation studies. Several books and guides discuss the application of these methods (e.g., Heyer et al. 1994; Wilson et al. 1996; Forman et al. 2003; National Research Council 2005; Manley et al. 2006; Clevenger and Huijser 2011; McDiarmid et al. 2012; Graeter et al. 2013), as do numerous articles in refereed journals (e.g., Enge 2006; Andrews et al. 2008). Examples of these methods are discussed further in Chapter 8 and Section 12.4.3.

4. *Establish Experimental and Control Sites.* Rather than the typical expectation of monitoring a single experimental site (i.e., treatment) alone, selecting two or three additional control (i.e., reference) sites for monitoring can be useful for scientific comparison and analysis; the control sites must be ecologically similar to the experimental site. Comparison to a single reference site can cause problems if the reference site turns out to be different from the treatment site (e.g., in habitat character or population densities).

Although monitoring a single site can provide information about what is occurring right there, monitoring multiple sites can provide a broader picture of whether the mitigation strategy is effective overall, across sites. In addition, where different structures or sites vary in efficacy, comparisons among multiple sites can reveal the factors that explain why some structures work better than others. Finally, where many monitoring locations are necessary to provide the appropriate level of information, or where monitoring encompasses large stretches of road, subsampling sites or structures can provide a more thorough assessment without adding more locations than time and money can support. Most projects typically do not have the necessary resources for the ideal level of monitoring. However, comparative monitoring among multiple sites is increasingly important as we gain an appreciation for the scope and severity of effects of roads on small animals.

5. *Document Site Conditions.* Many external factors can affect the results of a monitoring study. These include physical, biological, and environmental variables, such as topography, soil characteristics, water quality and quantity, vegetation composition, artificial light, and noise (Chapter 9). Relevant factors should be quantified to analyze their effects.

6. *Acquire Pre-existing Data on the Study Site(s).* Based on the identified species and habitat factors, it is important to obtain any valid and available pre-existing data, such as previous scientific reports and records on the species of interest and on the adjacent habitats. It is also necessary to record the specific characteristics of the road, such as features of the ROW, road width, number of lanes, traffic volume, crossing structure type and dimensions, and structure approach features.

7. *Minimize Collection Bias Among Survey Technicians.* Ideally, from a biological standpoint, monitoring should occur over multiple generations of the target species. As such, monitoring programs should be designed so that data can be collected by different surveyors in a standardized manner, thus minimizing bias (Box 12.2). Differences among observers can arise from varying abilities to find animals or signs of animal presence, inconsistent criteria used to identify species, different interpretations of written protocols, or inconsistencies in making judgment calls. In particular, in volunteer monitoring programs, observer differences are known to be a major source of variation (e.g., Sauer et al. 1994). When the data collectors of pre- and post-mitigation datasets are different, the variability among observers can bias the interpretation of monitoring data.

8. *Schedule Appropriate Timelines and Data Collection Efforts.* The extent of road effects on wildlife or the efficacy of a mitigation structure cannot be detected with a single survey during each time period. For this reason, repeat surveys are critical to estimate variation or identify seasonal patterns of activity prior to establishing monitoring protocols. The sampling intensity and duration should increase with increasing levels of natural variability and complexity of the scientific question or the targeted ecosystem. If variation is moderate to high, power analysis (e.g., Gibbs et al. 1998) or computer simulation (e.g., Rhodes and Jonzén 2011) can then be used to determine the optimal number and seasonal timing of surveys. In practice, trade-offs in the actual frequency of data collection can depend as much on resource availability (i.e., both time and money) as on statistical defensibility requirements. Monitoring provides long-term cost savings by identifying adaptive management opportunities that optimize functioning of mitigation measures. Therefore, when resources are insufficient to properly monitor a project, implementation should be delayed until the necessary resources are acquired to maximize cost effectiveness.

9. *Develop a Uniform Data Entry and Management System.* Standards for data entry forms and database management software packages can be designed and programmed to avoid inconsistencies in the data that can occur when multiple people are collecting and entering data. This uniformity improves transferability of data and reduces time needed to prepare data for analysis.

10. *Determine the Personnel and Equipment Needed.* The number and type of personnel needed

- Document every aspect of the monitoring
  process, including the time and spatial scale
  of surveys, what specific data are to be
  collected, what equipment is used, and how
  data are entered and processed. Protocols
  are adequate if someone else can follow
  the instructions and duplicate the data of
  the initial surveyor. Having two or more
  individuals test written protocols is useful
  for validating the protocols and identifying
  aspects that might be vague or confusing.
- Whenever possible, use previously established protocols for surveying a particular
  taxon. Protocols should include a process
  for qualifying surveyor abilities specific
  to target species. Several standard protocols for monitoring small animals have
  been published (e.g., Heyer et al. 1994;
  Wilson et al. 1996; McDiarmid et al. 2012;
  Graeter et al. 2013).
- Repeat testing of survey protocols (with
  the same or multiple observers) is useful
  for measuring the extent of variation to be
  expected from a survey technique. Taking
  the time to compare multiple survey techniques in the initial stages of monitoring
  can enhance long-term efficiency and data
  quality.

and amount of training and oversight required
depends on the tasks required and the qualifica-
tions and experience of available recruits. Along
with personnel and administration needs, the
costs of equipment and supplies should be projected for the duration of the study.

11. *Develop a Budget.* Reconciling the budget to meet
    the above needs within the resources avail-
    able may require scaling back some monitoring
    activities by prioritizing data collection efforts,
    as discussed above. However, also as discussed
    above, for maximum cost effectiveness, addi-
    tional resources should be acquired to satisfy the
    objectives of the monitoring plan.

12. *Build Flexibility and Adaptability into the Budget.*
    Many unexpected circumstances can occur
    over the course of a monitoring project. For this
    reason, it is important to build flexibility and
    adaptation into the budget to address any uncer-
    tainties that may occur during the life cycle of
    the project. Some examples include unplanned
    road construction and maintenance activities,
    extreme weather events, or human impacts (e.g.,
    vandalism, theft). When these issues arise, hav-
    ing included contingency funds into the budget
    at the outset will greatly improve opportunities
    to address these issues and thus reduce their
    cumulative impacts.

12.4. Pre- and Post-mitigation Monitoring

In order to conduct evaluations of the performance of
mitigation measures, pre-mitigation data must be collected to establish a baseline for conditions prior to the
mitigation. These pre-mitigation data in turn allow for
the quantification of changes in road effects after mitigation. However, for a true, scientifically defensible
evaluation, it is imperative to standardize data collection protocols and methods used in both monitoring
phases. Further, in many situations, anthropogenic
effects or alterations to ecological infrastructure (e.g.,
land use changes) have already occurred before the road
is built or widened. Without pre-mitigation monitoring, and with only post-mitigation monitoring, certain
effects may be attributed to the road, even though the
existing alterations to the landscape may have already
been contributing to impacts before new construction
occurred. Additionally, collecting data during construction, particularly in the absence of pre-construction
data, may help to address some issues adaptively before
construction is complete (Section 11.2).

12.4.1. Importance of Pre-mitigation Monitoring

Pre-mitigation monitoring is key to identifying which
issues may benefit from mitigation (what problems
are observed that can be fixed, and what other problems already exist that cannot be mitigated). Without
pre-mitigation monitoring, it may not be possible to
fully assess the effectiveness of the mitigation measure. In any case, it is important to have realistic goals

that relate to the specific conditions and objectives of the project. For example, in the instance of widening a rural, two-lane road to a four-lane road with a median, traffic density or speeds may increase, and then wildlife avoidance behavior may be more likely. Avoidance will be much more difficult, and potentially impossible, to mitigate (Section 4.6.1). When an animal exhibits deterrence behavior as a result of the road and associated traffic, it is quite possible that culverts or fencing will not be enough to encourage that animal to cross. In general, without an understanding of the pre-mitigation conditions, it may not be possible to ascertain what measure is needed to effectively mitigate the problem. In the instance where a new road is planned, pre-mitigation monitoring can provide valuable information on resident species, populations, and movements, which can help identify appropriate mitigation measures and specific design features related to the road project that can ensure ecological infrastructure will be maintained.

12.4.1.1. Case Study: Using Pre-Construction Monitoring to Guide the Placement and Design of Mitigation Measures along a North Carolina Highway

Daniel J. Smith

During the summer months, the Outer Banks of coastal North Carolina, United States, is a primary tourist destination for residents of the Mid-Atlantic states and is also a frequent site for landfall of hurricanes. A primary transportation corridor to and from the Outer Banks is US Highway 64. This road represents one of two coastal evacuation routes and has been targeted for widening from two to four lanes to improve traffic flow during emergencies. The road bisects a network of federal, state, and privately managed conservation areas that provide habitat for black bears (*Ursus americanus*), red wolves (*Canis rufus*, a federally endangered species), migratory birds, a diverse assemblage of herpetofauna, and numerous other species.

Comprehensive wildlife surveys were conducted from April 2009 to July 2010 to assess potential impacts and to make recommendations for crossing structures and other measures to reduce adverse effects along a 21 km section of the road proposed for widening (Smith 2012). Surveys consisted of road cruising (driving surveys to record all vertebrate species; three times per week), track beds (walking surveys within the ROW to record tracks in sand bed of medium to large vertebrates, including mammals, turtles, snakes, turkeys; twice a week), and camera traps (images to record all medium to large vertebrates; checked weekly [Vaughan et al. 2011]). This use of complementary techniques provided data on successful and unsuccessful road crossings. To determine species presence and potential road avoidance, mark-recapture studies for small mammals and herpetofauna were performed at roadside (n = 7 arrays, within ROW) and control (n = 4 arrays, at least 500 m from the road) locations in differing habitat types adjacent to the highway corridor. Specifically, mark-recapture arrays consisted of 30 m silt fences with 4 screen funnel traps, 6 pitfall traps on control sites, and 3 pitfall traps (habitat side only) on roadside sites. All were checked daily for three consecutive days per week. Lastly, existing telemetry data were available on red wolf movements in proximity to the road (Vaughan et al. 2011).

Roadkill surveys produced 27,877 individuals of 113 species. From 31 track bed stations, 18 different species or taxa were recorded from 7,477 sets of animal tracks. From the trapping arrays, 855 individuals were captured in control arrays, including 13 amphibian species (720 amphibians), while the roadside traps captured 909 individuals including 19 amphibian species (767 amphibians). Controls produced 180 captures per trap and roadsides produced 110 captures per trap, so on a per-trap basis, the ratio of total captures for control to roadside sites was about 1:1.1 for frogs and 1.4:1 for salamanders. For reptiles, 16 species were detected in both control and roadside sites, and the abundance (number of captures) was also similar: 78 at control sites and 79 at roadsides. This included more lizards at control sites (55 vs. 34 at roadsides), but more snakes and turtles at the roadside sites (24 snakes at roadsides vs. 15 at controls; 21 turtles at roadsides vs. 8 at controls). For small mammals, we also captured 9 mice and rats and 47 shrews and moles at control sites and 31 mice and rats and 33 shrews and moles at roadside arrays.

Spatial analysis of field data revealed 15 areas (ranging in road length from 100 m to 2,800 m) of significant wildlife activity. Results of field surveys and landscape analysis were used to determine candidate locations, structure types, and design specifications for wildlife crossings based on site characteristics and target species requirements. Specifically, recommendations included construction of 8 large animal crossing structures and 13 small animal crossing structures. Planning for road improvements for US Highway 64 also included the analysis of potential ecological effects of projected sea-level rise based on recent climate change models. As such, the improvements will include elevation of the road, which, along with the planned crossing structures, will allow for movement of species assemblages as the area transitions from freshwater wetlands to estuarine systems. This project was successful due to active participation and review by a diverse group of stakeholders representing multiple levels of government, private industry, and environmental advocates. It serves as an excellent example of comprehensive monitoring and successful integration of long-term transportation and conservation planning.

12.4.2. When Pre-mitigation Monitoring Is Not Possible

While post-construction monitoring may have been included in the project budget, rarely are pre-construction data collected for comparison. In the absence of these data, the focus of analysis shifts to the function of the mitigation measures over time, and in relation to areas without mitigation. For example, this may include trends in the use of a particular wildlife crossing over time.

12.4.3. How to Implement Pre- and Post-mitigation Monitoring

When considering monitoring for a transportation project, it is important to know the goals of the mitigation efforts. Monitoring for both pre- and post-mitigation should be designed as one project or program that addresses a particular question or goal, both before and after construction. For example, if the goal is reduction of road mortality, the monitoring protocols and methods should fit this goal. If the goal is to increase habitat connectivity, the protocols should address this goal. Ideally, the mitigation should address both of these goals.

12.4.3.1. Evaluating Reductions in Road Mortality
Of all the effects of roads on small animals, road mortality is the easiest to monitor. On-road data may in-

clude location and date of observations of both alive (after being hit by a vehicle) and dead animals, with the combined total being the number of wildlife-vehicle collisions (WVCs), as well as identification of species (or species grouping), age class, sex, direction of movement, and behavior. In pre-mitigation monitoring, tallies of road mortality can assist in first recognizing the problem and then finding ways to provide mitigation solutions. For example, road mortality numbers for a particular species that are higher than expected may justify and secure funding from transportation agencies to provide mitigation.

While some transportation agencies collect ongoing WVC data for large animals that are a safety hazard to motorists, data collection for smaller animals is typically not standard. Encounters of small animals on roads are recorded either through systematic surveys conducted by biologists (e.g., Langen et al. 2007; Section 8.2.1.4), or opportunistically in personal or organized citizen science databases (Chapter 5). Road mortality surveys for smaller animals, systematic or opportunistic, may be mandated if a certain species represents a safety hazard or is on a state or federal threatened or endangered species list, and is also heavily impacted by roads.

It is important to note that if a particular species is not found during road wildlife surveys, it does not necessarily mean it is absent from the area. It may not have been found because the road survey was limited in scope, the species in question may avoid use of roads and open road and roadside habitats, or a species that was once common in the area is now rare. The last situation is common where significant landscape alterations and habitat fragmentation has occurred. Historical records (e.g., species occurrences, land cover) can be used to help ascertain which species were, and still may be, present in an area.

At the most basic level, collecting road mortality data simply consists of identifying and counting dead animals. As long as counts are standardized (Sections 8.2 and 12.3), raw counts of dead animals can serve as the basis for evaluating the effects of mitigation on mortality. A typical procedure for collecting mortality counts is to walk relevant stretches of road and count dead animals on the pavement and the road shoulder. Numbers of dead animals per unit time and per unit of road length for each species can then be compared pre- and post-mitigation. Alternatively, if pre-mitigation road mortality data are not available, mortality counts can be compared between mitigation (i.e., treatment) sites and reference (i.e., control) sites.

As animal movement generally depends on seasonal and interannual changes in temperature and precipitation (Chapter 2), mortality counts can vary over the course of any given year, challenging the detection of real effects of mitigation on mortality. Ensuring that surveys are carried out during the same time of year and in similar weather conditions can help reduce this variability (Section 8.2.1) and make mitigation efficacy easier to quantify. Surveying at optimal times requires some background knowledge on when different kinds of species are most likely to move, and therefore most likely to be killed while moving (Chapter 2). In addition to treating seasonal and weather conditions consistently among years, effective monitoring will usually require repeat surveys within each seasonal movement period (e.g., breeding, juvenile dispersal). The required number of surveys will depend on variability in counts from one survey to the next (i.e., greater variability will require more surveys). In most cases, a minimum of three to five surveys per seasonal movement period will provide appropriate data to assess how counts are changing over time.

Roadkill counts are potentially useful as an index of relative mortality, but they should never be taken as absolute estimates of mortality, even for the specific days of a survey (Section 3.2.1). Small animal counts are generally underestimated because carcasses may be rapidly removed by scavengers or knocked off the road by vehicles, and because injured animals may survive long enough to move off the road before dying (Guinard et al., 2015). For example, Hels and Buchwald (2001) concluded that standard foot surveys would only locate 7–67% of dead amphibians; this wide range of variation can be attributed to taxonomic differences in removal rates. Further, it has been speculated that predation rates may increase near mitigation features where large numbers of animals are directed along or toward a specific path. To our knowledge, this has not been investigated adequately to determine whether the predation rates are increased at all, or are substantial enough to counteract the benefit of a given mitigation feature (e.g., raccoons depredating turtle nests, Crawford et al. 2014). Nevertheless, predation may also be

an important consideration for both pre-mitigation design as well as post-mitigation management. Finally, it is important to keep in mind that depending on the mitigation features employed, road mortality data may only tell part of the story. For example, the simplest method to evaluate effectiveness of a barrier fence may be to conduct roadkill surveys within the fenced area. However, these data may only reveal *if* animals can trespass the fence, not *how* they trespass the fence. Techniques provided in Section 12.4.3.4 may help answer the question of how the barrier fences are being breached, and thus, what options may exist to correct this problem.

12.4.3.2. Extrapolating Roadkill Counts to Population-Level Effects

Given the inherent biases in roadkill counts of individual animals, data from these surveys can provide a relative estimate of mortality rates. The ideal way to assess this relative change in mortality and the effectiveness of the mitigation feature is to compare post-mitigation mortality levels to pre-mitigation levels. However, pre-mitigation data are often not available and roadkill data are the only sources of information for modeling population-level effects. Studies that simply measure the number of successful crossings by wildlife or the change in wildlife mortality will not adequately assess whether the mitigation measure has resulted in a change in population abundance or persistence (van der Ree et al. 2007). Tallies of post-mitigation wildlife mortality will not provide information on the impact, whether positive or negative, to the population. Whether modeling population-level effects before or after construction, an understanding of both the natural history of the animal (e.g., age at sexual maturity, adult survival, longevity) and their detection rates should be incorporated into models to determine whether the post-mitigation mortality levels are sustainable for the population. Ideally, the pre-mitigation population size or density would also be available for inclusion in the assessment of post-construction effectiveness.

There are two measures of population size that may be used when evaluating the effectiveness of the mitigation measure. The first is the biological population, and the second is the group of individuals within the impact area of the road and mitigation measure. In the first instance of biological populations, the critical parameter is typically either population size (N) or change in population size from one year to the next (λ). In this case, the goal is to stabilize population size, or to establish annual growth in the population (i.e., λ > 1) where reductions are known to have occurred. Further, for species regulated by metapopulation dynamics (Section 8.3), the goal may be to maintain natural movements (immigration and emigration) among separate subpopulations. That being said, given the scope, time, and resources available for a monitoring project, it can be difficult or nearly impossible to quantify the biological population size accurately.

For the second measure, the spatial extent of the population to be assessed should reflect the impact area of the mitigation measures. As such, for fully aquatic species (e.g., fish or some amphibians), aquatic habitats provide a defined space in which to measure the efficacy of mitigation measures. For semi-aquatic and terrestrial species, the spatial extent of the population should be based on the species' maximum dispersal capabilities. For most small animals, the spatial extent is likely between a few hundred meters and a few kilometers. Frogs, salamanders, small snakes, and small mammals may have movement distances toward the low end of this range, whereas medium to large snakes, turtles, and medium-sized mammals may exhibit longer movements. Despite the vast amount of research that occurs disproportionately in North America, specific information on maximum dispersal distances is still not available for many species in published scientific journal literature.

12.4.3.3. Evaluating Improvements to Habitat Connectivity

Measuring and monitoring movement can be more difficult than assessing mortality, but assessing movement is still feasible. These types of assessments require documentation of crossing attempts, crossing success, and avoidance rates. Certain species may avoid roads completely or may be unable to successfully cross roads in the absence of suitable crossing structures (Chapters 2 and 4). Thus, determining the efficacy of a crossing structure requires measuring the change in animal crossing rates after the mitigation measure is implemented (e.g., van der Grift and Schippers 2013). Whereas individual-level changes in movements across the road can sometimes be challenging

to assess, monitoring population-level contributions of these individual movements may border on impossible. As small numbers of long-distance movements may be sufficient to recolonize extirpated sites and maintain genetic variation, it is often impossible to differentiate between low frequency movements and no movements at all. Coupling multiple sampling techniques can assist in elucidating how individual behaviors and migration rates across roads influence population genetics (Riley et al. 2006). Ultimately, recolonization or restoration of genetic variation can result from effective mitigation measures, but these changes may require decades to become apparent or may be difficult to quantify experimentally. Techniques exist for retrospective population-level assessments (Hels and Nachmann 2002; Keller and Largiadér 2003; Becker et al. 2007). However, as with mortality assessments, the ideal way to assess this relative change in connectivity and effectiveness of the mitigation feature is to compare post-mitigation crossing rates to pre-mitigation rates (see also Box 12.3). Where pre-mitigation data are not available, crossing rates can be compared between the mitigation site (i.e., treatment) and a similar site or sites where no mitigation was implemented (i.e., control[s], Section 12.3).

12.4.3.4. Techniques to Monitor the Effectiveness of Mitigation Measures

There are several active and passive monitoring techniques that are appropriate for determining levels of use or effectiveness of mitigation measures (see Section 8.2.1). Active methods are more time consuming and labor intensive, but yield finer-scale spatial and temporal data. Due to costs and time investments, active approaches may not be feasible for all monitoring projects. As such, other passive and less labor-intensive monitoring techniques may be more applicable. Regardless, for both active and passive methodologies, the resulting data are most useful when comparisons can be made between pre- and post-mitigation data or among multiple locations (i.e., treatment vs. control). Further, monitoring the mitigation measures and collecting data, both during and after construction, can also identify deficiencies (e.g., insufficient height in fencing, avoidance of a type of crossing structure by certain species found in the adjacent habitat) that may be addressed with adaptive management (Sections 11.2.1

Box 12.3. Before-After-Control-Impact (BACI) Analysis

Before-After-Control-Impact (BACI) designs are particularly effective for answering questions related to mitigation effectiveness (Roedenbeck et al. 2007). In concept, this design collects data before (B) and after (A) the mitigation measures are in place at both unaffected control sites (C) and the treatment, or impact (I) sites (i.e., post-mitigation sites in this example) for purposes of comparison. If the differences between control and impact sites are greater after the impact than before, it suggests that the observed change is a result of the intervention measure (Roedenbeck et al. 2007), provided that additional variables are also measured to properly account for variation in nontarget ecological effects. On the other hand, if the differences between control and impact sites remain the same after the intervention (e.g., installing mitigation measures), it would imply that any differences between sites are independent of the intervention. Before and after measurements at impact sites alone, or similarly, measurements at control and impact sites only after the intervention, cannot properly assess the changes associated with that intervention. Further, if the goal of the monitoring effort is to assess changes in population abundance or persistence, studies must incorporate information such as detection rates, survival, reproduction, and population size or density as part of the design; if not, no amount of BACI studies will provide population-level information.

Limited resources make BACI designs atypical in most transportation projects. However, where feasible, these designs can be immensely useful. As in scenarios where pre-mitigation data are not available or such collection is not possible (Section 12.4.2), the alternative to a BACI design is to monitor the function and performance of the mitigation measure over multiple years.

Essential Elements of a BACI Design

Collaboration and Commitment among Stakeholders. A BACI design requires a proactive and engaged long-term collaboration and commitment between researchers and transportation agencies for several reasons. Monitoring needs to be conducted prior (ideally several years) to the road construction phase to gather information before the impact; thus, there needs to be a multi-year commitment of resources. Extensive collaboration is also important because multiple sites need to be monitored in order to have both control and impact sites, which may often involve different landowners and sites away from the road. Educating stakeholders about the significant value that is added with a BACI design can help with generating the time and resources required.

Replication. For any study design including BACI, replication of both control and impact sites is extremely valuable for determining the amount of variation among sites. This amount of variation can also be minimized by controlling for confounding factors related to spatial variation among sites, such as differences in vegetation or topography. For example, if an experiment tests what type of crossing structure (e.g., concrete box or metal culvert) is more effective, a minimum number of replicates of each treatment should be determined (ideally at least three), but the number of replications should be based on an a priori statistical assessment. In order to properly replicate and compare impact sites, transportation agencies and other stakeholders must agree to construct different passage types at certain locations, even though not originally planned. Agencies are more easily convinced if the mitigation treatment types selected meet the minimal safety and hydraulic components required at all of the selected sites.

Advantages of a BACI Design for Small Animals

A BACI study design, specifically using control and impact sites, helps to manage unwanted effects caused by temporal and spatial variability, which is particularly valuable for small animals since many taxa (e.g., anurans) can exhibit dramatic interannual variation in abundance. Further, from a cost perspective, monitoring designs that include multiple sites, particularly multiple treatment sites, can be less expensive for smaller animals because applicable structures may be smaller and length of fencing may be shorter.

and 12.7). Certain techniques described below may be more appropriate for evaluating the effectiveness of a structure as opposed to other mitigation measures (e.g., barrier fencing). In some cases, simply inspecting the mitigation areas periodically and walking around the mitigation measure can reveal areas for improvement. For example, walking along a barrier or guide fence may reveal that animals are burrowing underneath it, which may necessitate the use of different fencing material, increasing the depth that the fence is buried, adjusting the position of the fence, or other modification.

DRIFT FENCES AND TRAPS. Drift fences with funnel and pitfall traps provide a low-tech approach to estimating road-crossing rates, and also provide insights on the directionality of movement and behavioral responses, such as avoidance of the road or a particular mitigation measure (see also Chapter 8). Arrays of drift fences with funnel and pitfall traps are commonly used in many areas of small animal biology (Enge 1997; Farallo et al. 2010), but certain designs are more applicable to estimating the barrier effect of roads (e.g., Gibbs 1998; Smith 2006, 2011). Drift fences (e.g., aluminum flashing, plastic woven silt fence) are sunk into the ground to block or direct animal movement such that animals moving along fences will fall into pitfalls or crawl into funnels where they can be counted, measured, and then released (e.g., Willson and Gibbons 2009). The combined use of funnel and pitfall traps addresses the individual shortcomings of each trap type in capturing different types of species (Enge 1997; Farallo et al. 2010). Drift fence and trap configurations should include an appropriate number of traps (see Willson and Gibbons 2009), with equal numbers of drift fences in treatment and control habitat areas in order to compare diversity and abundance (requires mark-recapture techniques; see below).

Placement and orientation of the arrays should be based on the size of the road and the amount of traffic. For example, on rural roads with very low traffic volumes and speeds, where surveyor safety is less of a concern, drift fences with funnel and pitfall traps should be placed within the ROW, near the road (e.g., within a few meters) to detect road crossing attempts (Figure 12.1a). Ideally, these drift fences should be set up in three ways:

1. Parallel to the road with traps on the habitat side of the fence to count animals moving toward, and potentially crossing, the road.
2. Parallel to the road with traps on the same side as the road to count animals that have already crossed the road.
3. In reference areas (located at a sufficient distance so that the target species is not influenced by the road) of habitat similar to that on either side of the road. These distances are based on movement data of target species, and this setup can help to produce a baseline of the number of captures expected independent of road effects.

On roads with higher traffic volumes and speeds, where surveyor safety can be a significant concern, drift fences and traps should be placed at the outer edge of the ROW (e.g., 10 m or more from the road). Because of the distance from the road, animals may move in irregular directions; this increases the uncertainty that an animal has actually crossed the road. In this case, employing mark-recapture techniques (see below) in addition to the arrays can identify animals that have crossed the road. These drift fences should be set up in two ways (Figure 12.1b):

1. Parallel to the road and directly across from each other; animals captured on the habitat side are released on the road side of the drift fence, and vice-versa.
2. In reference sites as described in 3 above.

MARK-RECAPTURE. Mark-recapture techniques, such as passive integrated transponder (PIT) tagging, can be employed separately or in conjunction with drift fences and traps. Either individual or batch marking can be beneficial to distinguish animals crossing the road from those that may have entered the road and turned back. Marking animals on the side of the road where they were first encountered allows for the measurement of both barrier effects (captures on the same side of the road, which may also indicate a behavioral avoidance response) and crossing rates (captures on the opposite side of the road), but will require daily sampling. PIT tagging is also a useful technique for evaluating crossing structure use; for example, tag readers have been built into passage structures to automatically monitor individual movement through fish passages (e.g.,

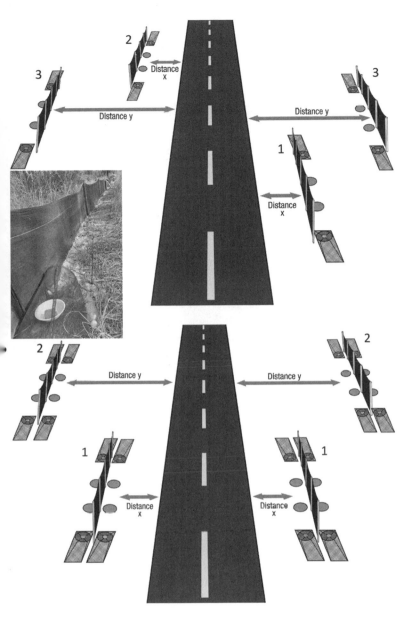

Figure 12.1. Two drift fence and trap designs (inset) for detecting road crossing attempts by small animals. (The inset applies to both configurations.) The configuration in (a) is for roads with very low traffic volume and speed (distance x is less than 3 m, distance y depends on the target species). The configuration in (b) is for roads with higher traffic volume and speed (distance x is 10 m or more, distance y depends on the target species). See 12.4.3.4 for additional detail on these configurations. *Credit: Daniel J. Smith (inset) and © Victor Young Illustration, adapted from conceptual drawings by David M. Marsh and Daniel J. Smith.*

Castro-Santos et al. 1996) and through culverts for desert tortoises (*Gopherus agassizii*; Boarman et al. 1998).

RADIOTELEMETRY. Radiotelemetry and Global Positioning System (GPS) tracking are frequently used where the monitoring goal is to quantify and compare movements surrounding the road both pre- and post-mitigation. However, these techniques can be time consuming and current options may be limited for some smaller animals due to size and weight of equipment, although microtechnology is rapidly progressing. In addition, radiotelemetry is inherently limited to small numbers of animals, as compared to marking techniques such as PIT tagging. However, radiotelemetry can provide valuable data on specific movement paths, which can allow for the differentiation between road crossing and road avoidance. For example, Shepard et al. (2008) demonstrated that massasauga rattlesnakes (*Sistrurus catenatus*) and two species of box turtles (*Terrapene* spp.) consistently altered their movement paths to avoid crossing roads in wetland habitats in Illinois. Lastly, direct observation of individual animals is possible with active monitoring via telemetry and GPS versus passive monitoring via PIT tagging.

REMOTE CAMERAS. Cameras can record movement of animals whether through a crossing structure or along directional or barrier fencing (e.g., Pagnucco et al. 2011, 2012). Digital video or still cameras (usually motion- and thermal infrared-activated) are generally accurate, but require greater up-front costs. However, there are cost savings in time and personnel because cameras allow remote monitoring and can be deployed specifically during periods of peak movement. In addition, camera images or recordings can provide valuable information on species behavior when encountering the ROW and the crossing structure, including reactions to structure size or shape, to different approach area characteristics, or to noise and vibrations. These findings can inform the need for modifications to structures and approach areas in order to improve their function for species that exhibit avoidance behavior (see additional discussion on behavioral responses to roads in Chapter 2).

TRACK SURVEYS. Track surveys allow for the identification of species or species groups moving through a specified area by detecting their footprints, tail tracks, and so forth. These surveys can provide data on multiple taxa simultaneously, along with assessments of relative movement frequency. Use of strategically placed track stations can enhance track surveys; these typically involve the placement of substrate at a particular area of interest (e.g., gypsum powder for small mammals, sand, or clay) or clearing of the vegetation and soil to facilitate the visibility of animal tracks. Track survey locations should be based on a specific question. For example, to answer whether animals are evading a barrier fence, placement of a track station at the ends of the fence, with or without a camera, can reveal whether animals are getting past it. Cameras used in addition to track stations or surveys can show how animals are circumventing the fence.

12.5. Benefits of Long-Term Monitoring

Long-term monitoring is valuable for any effort to understand ecological systems, especially for small animals such as amphibians and reptiles whose habitat use and movement patterns are seasonally and interannually variable (Bernardino and Dalrymple 1992; Smith 2006; Chapter 2). Most monitoring efforts tend

to be short term (i.e., one to two years) and these can be beneficial for assessing short-term effectiveness of a mitigation measure such as reduction of roadkill or improvements in connectivity. However, in order to assess the response to a mitigation measure by an assemblage of species or populations, long-term monitoring is necessary.

Long-term monitoring generally means multiple years (at least five, ideally many more) to discern differences in animal movement and population dynamics that may vary with temperature, rainfall, or generations. As an example, if monitoring is conducted during a single drought year, it could potentially omit significant movements by amphibian populations present in the study area (Smith 2006). The benefits of long-term monitoring apply equally to pre- and post-mitigation. Multiple years of pre-mitigation monitoring provide both a much better picture of current conditions and a better foundation for comparison to post-mitigation conditions. Post-mitigation monitoring should include measuring animal use of mitigation structures and should continue long enough to establish whether the mitigation measures have been effective. This monitoring addresses not only changes in animal use over time but also changes in the public infrastructure (e.g., traffic volume and speed, structural integrity and desired functionality of mitigation features). Monitoring for multiple years allows for improvements to be implemented as problems arise; this is adaptive management (Section 12.7), which greatly increases cost effectiveness, minimizes planning and oversight burdens, and reduces the likelihood of high visibility problems. Long-term monitoring of wildlife crossings can improve understanding of different responses to roads and crossing structures by various species and can increase the effectiveness of wildlife crossing designs (Sections 12.5 and 12.6).

12.6. Long-Term Monitoring: Lessons from Banff National Park, Alberta, Canada

The longest, continually running monitoring program (1996 to present) that assesses the performance of mitigation measures designed to address highway impacts on wildlife is associated with the Trans-Canada Highway (TCH) in Banff National Park, Alberta, Canada (Clevenger et al. 2009; Clevenger 2012). The program's intent is to specifically monitor and evaluate solutions designed to address the conflicts between highways (and traffic), wildlife conservation, and habitat connectivity. The research conducted on the wildlife crossings in Banff, a model of worldwide importance, has generated an enormous amount of information, including over 50 scientific papers, books, and reports (Clevenger 2012).

The TCH is the major transportation corridor that runs through the Bow River Valley west of Calgary, Alberta, and bisects Banff and Yoho National Parks, extending 76 km east of the park's western boundary at the Alberta–British Columbia border. Safety concerns first identified in the 1970s resulted in widening the TCH from two to four lanes (Clevenger 2012). The area contains significant populations of large ungulates and carnivores, such as elk (*Cervus elaphus*), moose (*Alces alces*), deer (*Odocoileus* spp.), grizzly (*Ursus arctos*) and black bears (*U. americanus*), and wolves (*Canis lupus*). Upgrades included exclusionary wildlife fencing for large animals and 54 wildlife crossing structures including 8 overpasses (Clevenger 2012).

The approach to monitoring and research in Banff has focused on practical applications in transportation and environmental management. Key contributions (from Clevenger 2012) that have advanced the discipline of road ecology and aided transportation agencies in planning and designing effective highway mitigation measures include the following findings:

- There is an acclimation period for wildlife to adapt to using wildlife crossings; therefore, monitoring should be long-term (Clevenger et al. 2009).
- Roadkills and successful road crossings by wildlife do not always occur in the same location over time (Clevenger, Chruszcz, et al. 2002).
- Preferences for wildlife crossing structure types and designs are species-specific (Clevenger and Waltho 2000, 2005; Clevenger et al. 2009).
- Human presence at wildlife crossings deters wildlife use (Clevenger and Waltho 2000).
- Wildlife crossing structures and fencing significantly reduce WVCs; in Banff, for example, WVCs were reduced by >80% for all large mammal species and 94% for ungulate species (Clevenger et al. 2001*a*).

- Crossing structures have benefitted grizzly and black bear populations (Sawaya and Clevenger 2010).
- Cameras are as effective as track pads in monitoring crossing structures for large mammal use, and are more cost effective over the long term (Ford et al. 2009).
- High levels of wildlife use were recorded at Banff crossing structures (Clevenger, Chruszcz, et al. 2002; Clevenger et al. 2009); as of April 2010, this tally included 220,000 detections of 11 species of large mammals (Clevenger 2012).
- The Banff crossing structures did not act as traps for prey of large mammalian predators (Ford and Clevenger 2010) and few documented cases exist of crossing structures serving as prey traps elsewhere for these taxa (Little et al. 2002).
- Research in Banff has shown that drainage culverts can act as habitat linkages for small and medium-sized mammals (Clevenger et al. 2001*b*).
- Guidelines for planning and measuring performance of wildlife crossings should include a wide range of temporal and spatial considerations and ecological goals (e.g., Clevenger 2005; Clevenger and Waltho 2003).
- Models that simulate wildlife movements based on empirical data (Clevenger and Wierzchowski 2006) or expert opinion (Clevenger, Wierzchowski, et al. 2002) can be used successfully for determining placement of mitigation structures.

The duration of the TCH monitoring program and the large number and types of mitigation measures located in a study area with abundant baseline ecological data has placed Banff on the leading edge of road ecology research (Clevenger 2012). Though the program's focus was primarily on large mammals, this research demonstrates the benefits of long-term monitoring, and represents a model approach equally applicable to small animals.

12.7. Adaptive Management

Adaptive management in ecology was first described by Holling (1978). The basic concept is simple; some kind of management action is taken, the results are monitored, and then the management is re-evaluated based on the results of the monitoring. An adaptive management approach balances management requirements with the need to improve our knowledge of the function of the system, which leads to better-informed and more effective management over the long term (e.g., McCarthy and Possingham 2007). Combined with science-based monitoring, an adaptive management strategy reduces the risk and uncertainty inherent with restoration (or mitigation) projects from the outset by encouraging flexibility and informed improvement of management plans, maintenance schedules, and design enhancements. Effective adaptive management can represent an extremely valuable marriage of science and conservation, because the results of learning through experimentation can be implemented directly to realize the expected benefit for management and conservation (e.g., Walters 1986). In addition, because the life cycle of a road project includes several phases (early planning and engineering, design, construction, and maintenance; Chapters 6 and 11), there are multiple opportunities to learn, adjust, and adapt from phase to phase.

12.7.1. The Application of Adaptive Management

Adaptive management stresses the need to adjust and change as research and monitoring provide new information about the performance (effect reduction) of the mitigation feature (e.g., Williams 2011). Two guiding principles are generally acknowledged: (1) there is uncertainty associated with ecosystem-level restoration (e.g., reducing wildlife mortality or restoring habitat connectivity); and (2) the most effective way to reduce uncertainties about how natural systems (i.e., ecological infrastructure) will respond to management actions (e.g., installation of wildlife crossings, manipulation of approach areas, addition of barrier or guide fencing, or other mitigation measures) is by learning directly from the implementation of these management actions (CERP-AMS 2006).

12.7.2. Developing an Adaptive Management Strategy

12.7.2.1. Overview of the Comprehensive Everglades Restoration Plan Adaptive Management Strategy

The central theme of this approach, "learning by doing," is to have planning and design integrated with on-going monitoring, assessment, and evaluation (CERP-AMS 2006). Conceptually, the adaptive environmental assessment process included three fundamental elements that functioned in a feedback loop: (1) addressing uncertainties by testing hypotheses; (2) linking science to decision making; and (3) adjusting implementation, as necessary, to improve the probability of success (RECOVER 2011). The general framework can be translated as the following steps in a monitoring program:

- Testing hypotheses = monitoring the mitigation measures.
- Linking science to decision making = making recommendations based on the results and analysis of the monitoring.
- Adjusting implementation = modifying the mitigation measure to improve performance.

12.7.2.2. Integrating Adaptive Management into Project Development

Throughout this book, we have stressed the importance of the various elements that comprise adaptive management in the context of roads. Specifically, we have presented the importance of planning wildlife requirements early on (Chapter 6) to ensure they are funded (Chapter 7), and to establish objectives and a solid plan for where (Chapter 8) and what kind of mitigation measures (Chapter 9) or retrofits (Chapter 10) should be constructed and maintained (Chapter 11). Upon implementation, these mitigation measures must be effectively monitored using a sound study design, and the results should be used to determine what adaptations are needed (this chapter). In general, much of the monitoring related to the mitigation measure may be accomplished with ongoing maintenance (Chapter 11), though some issues may require specific focus on species use or avoidance. The importance of wildlife connectivity should be identified in the preliminary engineering phase (Section 6.3.1), even though the preliminary design features and project cost estimates are not developed until the detailed design phase. In this preliminary engineering phase, the explicit recommendations and commitments (whether binding through NEPA or other legal processes or voluntary; Section 6.3.1) should specify a monitoring program in detail. Funding for the monitoring should be included

Box 12.4. Funding Mechanisms: An Example from Maryland

In Maryland (and some other states), the budget for road projects includes money for monitoring the effectiveness of environmental compliance after construction. As the design process ends, this money is transferred to an account established to fund environmental monitoring projects after road construction is complete. The project for monitoring impacts on wildlife can be differentiated from other projects in the overall environmental monitoring account.

The Maryland State Highway Administration usually establishes a five-year period after construction to monitor design features related to environmental compliance in order to determine their effectiveness in meeting goals. Monitoring is performed by wildlife experts under the direction of the highway design office responsible for regulatory permit coordination. When monitoring results in recommendations to perform needed maintenance, an account is designated to provide funding for repairs or enhancements for that specific project.

Unfortunately, the funding that Maryland requires for adaptive management as part of the road project typically is not provided in other states. Monitoring and needed retrofits then must be funded by other sources in the transportation agency's budget—either as separate projects in the work program (like the retrofit project described in Section 10.8.3) or from general environmental funds. Additional information on funding and other assistance can be found from project partnering as described in Section 12.7.3.1 and from the resources described in Chapter 7.

in the preliminary cost estimate, as in the Maryland example described in Box 12.4. In addition, the Environmental Assessment (EA) process is essentially a form of adaptive management, and thus provides another way to integrate it into a project.

12.7.3. State Transportation Agency Practices

In addition to Maryland's monitoring strategy (Box 12.4), other state transportation agencies have implemented progressive policies and systematic programs to voluntarily evaluate and improve ecological infrastructure across existing roads.

- Arizona State DOT was performing a statewide inventory and evaluation for existing structures having the potential for being modified for use by large wildlife. For structures in areas not scheduled for major upgrades, the agency planned low-cost retrofits, such as modifying existing fencing to exclude elk from roads, and using natural topography to allow animals trapped within the fenced ROW to exit into adjacent habitat instead of constructing specialized escape devices (e.g., Figure 12.2; Eilerts and Nordhaugen 2010).

- As a result of an executive order, "Protections and Connections for High Quality Natural Habitats," Washington State DOT (WSDOT) has completed cost estimates for nine proposed retrofit projects. The project locations were identified primarily from deer and elk carcass removal data, and from least-cost models for rare forest carnivores (e.g., fisher, *Martes pennanti*, and marten, *Martes americana*). The projects employed directional and barrier fencing, escape devices, double cattle guards at intersections of side roads, and arched culverts or single-span bridges serving as wildlife passage structures under the road (with existing bridges used where possible). WSDOT also co-leads a multi-organization working group that is producing a statewide habitat connectivity assessment. The group has selected focal species and assembled draft GIS models that utilize least-cost and circuit theory modeling methods, which will be used to develop priorities for future retrofit projects (McAllister 2010, Chapters 8, 10).

- New York State Department of Transportation initiated a project to identify priorities and develop multi-agency strategies to improve environmental conditions on its roads. The agency organized internal teams and partnered with external resource agencies and environmental organizations to

Figure 12.2. Example of a modification of existing fencing to exclude large ungulates and other mammals from roads. The creation of an earthen ramp allows animals trapped within the fenced right-of-way to exit into adjacent habitat. *Credit: Daniel J. Smith.*

develop guidance on best practices, which can be implemented during road maintenance activities. In 2001, senior environmental specialists were assigned as full-time environmental coordinators to each of their 11 regional maintenance groups. These coordinators were able to incorporate environmental best practices, such as (a) invasive plant control, (b) water-level control and water quality improvement, (c) measures to reduce turtle mortality and deer-vehicle collisions, (d) living snow fences (see Chapter 11), (e) osprey nesting enhancements, (f) alternative mowing strategies to enhance grassland songbird nesting habitat, (g) migratory bird protection on bridges, (h) herbicide education, and (i) small petroleum spill abatement. States without these policies might be persuaded to adopt them if the policy can be promoted and developed (as most of these are) as a means to streamline compliance with other agency requirements and initiatives that govern transportation projects (Williams 2003).

12.7.3.1. Establishing Constructive Partnerships
Partnerships are beneficial for financial as well as in-kind support in conducting monitoring studies. Most research efforts that involve performing pre- and post-mitigation monitoring in an adaptive management context typically include affected wildlife resource and land management agencies as well as the transportation agencies. Nongovernmental organizations (NGOs) that have specific interest in the conservation of potentially affected species are typically involved as well. Citizen support is also important and can be anything from volunteer assistance with data collection to political and public support from stakeholder groups (Chapter 5). In an adaptive management process, true collaborative partnerships can reduce potential conflicts, provide coordination and oversight, help with permit acquisition, streamline the process of conducting the work, and provide support for recommendations from the study. Partnering with the resource and permitting agencies is an important component of comprehensive voluntary evaluation programs developed in some states.

12.7.3.2. Applying Lessons Learned from Monitoring in Adaptive Management
It is important to establish transportation-related research questions to ensure that the research produces deliverables that are useful to and practical for transportation planners and engineers. Resources such as this book and other manuals and guidelines cited throughout this volume have been compiled because most wildlife ecology literature—even research specifically pertaining to transportation issues—is unknown or not readily accessible to (or written for) engineers and planners. Moreover, the lessons learned from monitoring may not be ecological. For example, the inspection of a crossing from an ecological viewpoint can reveal design and construction flaws that undermine the usefulness of the structure (see also Sections 6.3.1, 6.7, and 10.4.3). If the monitoring study has not been specifically commissioned by the transportation agency to provide information on the effectiveness of the wildlife crossing structure, then it is worth approaching the agency as a partner. The transportation agency may consider it a bargain to supply additional funding for performance monitoring to an already funded study, rather than having to initiate a new monitoring study.

12.8. Key Points

- Monitoring related to transportation projects should have two purposes: (1) to identify location and severity of impacts of roads and traffic on wildlife, and (2) to evaluate performance of mitigation measures intended to remedy the impacts.

- Goals and objectives of monitoring should pinpoint key issues and define knowns and unknowns associated with the road, species of concern, and adjacent habitat areas, thus providing a framework of expected and measurable outcomes.

- Monitoring provides long-term cost savings by identifying adaptive management opportunities that maximize functioning of mitigation measures. Therefore, when resources are insufficient to monitor a project properly, project implementation should be delayed until the necessary resources are acquired to maximize cost effectiveness.

- Both pre- and post-mitigation monitoring are important to assess mitigation performance; pre-mitigation data must be collected to establish a baseline for conditions prior to the mitigation and will inform quantification of changes in post-mitigation road effects. For a true, scientifically defensible evaluation, it is imperative to use the same data collection protocols and methods in both monitoring phases.

- The goals of mitigation efforts are important when considering monitoring for a transportation project; both pre- and post-mitigation monitoring should be designed as one program that addresses a particular question or goal (e.g., reduction of road mortality, or improvements in habitat connectivity, or preferably both).

- Roadkill counts can provide an index of relative mortality, but should never be taken as absolute estimates of mortality, even for the specific days of a survey. Monitoring programs should be designed so that data can be collected by different individuals in a standardized manner, thus minimizing bias.

- There are several active and passive monitoring techniques that are appropriate for determining levels of use or effectiveness of mitigation measures. Active methods are more time consuming and labor intensive but yield finer-scale spatial and temporal data; passive and less labor-intensive monitoring techniques may be less costly but yield less detailed data.

- Long-term monitoring is particularly valuable for small animals such as amphibians and reptiles whose habitat use and movement patterns are seasonally and interannually variable. Long-term monitoring of wildlife crossings can improve understanding of different responses to roads and crossing structures by various species and can increase the effectiveness of wildlife crossing designs.

- Adaptive management places value on learning about the effectiveness of management measures by monitoring its outcomes, and provides the opportunity to learn, adapt, and adjust mitigation in all phases of a road project.

LITERATURE CITED

Andrews, K. M., J. W. Gibbons, and D. M. Jochimsen. 2008. Ecological effects of roads on amphibians and reptiles: a literature review. Pages 121–143 in Urban herpetology. J. C. Mitchell, R. E. Jung Brown, and B. Bartholomew, editors. Herpetological Conservation Vol. 3. Society for the Study of Amphibians and Reptiles, Salt Lake City, Utah, USA.

Becker, C. G., C. R. Fonseca, C.F.B. Haddad, R. F Batista, and P. I. Prado. 2007. Habitat split and the global decline of amphibians. Science 318:1775–1777.

Bernardino F. S., Jr., and G. H. Dalrymple. 1992. Seasonal activity and road mortality of the snakes of the Pa-hay-okee wetlands of Everglades National Park, USA. Biological Conservation 62:71–75.

Boarman, W. I., M. L. Beigel, G. C. Goodlett, and M. Sazaki. 1998. A passive integrated transponder system for tracking animal movements. Wildlife Society Bulletin 26:886–891.

Castro-Santos, T., A. Haro, and S. Walk. 1996. A passive integrated transponder (PIT) tag system for monitoring fishways. Fisheries Research 28:253–261.

CERP-AMS (Comprehensive Everglades Restoration Plan Adaptive Management Strategy). 2006. Comprehensive Everglades restoration plan: adaptive management strategy. Steering Committee and Writing Team. US Army Corps of Engineers, Jacksonville, Florida, and South Florida Water Management District, West Palm Beach, Florida, USA.

Clevenger, A. P. 2005. Conservation value of wildlife crossings: measures of performance and research directions. GAIA 14:124–129.

Clevenger, A. P. 2012. 15 years of Banff research: what we've learned and why it's important to transportation managers beyond the park boundary. Pages 409–423 in Proceedings of the 2011 international conference on ecology and transportation. P. J. Wagner, D. Nelson, and E. Murray, editors. Center for Transportation and the Environment, North Carolina State University, Raleigh, North Carolina, USA.

Clevenger, A. P., B. Chruszcz, and K. E. Gunson. 2001a. Highway mitigation fencing reduces wildlife-vehicle collisions. Wildlife Society Bulletin 29:646–653.

Clevenger, A. P., B. Chruszcz, and K. E. Gunson. 2001b. Drainage culverts as habitat linkages and factors affecting passage by mammals. Journal of Applied Ecology 38:1340–1349.

Clevenger, A. P., B. Chruszcz, K. Gunson, and J. Wierzchowski. 2002. Roads and wildlife in the Canadian Rocky Mountain Parks: movements, mortality and mitigation. Final Report (October 2002). Parks Canada, Banff, Alberta, Canada.

Clevenger, A. P., A. T. Ford, and M. A. Sawaya. 2009. Banff wildlife crossings project: integrating science and education in restoring population connectivity across transportation corridors. Final Report. Parks Canada Agency, Radium Hot Springs, British Columbia, Canada.

Clevenger, A. P., and M. P. Huijser. 2011. Wildlife crossing structure handbook: design and evaluation in North America. Report No. FHWA-CFL/TD-11-003. Federal Highway Administration, Washington, DC, USA.

Clevenger, A. P., and N. Waltho. 2000. Factors influencing the effectiveness of wildlife underpasses in Banff National Park, Alberta, Canada. Conservation Biology 14:47–56.

Clevenger, A. P., and N. Waltho. 2003. Long-term, year-round monitoring of wildlife crossing structure and the importance of temporal and spatial variability in performance studies. Pages 293–302 in Proceedings of the 2003 international conference on ecology and transportation. C. L. Irwin, P. Garrett, and K. P. McDermott, editors. Center for Transportation and the Environment, North Carolina State University, Raleigh, North Carolina, USA.

Clevenger, A. P., and N. Waltho. 2005. Performance indices to identify attributes of highway crossing structures facilitating movement of large mammals. Biological Conservation 121:453–464.

Clevenger, A. P, and J. Wierzchowski. 2006. Maintaining and restoring connectivity in landscapes fragmented by roads. Pages 502–535 in Connectivity conservation. K. Crooks and M. Sanjayan, editors. Cambridge University Press, Cambridge, UK.

Clevenger, A. P., J. Wierzchowski, B. Chruszcz, and K. Gunson. 2002. GIS-generated, expert based models for wildlife habitat linkages and mitigation planning. Conservation Biology 16:503–514.

Crawford, B. A., J. C. Maerz, N. P. Nibbelink, K. A. Buhlmann, T. M. Norton, and S. E. Albeke. 2014. Estimating the consequences of multiple threats and management strategies for semi-aquatic turtles. Journal of Applied Ecology 51:359–366.

Eilerts, B. D., and S. E. Nordhaugen. 2010. Looking to the future with retrofit options from lessons learned. Pages 25–29 in Proceedings of the 2009 international conference on ecology and transportation. P. J. Wagner, D. Nelson, and E. Murray, editors. Center for Transportation and Environment, North Carolina State University, Raleigh, North Carolina, USA.

Enge, K. M. 1997. Use of silt fencing and funnel traps for drift fences. Herpetological Review 28:30–31.

Enge, K. M. 2006. References on herpetological survey techniques. Florida Fish and Wildlife Conservation Commission, Gainesville, Florida, USA.

Farallo, V. R., D. J. Brown, and M.R.J. Forstner. 2010. An improved funnel trap for drift fence surveys. Southwestern Naturalist 55:457–460.

Ford, A. T., and A. P. Clevenger. 2010. Validity of the prey trap hypothesis for carnivore-ungulate interactions at wildlife crossing structures. Conservation Biology 24:1679–1685.

Ford, A. T., A. P. Clevenger, and A. Bennett. 2009. Comparison of non-invasive methods for monitoring wildlife crossing structures on highways. Journal of Wildlife Management 73:1213–1222.

Forman, R.T.T., D. Sperling, J. A. Bissonette, A. P. Clevenger, C. D. Cutshall, V. H. Dale, L. Fahrig, et al. 2003. Road ecology. Science and solutions. Island Press, Washington, DC, USA.

Gibbs, J. P. 1998. Amphibian movements in response to forest edges, roads, and streambeds in southern New England. Journal of Wildlife Management 62:584–589.

Gibbs, J. P., S. Droege, and P. Eagle. 1998. Monitoring populations of plants and animals. BioScience 48:935–940.

Graeter, G. J., K. A. Buhlmann, L. R. Wilkinson, and J. W. Gibbons, editors. 2013. Inventory and monitoring: recommended techniques for reptiles and amphibians. Technical Publication IM-1. Partners in Amphibian and Reptile Conservation, Birmingham, Alabama, USA.

Guinard, E., R. Prodon, and C. Barbraud. 2015. A robust method to obtain defendable data on wildlife mortality. Pages 96–100 in Handbook of road ecology. R. van der Ree, D. J. Smith, and C. Grilo, editors. John Wiley and Sons, Hoboken, New Jersey, USA.

Hels, T., and E. Buchwald. 2001. The effect of road kills on amphibian populations. Biological Conservation 99:331–340.

Hels, T., and G. Nachman. 2002. Simulating viability of a spadefoot toad Pelobates fuscus metapopulation in a landscape fragmented by a road. Ecography 25:730–744.

Heyer, W. R., M. A. Donnelly, R. W. McDiarmid, L. C. Hayek, and M. S. Foster. 1994. Measuring and monitoring biological diversity: Standard methods for amphibians. Smithsonian Institute Press, Washington, DC, USA.

Holling, C. S. 1978. Adaptive environmental assessment and management. John Wiley and Sons, London, UK.

Keller, I., and C. R. Largiadér. 2003. Recent habitat fragmentation caused by major roads leads to reduction of gene flow and loss of genetic variability in ground beetles. Proceedings of the Royal Society of London, Series B, Biological Sciences 270:417–423.

Langen, T. A., A. Machniak, E. Crowe, C. Mangan, D. Marker, N. Liddle, and B. Roden. 2007. Methodologies for surveying herpetofauna mortality on rural highways. Journal of Wildlife Management 71:1361–1368.

Little, S. J., R. G. Harcourt, and A. P. Clevenger. 2002. Do wildlife passages act as prey-traps? Biological Conservation 107:135–145.

Manley, P. N., B. van Horne, J. K. Roth, W. J. Zielinski, M. M. McKenzie, T. J. Weller, F. W. Weckerly, and C. Vojta. 2006. Multiple species inventory and monitoring technical guide. General Technical Report WO-73. US Forest Service, Washington, DC, USA.

McAllister, K. R. 2010. Washington's Habitat Connectivity Highway Retrofit initiative. Pages 363–365 in Proceedings of the 2009 international conference on ecology and transportation. P. J. Wagner, D. Nelson, and E. Murray, editors. Center for Transportation and Environment, North Carolina State University, Raleigh, North Carolina, USA.

McCarthy, M. A., and H. P. Possingham. 2007. Active adaptive management for conservation. Conservation Biology 21:956–963.

McDiarmid, R. W., M. S. Foster, C. Guyer, J. W. Gibbons, and N. Chernoff. 2012. Reptile diversity: Standard methods for inventory and monitoring. University of California Press, Berkeley, California, USA.

National Research Council (NRC). 2005. Assessing and managing the ecological impacts of paved roads. National Academies Press, Washington, DC, USA.

Pagnucco, K. S., C. A. Paszkowski, and G. J. Scrimgeour. 2011. Using cameras to monitor tunnel use by long-toed salamanders (Ambystoma macrodactylum): an informative, cost-efficient technique. Herpetological Conservation and Biology 6:277–286.

Pagnucco, K. S., C. A. Paszkowski, and G. J. Scrimgeour. 2012. Characterizing movement patterns and spatio-temporal use of under-road tunnels by long-toed salamanders in Waterton Lakes National Park, Canada. Copeia 2012: 331–340.

RECOVER (Restoration Coordination and Verification). 2011. Comprehensive Everglades restoration plan: adaptive management integration guide. Restoration Coordination and Verification, US Army Corps of Engineers, Jacksonville, Florida, and South Florida Water Management District, West Palm Beach, Florida, USA.

Rhodes, J. R., and N. Jonzén. 2011. Monitoring temporal trends in spatially-structured populations: how should sampling effort be allocated between space and time? Ecography 34:1040–1048.

Riley, S.P.D., J. P. Pollinger, R. M. Sauvajot, E. C. York, C. Bromley, T. K. Fuller, and R. K. Wayne. 2006. A southern California freeway is a physical and social barrier to gene flow in carnivores. Molecular Ecology 15:1733–1741.

Roedenbeck, I. A., L. Fahrig, C. S. Findlay, J. E. Houlahan, J.A.G. Jaeger, N. Klar, S. Kramer-Schadt, and E. A. van der Grift. 2007. The Rauischholzhausen agenda for road ecology. Ecology and Society 12(1):11. http://www.ecologyandsociety.org/vol12/iss1/art11/. Accessed 9 September 2014.

Ruediger, B., and J. Lloyd. 2003. A rapid assessment process for determining potential wildlife, fish and plant linkages for highways. Pages 206–225 in Proceedings of the 2003 international conference on ecology and transportation. C. L.

Irwin, P. Garrett, and K. P. McDermott, editors. Center for Transportation and the Environment, North Carolina State University, Raleigh, North Carolina, USA.

Sauer, J. R., B. G. Peterjohn, and W. A. Link. 1994. Observer differences in the North American breeding bird survey. The Auk 111:50–62.

Sawaya, M. A., and A. P. Clevenger. 2010. Using non-invasive genetic sampling methods to assess the value of wildlife crossings for black and grizzly bear populations in Banff National Park. Pages 42–55 in Proceedings of the 2009 international conference on ecology and transportation. P. J. Wagner, D. Nelson, and E. Murray, editors. Center for Transportation and Environment, North Carolina State University, Raleigh, North Carolina, USA.

Shepard, D. B., A. R. Kuhns, M. J. Dreslik, and C. A. Phillips. 2008. Roads as barriers to animal movement in fragmented landscapes. Animal Conservation 11:288–296.

Smith, D. J. 2006. Ecological impacts of SR 200 on the Ross Prairie ecosystem. Pages 380–396 in Proceedings of the 2005 international conference on ecology and transportation. C. L. Irwin, P. Garrett, and K. P. McDermott, editors. Center for Transportation and the Environment, North Carolina State University, Raleigh, North Carolina, USA.

Smith, D. J. 2011. Cost effective wildlife crossing structures which minimize the highway barrier effects on wildlife and improve highway safety along US 64, Tyrrell County, NC. Final Report. Report No. FHWA/NC/2009-26. North Carolina Department of Transportation, Raleigh, North Carolina, USA.

Smith, D. J. 2012. Determining location and design of cost effective wildlife crossing structures along US 64 in North Carolina. Transportation Research Record 2270:31–38.

van der Grift, E. A., and P. Schippers. 2013. Wildlife crossing structures: can we predict effects on population persistence? In Proceedings of the 2013 international conference on ecology and transportation. http://www.icoet.net/icoet_2013/documents/papers/ICOET2013_Paper101A_VanderGrift_et_al.pdf. Accessed 19 September 2014.

van der Ree, R., E. van der Grift, C. Mata, and F. Suarez. 2007. Overcoming the barrier effect of roads: how effective are mitigation strategies? An international review of the use and effectiveness of underpasses and overpasses designed to increase the permeability of roads for wildlife. Pages 423–431 in Proceedings of the 2007 international conference on ecology and transportation. C. L. Irwin, D. Nelson, and K. P. McDermott, editors. Center for Transportation and the Environment, North Carolina State University, Raleigh, North Carolina, USA.

Vaughan, M. R., M. J. Kelly, C. M. Proctor, and J. A. Trent. 2011. Evaluating potential effects of widening US 64 on red wolves in Washington, Tyrrell, and Dare Counties, North Carolina. Final Report. VT-NCDOT Contract No. 09-0776-10. Virginia Tech University, Blacksburg, Virginia, USA.

Walters, C. J. 1986. Adaptive management of renewable resources. McGraw-Hill, New York, New York, USA.

Williams, B. K. 2011. Adaptive management of natural resources—framework and issues. Journal of Environmental Management 92:1346–1353.

Williams, K. 2003. Environmental stewardship in NYSDOT highway maintenance. Pages 591–595 *in* Proceedings of the 2003 international conference on ecology and transportation. C. L. Irwin, P. Garrett, and K. P. McDermott, editors. Center for Transportation and the Environment, North Carolina State University, Raleigh, North Carolina, USA.

Willson, J. D., and J. W. Gibbons. 2009. Drift fences, coverboards, and other traps. Pages 229–245 *in* Amphibian ecology and conservation: A handbook of techniques. C. K. Dodd Jr., editor. Oxford University Press, Oxford, UK.

Wilson, D. E., F. R. Cole, J. D. Nichols, R. Rudran, and M. S. Foster. 1996. Measuring and monitoring biological diversity: standard methods for mammals. Smithsonian Institution Press, Washington, DC, USA.

Practical Example

2

MARK FITZSIMMONS AND
ALVIN R. BREISCH

Design and Effectiveness of New York State's First Amphibian Tunnel and Its Contribution to Adaptive Management

Introduction

As detailed in previous chapters, transportation engineering and planning processes tend to focus primarily on providing the public infrastructure that accomplishes the movement of people and goods between locations in a safe, efficient, and cost-effective manner. Ideally, some sense of maintaining ecological infrastructure enters into the planning and design process, but the bottom-line objectives are safety, efficiency, and cost effectiveness. Unfortunately, the successful achievement of the last two objectives frequently comes at a high cost to wildlife when public infrastructure affects the ecological infrastructure (see also Chapters 6 and 7).

In 1999, the Albany County Office of Natural Resources initiated a project on a section of county highway (see description below) in rural Albany County, New York State (NYS), United States to incorporate a combined tunnel and barrier-drift fence (guide or directional fence) system into a highway reconstruction project. The effort was in response to the observed incidence of high amphibian road mortality associated with movements between breeding, foraging, and hibernating sites. The project, the first of its kind to be implemented in NYS, involved the cooperation of the Albany County Department of Public Works and NYS Department of Environmental Conservation. We present here a critical review of the project's design, construction, and maintenance requirements, as well as possible positive and negative impacts on the target species.

Description of Project Area

The project site is an approximately 90-m (300-ft) segment of a 2-lane, ~7 m (22 ft) wide county highway traversing an area of extensive wetland, much of which is located in Black Creek Marsh State Wildlife Management Area. The topography of the surrounding wetland is generally flat on both sides of the highway and for several kilometers north of the project site (Figure PE 2.1). South of the project area, the roadway has an approximately 8% upgrade with relatively steep downward slopes extending east of the road corridor and variable upward slopes extending to the west. This road segment has served as a control area for monitoring road mortality beyond the limits of the project (Figure PE 2.2).

The south border of the project area (i.e., the dividing line between project and control areas) is defined by Black Creek, a permanent water course with extremely high seasonal variability in depth and flow. The habitat adjacent to the control area immediately south of the tunnel site and east of the road corridor is shrub swamp; west of the control road segment is upland with mixed hardwood-conifer overstory, a kettlehole bog with permanent open water, and sparsely distributed private residences. The kettlehole supports substantial and diverse amphibian breeding activity. This segment of the subject highway has served as a control for monitoring road mortality given the high degree of amphibian movements to and from breeding habitat and absence of mitigation measures. Further, continuous

Figure PE 2.1. Road segment where tunnel project was planned for installation. *Credit: Mark Fitzsimmons.*

Figure PE 2.2. Control road segment immediately south of project site. *Credit: Mark Fitzsimmons.*

fencing would be difficult to construct in this area because of the presence of several driveways.

Pre-installation Amphibian Monitoring

Pre-project indications of amphibian species diversity, abundance, and road mortality were based on almost 20 years of observations made by the authors at various sites in Albany County. Monitoring of the project site and the surrounding area (within a 1.2 km [0.75 mi] radius or ~4.5 sq km [~1.75 sq mi]) yielded records of 20 amphibian species (12 salamanders and 8 frogs), including 5 species that are locally rare or have special status; this is an extremely high diversity for such a localized area and notably

higher than other sites we monitored. Another unique aspect of the project area distinguishing it from other sites demonstrating high amphibian road crossing is that it traverses two areas with generally similar habitats as opposed to separating upland from breeding habitat. Thus, amphibian movements across the highway were apparent for most months of the year, and were driven by not only breeding behaviors, but also foraging, and other activities. In fact, February is the only month of the year in which at least some amphibian activity on road surfaces has not been observed since monitoring began.

Site Selection, Preliminary Planning, and Design Considerations

As mentioned above, three main factors were paramount in guiding the decision on site selection: significantly high amphibian diversity, high abundance, and correspondingly high road mortality. However, a number of other factors also contributed to preference for this particular site over other possible candidates:

- Resident amphibians inhabiting the immediate vicinity of the project site included populations of locally rare and special status species. It is noteworthy that the occurrence of northern red salamanders (*Pseudotriton ruber ruber*) in the project area is at the absolute northern extent of its geographic range. The habitat surrounding and including the project supports wood turtles (*Glyptemys insculpta*), a designated species of greatest conservation need in New York, as well as a New York State designated species of special concern and species of regional conservation concern in the Northeast. Historically, bog turtles (*G. muhlenbergii*), a New York State endangered species that is also federally designated as threatened under the Endangered Species Act, occupied the adjacent wetlands but has not been found there since 1957.
- The majority of the land adjacent to the project site as well as the adjacent control area is protected as within the Black Creek Marsh State Wildlife Management Area, thus precluding development to the habitat supporting target amphibians and non-target species.
- Black Creek is an approximately 3 m wide, permanent stream with seasonally variable depth and

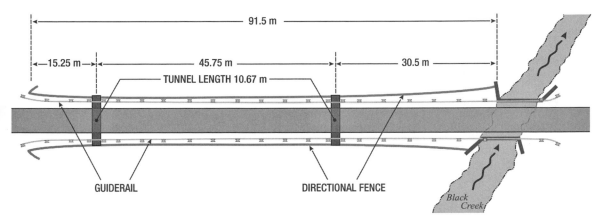

Figure PE 2.3. Site plan of tunnels and directional fence system. *Credit: © Victor Young Illustration, adapted from the original by John E. Merrill Jr.*

flow that serves as the southern boundary of the project site, separating it from the control area in which comparative counts are made.

- The road surface is perched well above adjacent habitat, which facilitated installation of tunnel structures with minimal excavation below the existing grade, minimizing changes to post-construction hydrology and reducing project costs.

- A pending highway improvement project presented an opportunity for substantial cost savings by incorporating installation of the tunnel structures into the road reconstruction and repaving.

Following site selection, project timing was given careful consideration. In order to minimize direct and indirect impacts related to construction activity, project installation was scheduled in the fall, thereby avoiding spring and early summer peaks in amphibian movements. Also, as previously noted, we made every effort to schedule project installation coincident with a planned highway improvement project, which significantly reduced construction costs.

In terms of project design, two main objectives were identified: (1) physically limiting movements by amphibians and nontarget species onto the road surface, and (2) directing movements of amphibians toward the mouth of a tunnel for safe passage under the highway. To accomplish these objectives over the 90± m (300 ft) segment comprising the project site, it was decided to install two tunnel structures

traversing the full width of the two-lane county highway road corridor. Finally, extending beyond and between the tunnel openings on each side of the highway is a permanent, wooden fence aligned parallel to the travel lanes (Figure PE 2.3).

Tunnel Structure: Materials and Installation

The tunnel structures are precast concrete box culverts with an outside dimension of 1.5 m (5 ft) square, inside dimension of 1.2 m (4 ft) square, and wall thickness of ~15 cm (6 in). The tunnels were delivered in three U-shaped sections with a separate cap. Each section measured 3.6 m (11.7 ft) in length and weighed approximately 4,309 kg (9,500 lbs)— 3,175 kg (7,000 lbs) for the U and 1,134 kg (2,500 lbs) for the cap (Figure PE 2.4). The structures were installed as an upright U that was partially filled with native substrate over sand and then capped. The resulting opening was 1.2 m (4 ft) wide with a floor-to-ceiling height varying from 0.3 to 0.4 m (12–18 in; Figures PE 2.5 and PE 2.6). The total 10.7 m (35 ft) tunnel length was sufficient to cross the full 6.7 m (22 ft) width of the 2 travel lanes plus shoulders and embankments. Two tunnel structures were installed approximately 46 m (150 ft) apart (Figure PE 2.3). Although not part of the project design, the post-construction elevation of the south tunnel was slightly higher (approximately 15 cm) than the north tunnel. This minor deviation resulted in differences in substrate, essentially yielding a "dry" (south) and a "wet" (north) tunnel.

a

b

c

Figure PE 2.4. Installation of precast concrete box culvert sections (a, b) and caps (c). *Credit: Mark Fitzsimmons.*

Although the underlying motivation for the tunnel project was road mortality of amphibians, the project was designed to accommodate a wide diversity of species, including reptiles and mammals, and this resulted in larger dimensions than other tunnel designs we reviewed at the time. Additionally, the larger tunnel mouths contribute to equalizing ambient light, temperature, and humidity between outside and inner tunnel conditions. In order to achieve this in smaller tunnel designs, slotted drains in the road surface have been added along the length of smaller tunnel designs (Chapter 9). However, in addition to bringing light, temperature balance, and moisture to the tunnel interior, the slotted drains also introduce road salt and sand from winter road treatments (see also Chapter 4 for a discussion of effects), which would have to be periodically flushed or otherwise cleaned out of the tunnels, thereby increasing maintenance costs. Additionally, one

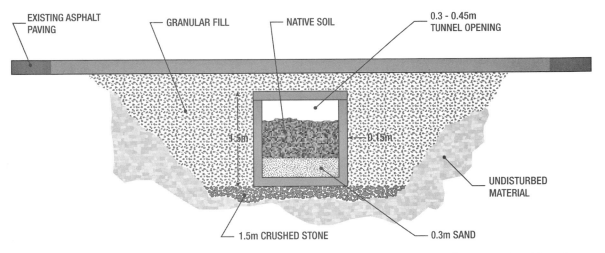

EXISTING ASPHALT
PAVING

GRANULAR FILL

NATIVE SOIL

0.3 - 0.45m
TUNNEL OPENING

1.5m

0.15m

UNDISTURBED
MATERIAL

1.5m CRUSHED STONE

0.3m SAND

Figure PE 2.5. Cross-section view of tunnel structure and specifications. *Credit: © Victor Young Illustration, adapted from the original by John E. Merrill Jr.*

Figure PE 2.6. Newly installed tunnel and directional fence. *Credit: Alvin R. Breisch.*

potential disadvantage of large tunnel openings is the risk of allowing access to predators into a confined setting that limits the ability of amphibian prey to escape; however, there is currently little evidence that suggests crossing structures act as prey traps. Placement of concealing cover objects (e.g., boards, flat rocks) along the length of the tunnel interior prior to capping offers some mitigation in this regard (Section 9.7.1.2).

Fence Materials and Installation

One of the most important design features of the project relates to the fence extending between and beyond the tunnel openings on each side of the road, parallel to the travel lane, for a distance of 90 m. The area between the fence and the road surface was backfilled and graded to stabilize the slope and prevent animals that circumvented the fence from being trapped. The fence was constructed of pressure-treated lumber and supported by 89 mm × 89 mm (4 in × 4 in) pressure-treated posts placed at 1.2 m (4 ft) intervals. A pressure-treated 38 mm × 140 mm × 3.6 m (2 in × 6 in ×12 ft) board was attached to each post on the roadside, and a 38 mm × 140 mm × 3.6 m (2 in × 6 in ×12 ft) board on top of a 38 mm × 286 mm × 3.6 m (2 in × 12 in ×12 ft) board was attached to each post on the wetland side (Figure PE 2.7). At the northern terminus of each guide fence a 1.8 m (6 ft) sharp angling of the fence or dogleg (Chapter 9; Figure PE 2.8) was added to direct

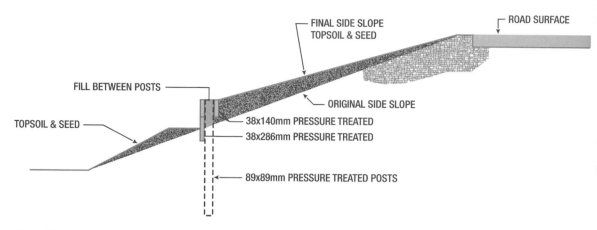

Figure PE 2.7. Cross-section showing fence in relation to road surface. *Credit: © Victor Young Illustration, adapted from the original by John E. Merrill Jr.*

animals away from the road and back toward a tunnel entrance. This feature was observed to work successfully with a variety of salamander and frog species.

Post-construction Monitoring

Due to manpower and funding limitations, intensive ongoing monitoring of the project following construction was not practical. However, periodic counts were made and various data were recorded for each observation, including date, time, weather conditions, species, location along the fence, direction of movement, and the distance from the fence structure when first observed. The fence was calibrated at 1 m intervals in order to record location of an animal along the fence.

The majority of this monitoring was conducted from early spring through early fall when weather and other conditions were appropriate for amphibian movement (either for breeding, foraging, or otherwise). Amphibian movement typically occurred at night, with temperatures above freezing and at least some precipitation or wet ground conditions. Monitoring during periods of peak movement (late March to late April) occurred approximately 4–6 times per year with each survey averaging 1.5–2 hours, depending on the duration of optimum weather conditions. Additional monitoring was conducted at other times of the year depending on weather conditions.

Results of the counts conducted while monitoring over a 6-year period following installation of the structure showed that approximately 2% of over

Figure PE 2.8. The sharp angle or "dog-leg" at the terminus of the fence. *Credit: Alvin R. Breisch.*

Figure PE 2.9. Red salamander (*Pseudotriton ruber*) moving along the fence. *Credit: Mark Fitzsimmons.*

300 individuals (representing 16 species) observed were roadkills within the bounds of the tunnel-fence system. This count compared to 40% of almost 600 individuals (representing 19 species) as roadkills in the adjacent control (unfenced) area on the same road extending immediately south from the project area. The 2% incidence of roadkills cited above within the project area included just eight individuals of four frog species and no salamanders. The reduction in roadkill following the structure installation indicated a positive effect of both the guide fence and tunnel system and overall success of the project goals.

Monitoring has also shown very few movements of live amphibians across the road in the project area, which attests to the effectiveness of the fence component of the project. Seventeen individuals of six species were observed alive on the road surface within the project area (five frog species and one salamander species). It should be noted that the majority of individuals were observed on the road moving along rather than perpendicular to the travel lane, suggesting that their point of origin may have been beyond the limits of the tunnel-fence system.

Monitoring also documented individuals of six salamander and three frog species using the tunnel system. In addition, a variety of nontarget species (three reptile and six mammal species) were confirmed. While no attempt was made to measure project effectiveness to any degree of statistical significance, the observations resulting from continued monitoring are strongly indicative that the mitigation structures were successful in physically impeding most amphibian movements onto or over the road surface along the 90 m length of the project site. Additionally, movements of amphibians and various incidentally observed species along the fence were successfully directed toward tunnel openings (Figure PE 2.9).

Figure PE 2.10. Examples of bowing of fence (a) and separation from tunnel structure (b). *Credit: Alvin R. Breisch.*

Adaptive Management: Design Lessons

Many of the adaptive management lessons we present here are site-specific considerations that contribute to the general field of road ecology. This information is not always easy to find in published sources in a synthesized manner. A significant problem during the planning stage is the disconnect between project engineering and the requirements of animals in the affected area. With the application of standard construction techniques, some of these issues could have been avoided. That being said, we have presented the issues we encountered as a ser-

vice to the readers of this book to further learn from our experiences.

For example, the directional fence should have been attached to the concrete tunnel structure and tied into the slope on road verges. Without these simple modifications to the project design, we observed bowing of the fence between the tunnels and separation of the fence from the tunnel structure. This movement of the fence resulted in gaps through which amphibians could gain access to the road surface (Figure PE 2.10).

Careful attention to localized hydrology is key to

the final characteristics of passage structure interior and openings (see also Chapter 9). Our project site included wetland areas on both sides of the highway, with a discernible flow gradient between them. The localized hydrology of the project site became problematic with high water levels, especially following snowmelt and large rain events. At such times, instead of the box culverts functioning as tunnels, the wet passage structure functioned essentially as a drainage culvert. The structure floor underwent significant scouring with much of the bottom substrate eventually being washed out. Ultimately, one of our structures was deepened to the point of permanent water within the tunnel rather than moist soil conditions as designed (Figure PE 2.11). Additionally, the approach zone to the tunnel opening gradually eroded to form a pool of standing water at the entrance. In spite of this change, egg masses of northern two-lined salamanders (*Eurycea bislineata*) were found at the entrance of the structure, and individuals of six amphibian species and a hatchling snapping turtle (*Chelydra serpentina*) were observed swimming into the tunnel.

Adaptive Management: Maintenance Issues

Various aspects of the project design process included consideration of keeping maintenance costs to a minimum. Nonetheless, certain ongoing maintenance requirements are inevitable and are important to the long-term effectiveness and success

Figure PE 2.11. The north or "wet" tunnel, just a few centimeters lower in elevation than the south or "dry" tunnel, was scoured by storm events and became a permanent water passage. *Credit: Mark Fitzsimmons.*

of this type of project (see also Chapter 11). Weed growth above and below the fence needs to be cut back at least once annually, and in some years twice. This vegetation management not only facilitates monitoring by removing sight obstructions, but also eliminates bridging of vegetation, particularly grasses, over the fence (Figure PE 2.12; Chapter 11). Unless the vegetation is cut back to ground level, many of the amphibian species can readily surmount the fence via this vegetation and make their way onto the road surface. Periodically, vegetation removal should also include encroaching shrubs and small tree saplings, which tend to hinder access for people conducting post-construction monitoring along the fence.

Another ongoing maintenance requirement involves filling and regrading animal burrows along the fence. We found that woodchucks (*Marmota monax*) are the most frequent culprit, but eastern chipmunks (*Tamias striatus*) and striped skunks (*Mephitis mephitis*) are also responsible for their share of burrowing and digging activity. Unless filled, the fence may become undermined, creating possible structural problems, and potentially allowing passage of amphibians under and behind the fence, where they eventually gain access to the road surface.

A final maintenance responsibility with this project involves refreshing calibration marks along the directional fence. In our case, the use of perma-

nent brass nails eliminated the need to repeatedly measure each 1 m calibration mark along the fence; however the use of "permanent" ink markers to indicate location codes was anything but permanent and had to be re-applied each spring before that season's monitoring commenced.

Impact of an Extreme Weather Event

Tropical Storms Irene and Lee deposited 28 to 30 cm of rain on the project area between 28 August and 8 September 2011. According to the Albany County Department of Public Works, the project road segment was inundated by as much as 10 cm (~4 in) of water above the crown of the road, yielding a depth of approximately 1.5 m (4.9 ft) from the road crown to the base elevation of the adjacent wetland. The road surface was flooded throughout the peak period of the storm and for almost a day after the rain had ended.

Upon inspection after the water receded, it was

Figure PE 2.12. Without periodic cutting and mowing, vegetation will form a bridge over the fence, allowing amphibians to access the road surface. *Credit: Alvin R. Breisch.*

Figure PE 2.13. Partial road and shoulder collapse and completely submerged tunnel opening following Tropical Storms Irene and Lee. *Credit: Alvin R. Breisch.*

found that portions of the shoulder and edges of the road had collapsed in the project area and that the mouth of each tunnel was completely submerged (Figure PE 2.13). At several locations along the fence, the road bank had subsided and the backfill behind it had eroded away (Figure PE 2.14). In addition, the inside of both tunnels had been scoured clean by the flood waters leaving no substrate in the tunnels, essentially resulting in water passages connecting the wetlands on each side of the road (Figure PE 2.15).

The aftermath of the back-to-back tropical storms had a significant impact on the structure and function of the tunnel and fence system. The design goal of moist, terrestrial substrate was replaced by permanent aquatic conditions with a constant but variable flow depending on spring snowmelt and

Figure PE 2.14. Subsidence of bank and erosion of backfill behind the fence created passageways for amphibians to trespass directly to road surface. *Credit: Mark Fitzsimmons.*

Figure PE 2.15. Formerly "dry" (south) tunnel after Tropical Storms Irene and Lee. Dotted line shows approximate level of tunnel floor before the storms. *Credit: Alvin R. Breisch, dotted line added by © Victor Young Illustration.*

precipitation. To some extent the aquatic tunnels will continue to serve as safe passage for more aquatic species of salamanders, frogs, and other incidental species; however, flow rates are frequently high enough to preclude amphibians and most other species from moving against the current. Furthermore, the remaining fence is in poor condition and while it will continue to keep some animals from gaining access to the road surface, it no longer functions with the same exclusionary success as the pre-storm structure. In fact, we now regularly find live and dead frogs and salamanders (including the regionally uncommon northern red salamander) on the road surface, a situation that rarely occurred prior to the tropical storms.

Repair of the resulting storm damage would require excavation of the entire road segment surrounding the tunnels as well as the road bank along the full length of the fence. This effort is beyond

the funding available to the Albany County Department of Public Works at this time, although smaller, sequential repair projects are being considered at the time of this publication.

Conclusions

This practical example demonstrates the importance of pre- and post-construction monitoring (Chapter 12), but also reflects the lack of available expert guidance in hydrogeology to address passage structure considerations for small animals at the time of project implementation. Had this volume been available to project engineers and the contractors responsible for installation during the construction phase, some of the problems may have been averted. Prior to the tropical storms, the fence created a highly effective system for reducing roadkill in the project area and the tunnels provided an effective under-the-road passageway for both aquatic and terrestrial animals. It should also be noted that this project was completed at a reasonable expense given that it was incorporated as part of a highway construction project in the area.

Despite these issues and even with implementation of potential solutions, the tropical storms presented problems beyond what could have been predicted or managed in the design and construction phases. Thus, this example also illustrates how even with the best of efforts, certain events beyond our control can affect a passage structure and the wildlife that may have relied on it. As such, the events described above support the need for adaptive management (Chapter 12), particularly in conjunction with ongoing post-construction monitoring, to continually maintain the effectiveness of a given passage structure.

13 — The Road Ahead

KIMBERLY M. ANDREWS

Many challenges face us in the field of road ecology, and it is time to catch up in our efforts with small animals. There is more to do than ever, but fortunately, we also have a growing contingency of experts and interested citizens. Together, we have the momentum to begin tackling this list of challenges. We can be both the professional innovators and advocates for the needs of the public and society. If we don't do it, who will?

In this book, we have attempted to present the current state of our knowledge. However, advances continue in this field every day, which provides much encouragement. That being said, we summarize here some of the primary persisting challenges (in no particular order of priority) which must be overcome to advance our effectiveness with the prevention, minimization, and mitigation of the effects of public infrastructure on ecological infrastructure. We encourage you to identify the areas that pique your interest and to dedicate yourself to moving us farther along the road.

13.1. Ecological Detection

Detection and Quantification of Ecological Effects

There are many ecological factors that influence ecosystem processes independently and interactively. Controlled experimental designs are necessary to collect useful data. However, scientific design often cannot incorporate all of the factors involved or the synergy of diverse variables. Although we will never be able to detect and quantify all ecological drivers, we can be aware of these limitations and purposefully design our studies around the biological parameters that are a priority. In turn, these results must be communicated with care to our peers and the public to avoid oversimplification.

Time Lags in Ecological Detection (Temporal Scales)

Not all ecological effects are immediately apparent and measurable. This lag is particularly true for events that are of a large scale (occurring across landscapes or the globe), those that operate on an evolutionary time frame (adaptive pressures), or those that affect long-term population viability, especially with long-lived animals (decreased reproduction or fecundity rather than immediate and visually apparent mortality). There is no way to avoid time lags or to make these effects more immediately apparent. However, we can diagnose which effects have delayed responses and measure early indicators. Additionally, predictive modeling exercises, when based on validated field data, can allow us to project trends over larger spatial areas and longer time frames.

Landscape-Level Variability (Spatial Scales)

In nature, no two locations are the same, even those hosting the same habitats and wildlife communities. The larger the spatial extent, the greater the amount of variability within the target system. As road networks influence landscapes and ecosystems, it is important to conduct research and plan new or modified public infrastructure in the context of large spatial scales. However, it can be logistically difficult to execute research, management, and planning practices at these scales, whether it is something as simple as having the time to collect data across all locations or something more complex, such as finding scientifically acceptable "replicate" field sites when sites across the landscape can vary considerably.

Individual, Population, and Community (Ecosystem Scales)

Organisms operate at different ecological levels of organization. The simplest way of understanding these is to study the effects of public infrastructure on individuals and extrapolate to population-level effects. However, the assumption of this extrapolation leaves much room for error. If our goal is to maintain or restore the functioning of the ecosystem, we are concerned with larger-scale processes, such as connectivity across landscapes among multiple habitats and metapopulations, retaining predator-prey interactions for intact wildlife communities and food chains, and ensuring population viability such that population growth is stable or increasing. While the mortality of a collection of individuals can be an indication of a problem for any of these larger-scale processes, the observed mortality could be occurring at low levels that may not threaten population viability. Road mortality data should be collected in conjunction with data on populations adjacent to roads so that mortality numbers can be assessed relative to numbers occupying the surrounding landscape. Modeling exercises can greatly assist with these problems provided that quality field and natural history data are used to parameterize those models.

Counting Individuals on Roads: Are These Numbers Current Trends, or a Legacy Effect?

Following some of the aforementioned reasoning, it can be misleading to simply quantify the number of animals observed on roads. If the number of animals killed is decreasing after the installation of underpasses, does it mean that fewer are crossing because more are using the passages, or does it mean that the population has declined and there are fewer individuals adjacent to the road to cross? When identifying mortality hotspots, is that where the priority mitigation locations should be or does it reflect a "legacy" effect where the individuals at initially prime habitat locations have already been selectively removed from the population? These potential

legacy effects are more of an issue for animals with high site fidelity, greater localized movements, and reduced probabilities of recolonizing neighboring habitats.

Emerging Issues

There are many emerging issues that can be exacerbated by the effects of roads. Climate change is necessitating shifts for some species in spatial or temporal activity patterns based on changing habitat conditions and resource availability. Intact ecological infrastructure will be necessary to allow adaptation, particularly for species that exhibit behavioral deterrence to hard, public infrastructure. Roads and development may increase population-level impacts where excessive amounts of individuals are killed as a result of urbanization. Additionally, roads create increased amounts of edges relative to habitat interiors. Habitat edges are disturbed environments that are susceptible to colonization and spread of invasive species and wildlife diseases. Very few data have been collected regarding how these issues may be influenced by the presence of various densities of roads or differing levels of urbanization. Climate change and increased edges are timely topics for research, though we may not be able to remedy their impacts proactively.

13.2. Engineering Design and Habitat Management

Beyond Mortality: Mitigating Habitat Effects

While there is still tremendous room for growth in testing wildlife crossing systems, we have made great advances in structural designs to keep animals off roads and to encourage crossing roads safely. However, we have made little progress in studying or quantifying the more diverse and potentially more severe effects on habitats. Further, we are still figuring out how to study or reduce other effects, such as the effects of barriers and habitat fragmentation, the formation of ecological traps from animal attraction to roadside habitats, the levels of environmental pollution and degradation adjacent to roadsides, and the extent to which these contaminants disperse via water, air, and food chains.

Limitations of Fencing and Passages

Directional fencing and passages are often suggested as the solution for all conflicts involving animals on roads. The purpose of these systems is to reduce mortality and restore connectivity by facilitating movement across the road. However, we are now aware there are many road effects beyond simple mortality and habitat connectivity. These effects are much harder to study, quantify, and solve with appropriate designs; therefore, their remediation has greatly lagged. Barrier effects that isolate animals on either side of the road cannot always be resolved with directional fencing and passage structures. This is particularly true if the target species exhibits avoidance behavior to roads or will not use culverts even when available.

Culverts also do little to minimize or reduce negative effects of roads on habitat quality (e.g., pollution from runoff, noise, or light). Lastly, fencing can cause its own suite of problems. For example, when it is applied as a barrier to keep nesting turtles off roads, nesting effort then concentrates near the fence, presenting an easy target for increased depredation. That loss of potential population recruitment may have an even greater effect than would occur from road mortality alone.

Mitigating Effects on Wildlife Communities versus Single Species

Ideally, we should design and implement road mitigation approaches that will benefit entire wildlife communities rather than just catering to one or more target species. As such, while directional fencing and culverts may work with many species, they are not a one-size-fits-all solution. For example, we are still learning how to prevent arboreal species from trespassing directional or barrier fence systems. Incorporation of an overhanging lip on the fencing structures can reduce trespass but may not prevent it. We can probably aim only for reductions in mortality in these situations rather than total prevention. However, if we can reduce the effects enough to maintain population viability, we need to accept some residual mortality as part of an overall successful effort.

Increased Monitoring Needs

There is a growing number of wildlife passage structures installed specifically for small animals. However, there are few such projects where monitoring has occurred and even fewer where monitoring continued for longer than one year. Typically, funding resources and logistics will only allow for post-construction mortality counts and will not allow for a robust assessment of population effects. Population genetics data for small animals generally are lacking, and are particularly uncommon in fragmented landscapes or relative to pre- and post-construction of roads. Further, where genetic effects have been studied after a road is in place, we lack baseline data on genetic exchange rates across the road. We assume connectivity is disrupted when we see animals being hit on roads, but rarely can we be certain that those mortality levels truly are disruptive to population persistence.

Need for Low-Cost, Tested Passage Designs

The good thing about small animals is that huge, expansive passage structures are not necessary. In fact, small animals may be hesitant to use such large structures. Small culverts can mean a smaller price tag with respect to individual structures. However, some instances require more numerous, closely spaced passages to accommodate low mobility species, which ultimately may be more costly. There are many companies that are producing small, affordable structures; preliminary results suggest that they may be effective, but long-term monitoring is necessary to identify the most effective structures and system configurations.

Adaptive Management and Flexible Roadside Maintenance Plans

Where funding support is in place for structure monitoring, maintenance, and roadside management, plans must be contractually specified prior to project construction and completion in order to guarantee these activities. However, structure performance, wildlife use, vegetation encroachment, human traffic patterns, and adjacent habitat development can all change over time. Thus, management and maintenance regimes and plans must be adaptive and allow flexibility to address these changes as they occur.

Trade-offs in Deciding Which Effects to Mitigate

There is great diversity in the ecological effects of roads and in the natural history of small animals. As such, it is impossible to create a mitigation measure that will address all ecological effects for all affected species, even with multi-component mitigation systems.

Typically, endangered or rare species are prioritized in small animal projects; however, while it is important to consider these declining or rare species, it is also important to accommodate proactively the species that are thought to be common. Further, identification and management of certain key species in turn can benefit many other species and maintain important ecological relationships (e.g., predator-prey interactions) and overall ecosystem functionality. Additionally, identification of key biological traits contributing to species vulnerability to road effects can then allow for the development of mitigation strategies that target the greatest number of species.

Trade-offs in Deciding Which Effects to Mitigate for Which Species

Not all effects of roads are preventable or reversible, and it is arguable that any can be negated fully. Once roads are in place, some effects cannot be mitigated at all while others cannot be mitigated simultaneously with others. A structure that mitigates one negative effect may counter or prevent the amelioration of another. Therefore, effects should be prioritized that are most likely to have severe consequences for ecosystem function. While mortality reductions are important, mortality is not always the most detrimental effect. Typically, mortality is at the top of the mitigation list, but this overlooks broader effects. We encourage investigators to question the *what* and *how* of the species and the ecological processes the road is affecting before automatically prescribing the standard remedy of underpasses and fencing.

13.3. Public Perceptions

Societal Attitudes

Society supports allocation of public resources to issues it cares about or that affect it directly. Most individuals are not directly affected by small animal conflicts with roads and will only be assertive in supporting these initiatives if we engage their interest. In particular, the human fear of snakes is all too common and many people are comforted by seeing harm done to these animals. We have much work to do to gain public support for smaller, less charismatic, or negatively perceived species. It is a challenge that may be one of the greatest

hurdles in developing considerations for small animals in road ecology.

Resource Limitation

The deficient public and political support for the resolution of small animal road conflicts then translates to financial limitations. Resources to address road conflicts continue to be limited despite growth in the field of road ecology. While available funding for conservation is a zero sum game (i.e., a limited pot of money can only be divided so far), we can increase support through communicating the added value of ecological infrastructure beyond the scientific community.

Driver Behaviors

Many small animals are hit on roads simply because people do not see them. However, many drivers also will not swerve to avoid them, as these animals do not pose a risk to them or their vehicles. Further, it is not uncommon for drivers to intentionally run over some animals. This pattern is pervasive worldwide with snakes but also occurs surprisingly often with turtles. Conversely, it is not uncommon for drivers, even those who are not professional biologists, to stop their car and move animals out of roads, in turn putting themselves in danger of being struck by a vehicle.

Safety Issue

The primary factor driving resource allocation for wildlife and roads is the risk to human safety and property. Safety is an issue in the instances of drivers swerving to either hit or miss animals. There are also anecdotal accounts of "frog slicks" formed by frogs that were hit during mass breeding migrations, and caused vehicles to lose control or slide off the road (M. Lannoo, Indiana University School of Medicine, personal communication). However, in general, it is true that with most small animal species, the safety concern is not as prominent as with large vertebrates. It is up to us to argue the many valid reasons why small animals should not be ignored.

Research Needs

There is a major information deficit regarding the general natural history of many small animals, and even less about how these animals are affected by urbanization and public infrastructure. Securing funding for road-related research projects can be challenging; many entities that typically fund research feel that these questions are the responsibility of transportation agencies to support and answer. However, transportation agencies do not typically have the means to fund true scientific research and have not typically prioritized the mitigation of impacts to small animal due to the low risk to human safety. When wildlife contracts are secured with transportation agencies, they are usually only for species with regulatory status and for short-term assessments (often for one year or less) immediately prior to or following a construction project. These restrictions are not conducive to implementing an effective wildlife assessment. Further, there is a noticeable geographic bias in the research, in that it has been conducted primarily in the eastern United States, Canada, and European countries. Within the United States, the southeastern region contains the overall highest diversity of small animals, a high road density, and therefore, the greatest abundance and severity of road conflicts. However, expansion of research on small animal conflicts with roads is sorely needed throughout the world.

Need for Partnerships

Partnerships are essential because the sum of our knowledge is greater than its parts. We can rapidly gain ground by consolidating and increasing communication of our knowledge bases, whether through verbal conversations or in written discussions such as in this book. Partnerships and frequent information exchange on what we learn and what we still do not know will only improve our ability to make sound management calls. We need to prioritize knowledge gaps so that we can work together to collect these data and experiences to inform and improve the design of long-term, adaptive, planning and monitoring regimes.

13.4. Closing Thoughts

We propose a shift to long-term management approaches in a world that typically executes short-term plans. Recognizing limitations, setting achievable project objectives, and reviewing performance-based measures will help us to realistically set goals and define measurable successes. While it may not be possible to prevent or fix all of the negative consequences of roads, proactive planning can minimize these high profile problems involving low profile animals.

Development of new roads has slowed in North America. However, there are still many opportunities to apply what we have learned in developing countries where the rate of new road construction is high, and in the western United States where the boom in energy development has required extensive new road construction. There is also a great deal of opportunity to implement mitigation in our existing road network within North America and across the globe. Existing maintenance schedules for bridge repairs open the possibility of adapting these to serve as underpasses for wildlife for minimal costs. Small sources of funding can be used to support wildlife mitigation measures that in turn can have large positive impacts. While the challenges are great, the road ahead is full of possibilities that can be realized with a little bit of creativity and persistence. And importantly, there are more and more of us that are traveling this road together. We hope that you will take some of the resources and knowledge presented here and will join us on this journey.

Index

Page numbers followed by f indicate figures and those followed by t indicate tables.